Michel Rouhana O.A.M.

CREATION OUT OF LOVE

vs

CREATION OUT OF NIHIL

A Luminous Bridge between Science and Religion

Creatio Ex Amore vs Creatio Ex Nihilo

An Essay in *Râzâtic* Theology

1st English Edition.
No part of this book may be reproduced without the written consent of the author.
Email: bounamichel@gmail.com
Mobile: Lebanon (+961 81 367247), Canada (+1 437-788 4837)

Nihil Obstat

الأباتي مارون أبوجوده

أب عام أنطونيّ

Rev. Abbot Maroun Abou-Jaoudeh,
Superior General of O.A.M.

Address: Monastery Saint Roch
            Dékouaneh, Beirut
            B.P. 55035
            Tel : +961 1 681455 / 6 / 7
            Email: sec.oam@gmail.com
            Site internet: www.antonins.org

ISBN 978-1-7781234-4-3

Cover: The "raindow" used by the Creator, in Genesis, as signature of his Pact with humans , represents the most tangible bridge connecting Science and Religion. (Designed by the Author).

Catalog data :

Rouhana, Michel C. (-1949 -).

Creation out of Love vs Creation out of Nihil: A Luminous Bridge Between Science and Religion [Printed text] / Michel C. Rouhana, O.A.M., Ph.D. - 1 volume (450 pages) : illustrated ; 17x24 cm.

This page contains on top: "Essay in *Râzâtic* Theology".

It contains bibliographic references and a thematic index.

English text. It contains transliterated texts from Hebrew and Syriac. It includes as well Syriac, Arabic, Hebrew, Greek, Latin, and German words and texts.

1. Christianity and Science. 2. Quantum theory. 3. Creation - Patristic teaching. 4. Syriac Fathers. 5. Mystics - Catholic Church. 6. Mary (saint; figures of the New Testament) - Theology. 7. Space-time. 8. Time - Religious aspect - Christianity. 9. Environmental ethics - Biblical teaching. I. Rouhana, Michel C. II. Title.

Dewey : 231.765 (23rd ed.) = Creation (Christianity)

Acknowledgment

I raise all my gratitude to the Holy Providence that led me in unfathomable ways until I encountered the Eternal Motherhood.

From now on, no full stop can follow the many thanks I give to my natural mother. Now, I have become more aware of the Eternal Motherhood that lies behind Jesus Christ's following words: "Blessed rather are those who hear the Word of God and obey it." He said it in response to the woman who told him: "Blessed is the womb that bore You and blessed are the breasts that nursed You." By this dialogue, the Incarnate Word set the first foundation of His Kingdom as Childhood, with the Eternal Motherhood standing behind all that we consider as the maternal womb that brings forth life. Now, I must give thanks to the Source before thanking the stream.

> Thank you, Holy Trinity
> Thank you, Holy Providence
> Thank you, Word of God

Thank you, Mary Mother of the Word of God. By you and your Son, humanity discovered the Motherhood and Childhood that Adam and Eve could never know.

Thank you, mom, family, school, Church, community, university, monastic life. Thank you, friendship, earth, water, nature, creation. I give thanks to all the females who have been part of my masculine priestly formation.

I also insist on thanking, by the Lord of Childhood, all those who assisted me, directly or indirectly, in bringing this book to its happy end, in particular, lately, Grod Nixon, Laura De Sanctis, Tony Dableh, Dr. Wisam Farjow and Harriette Mastert.

N.B. "I am seeking an imagination that can match mine."

I dedicate this book to

Pope Francis

The face of Childhood and Maternity

"Love" is the Triune God

He creates out of Himself (*Ex Amore*)

By Himself and in Himself.

Outside of Him is the chaos, the absurdity,

without Him no being could have existed,

And in Him, nothing can have an end.

Michel C. Rouhana O.A.M.

PREFACE

To reconcile Science and Religion remains for us Christians an inexhaustible and relevant topic. To dare to venture into the subject assumes an unlimited number of risks lying between the deification of Science and the profanation of Religion, passing through the softening of rationality or the hardening of affectivity.

Your contribution, dear Brother in Christ, to this journey of rapprochement between these two realities situated at the crossroads of the Divine and the Human is precious since it has been able to combine humility with depth: the humility of the learned man who seeks to scrutinize the world of knowledge without pretending to possess it, and the depth of the man of faith who bows before the Lord for spiritual enlightenment without, however, despising his intelligence.

Thus, with the original and innovative theological-scientific contribution of your book, the fruit of lengthy research and serious academic supervision, the challenge has been won!

I am not trying to outbid the testimonials of the scholars who accompanied you on this very distinct and thorny path of reconciliation between Science and Religion, but I want to tell you that your work did not catch me off guard! In fact, since I knew you in the novitiate, you were already diving, as St Ephrem says, in the voluminous books of our Maronite liturgy in search of a particular 'pearl' that is revealed to us today through your present work.

Dear Brother, I congratulate you on this research, and I note that I do not find any impediment to publishing this book. May our Lord, by His infinite Graces, transform it into a tool for Peace and Love between humans, serving to broaden the readers' knowledge and to deepen their faith.

Father Abbot Daoud Reaidy
Superior General of O.A.M.

TESTIMONIALS

Prof. Emeritus Dennis O'Hara Ph.D. Environmental Theology (Saint Michael's College, University of Toronto)

... The strength of this work is derived from the creative imagination of the author as well as his unique appropriation of cosmic and quantum physics, Eastern and Western culture, theological and liturgical texts, ecclesiastical art, and philosophy. It is through the blending of these various elements, with the guidance of the Spirit, that some truly unique and brilliantly insightful speculations are presented to the reader. The author's exploration of *Zimzum* is particularly rich and offers an understanding of immanence/transcendence that is playfully provocative.

Prof. Emeritus the late Paul Feghali (Mgr.), Ph.D. Philosophy and Biblical studies

... The kenosis of God revealed by the Incarnation is the main idea of this work, and the Creation out of Love and in God (*Creatio Ex Amore et In Deum*) is its key... God leaves a place in His suffering for the sufferings of those who love Him so that they participate with their sufferings in the continual redemption of the human race as well as of the universe... Childhood, Simplicity, and Silence, according to our author, are indispensable conditions (*sine qua non*) to advance in the search for the Truth, the only Truth, whether it is sought by Science or by Religion. (Translated from French by the author)

Prof. Emeritus the late Georges Rahme O.A.M., Ph.D. Teilhardian philosopher, former dean of the Faculty of Letters at the Lebanese University.

... It is a subject so daring but interesting in its originality despite its difficulty. Moreover, our philosophy and our theology need to be enriched by such research... The author manages to say that if the Sun and its Fire are inaccessible to the human eye, nevertheless God is, thanks to the Incarnation... Amazing! It is also the experience of Saint Ephrem who had the grace to have "the Fire in the Eye", as mentioned in Chapter II... The Light is there, no doubt, but also the Eucharist, that *"Râzâ malyâ râzê"*, that purifying Ember (*Gmurtâ mḥasyânitâ*) which purified the lips of the prophet Isaiah and purifies ours today. It is the Pearl of great price (*Margânitâ*) of the parable, the Mirror that makes the human 'side' of the Icon of the Church see, dynamically and continuously, the divine one...

Ex Amore, by itself, gives meaning to the cosmic dimensions that seek to explain the "quantic phenomenon" of all the scientists listed in Chapter V...

... I want to end my appreciation by highlighting the anthem of Jacob of Sarug, quoted in footnote 79, which, according to the author, should be referred to Saint Ephrem because, as he says, nobody was sensitive to the 'Marian *Râzâ*' like him. This hymn is also so personally dear to me because it affirms that, for our Syriac Church, the role of Mary in the creation is outstanding. For me, Mary is '*co-redemptrix*' and '*co-creatrix*'; otherwise, we would not have been created.

May Mary, through her intercession before her Son, keep you, dear Michel, in grace so that you may continue your research at the service of the Church and humanity. (Translated from French by the author)

Prof. Youssef El Hajj, Ph.D. in nuclear physics. Dean of the Faculty of Sciences in NDU University, Lebanon. (co-director of the author's doctoral thesis 2008 - 2011)

"In the last days, God says, I will pour out my Spirit on all people. Your sons and daughters will prophesy." (Acts 2:17)

... I think this achievement is one of the most complimentary. In fact, the idea is very daring. Is it about creating a hermeneutics from very modern and very, very cutting-edge scientific concepts?... This is an extremely bold idea and, at the same time, very dangerous.... Father Rouhana accepted the stakes, accepted the challenge, and I believe he rose to the level of the challenge. There is no doubt about it ... There is certainly a lot of very interesting insight in this thesis, not only for the uninitiated reader but also for scientists. (Translated from French by the author).

ABBREVIATIONS

of the Ephremian corpus in universal use

After Feghali		After Brock	After Bou Mansour
C Ex	Commentary on Exodus	*CExo*	*GETon*
C Gen	Commentary on the book of Genesis	*CGen*	*GETon*
C Nis	The Nisibeian Hymns	*Nis*	*CNis*
CSCO	*Corpus Scriptorum Christianorum Orientalium*	*CSCO*	*CSCO*
EC Syr	Commentary of the Gospel: Concordant or *Diatessaron*, according to the Syriac text.	*CDia*	
H Az	Hymns on *Azymes*	*Azy*	*Az*
Hc Haer	Hymns against Heretics	*Hae*	*CH*
Hc Jul	Hymns against Julian the Apostate		*CJul = Parad . Crucif = Az*
H Cruc	Hymns on the Crucifixion	*Cru*	*Az*
H Eccl	Hymns on the Church	*Ecc*	*Eccl*
H Epiph	Hymns on the Epiphany	*Epi*	*Epi*
H Fid	Hymns on Faith	*Fid*	*HdF*
H Jejun	Hymns on Fasting	*Jej*	*Jejun*
H Nat	Hymns on Nativity	*DeNat*	*Nat*
H Par	Hymns on Paradise	*Par*	*Parad-Eccl (et CJul)*
H Res	Hymns on Resurrection	*Res*	*Resurr*
H Virg	Hymns on Virginity	*Vir*	*Virg*

As far as we are concerned, we will follow the favorite abbreviations of Feghali, but for those of the quotes, we will respect theirs.

For the biblical books we opted for the abbreviations of the Roman Catholic Holy Bible. *Cf.* https://www.catholicdoors.com/bible/abbrev.htm

LINGUISTIC PRELIMINARY

We purposely decided on this preliminary, at the beginning of this book, to address matters that might block the desire to pursue reading. No doubt, this book is more complex than any other of its kind because on top of the input of post-modern western theologians, philosophers, Nobel-prize-winning physicists, and astronomers, it serves as a bridge, thanks to the pattern of Light, between Biblical revelation, the Kabbalah School, Syriac Antiochian theology, and physical sciences, specifically quantum physics. If we anticipate a shock, it is because of the leaps the mind of the reader will be solicited to make *vis-à-vis* some phenomena outside the box of traditional patterns. Accordingly, and to make it easier to grasp the leading ideas of this monograph, we found it convenient to introduce these keywords to you, beloved readers.

Ex Amore and *Ex Nihilo*

By using the title "Creation out of Love" (*Ex Amore*) instead out of 'Nothing' (*Ex Nihilo*), we do not mean to contradict the Christian dogma of "*Creatio Ex Nihilo*" based on the unique biblical reference found in the second Book of Maccabees, 7:28. It is instead a reconceptualization of the latter, based on new phenomenological and scientific discoveries which have proved that *Nihil*, philosophically and scientifically speaking as well, is but an absurdity. On the one hand, Philosophy teaches that as long as 'Being' exists, *Nihil*, which is the total absence of 'Being', makes no sense. On the other hand, scientists prove that there is no void in the universe, no emptiness, either at the macro level or at the micro level. Energy, gravity, electromagnetic fields, bosons, etc., fill the universe.

"Creation out of Love" is a theological proposal that fills the gap between the Biblical *Nihil* and the scientific space-time exhaustively occupied by 'seen' and 'unseen' substances. What kind of "occupants" will we be talking about? Can they originate from behind Max Planck's wall?

We are adamant to prove that yes, this is the case. Once 'Being' is considered Love itself, every existence should be coming out of 'Him', the Triune One, by 'Him', and in 'Him'. The 'Primordial Particle' that stands behind the Big Bang, hence behind all matter, is left by 'Him' after a specific centrifugal withdrawal (kenosis), allowing the expansion of the universe and, by the same token, space-time to take place. It is the major leap, out of the box, that this book intends to help humanity make.

Râzâ as a concept

It is a Persian root word originating from the Book of Daniel (Dan 2:18), from a specific portion written in Aramaic and not considered canonical by everyone. It became a key term within Syriac-Aramaic literature and liturgy. Written with a small "r", it designates a signifier or, more precisely, a dynamic symbol, a hierophanic revealer different from what the frustrating Greek concept "mystery" can mean.[1] When it comes to the Christian salvific Economy and sacraments, the Syriac liturgy uses this term frequently in its texts. 20th century translators confused it with the concept "mystery" as used in Saint Paul's letters. We will show that Saint Ephrem used it differently to serve his Christological teaching.

As a concept, *râzâ* is the root word behind the qualification *râzâtic* used in our monograph. It represents a vital discovery for our project of reconciliation between Science and Religion. We will compare it to Planck's constant and, more specifically, to what is considered the primordial particle of light determined by Einstein: the quantum of energy, the photon ($E = h\nu$). We consider it a primordial linguistic element and desire to propose this concept to humanity, in all its languages, as a substitute for the word "mystery" (*Mysterion*) as used in Christianity. According to Saint Ephrem, we will be seeking a Christianity without mysteries. Mysteries no longer fit Christianity, especially after what Jesus declared, saying: "No longer do I call you servants, for a servant does not know what his master is doing. However, I have called you friends because everything I have learned from My Father I have made known to you".[2] He left it to us then to seek the Luminous Eye (*'aynâ šafyâ*) capable of 'seeing' and the pure Heart (*Lebâ dakyâ*) capable of 'understanding'.

The famous Syriacist Sebastian Brock, in his book The Luminous Eye, says precisely, that "Inherent to *râzâ* there is what Ephrem calls the hidden power (*haylâ kasyâ*)." This power is potentiality, force, living-loving energy... It is the core of the difference between this concept and the one called "mystery". (*Cf.* p. 57, § 2.2.1.)

Râzâ as substantive (the *râz*)

This substantive form, "the *râz*", is formed by replacing the last letter '*â*' of the Syriac word with the definite article "the", as Syriac dictionaries explain. Consequently, it becomes obvious that the root of the Ephremian concept is *râz*. That is why we will write this noun *râzâ* when it is transliterated and as the *râz*

---

1   Brock, Sebastian. *The Luminous Eye, The Spiritual World Vision of Ephrem the Syrian.* Cistercian Publications, Kalamazoo, Michigan, 1992, p. 46. From now on, we will refer to this work by the letters *TLE*.

2   Jn 15:15.

when using it as an English word.

While in the French language, we had to give it a gender (masculine), in English we do not have this problem, so we can keep the unspecified pronoun "it" unless referring to sacred or divine persons.

When it comes to its written form for pronunciation, although the English language contains the sound (ŏ) for the letter "a" as in the word (ball), we will continue using "â" to suggest the sound (ŏ) of the Syriac occidental tradition (as in, *e.g.*, the Maronite). We shall also include the diacretic to establish a unified notation for this word in all European languages. Chaldeans and others of the Syriac oriental tradition can pronounce it as the French "a" according to their tradition.

Two of the most famous Syriacists, Edmund Beck and Tanios Bou Mansour, did the same by giving the reader the free choice of pronunciation. However, for us, although we are Maronite, we opted for the French pronunciation of the letter "a" instead of "ŏ" because of the Aramaic accent used by Lord Jesus and the disciples. The Gospel proves it in two places: *"Elî, Elî, lma Shbaktan?"* and *"Tabitha kumî"*. Saint Paul did the same in 1 Cor 16:22, where he wrote: *"Maranatha"*. The Syriac plural of *râzâ* is pronounced *râzê*. We will keep it this way for uniformity with its French use.

The verb to *râzify*

We find it written in two forms: one referring to the past *râz* (ܐܪܙ, ܐܪܝܙ, ܝܪܙ), and the other to the atemporal present *étérêz* (ܐܬܪܙܝ – ܐܬܪܐܙܝ), as the Syriac-English dictionary makes clear.[3] This verb and its grammatical applications derive from the root *râz*, just as, for example, the verb 'to love' derives from 'love'. *Râz* and *étérêz* are the two forms of the Syriac verb often encountered and widely explained by dictionaries through the nuances of their applications. However, as we notice throughout the various sources, the Syriac verb *râz* is transitive. Its meaning is comparable with the verbs 'to reveal', 'to declare', 'to teach', 'to pass a secret', and 'to initiate someone'. To build a transitive verb out of the substantive form of a word makes the style more direct and, consequently, more significant. The verb "to *râzify*" that we have forged to serve the vocabulary of the Incarnation and Redemption suggests an act behind which the subject of the verb and its direct object switch roles. Thus, Mary might become the direct object of the birth of her Son Jesus, instead of being its subject.

Râzâ or the *Râz* as a sign, a signified, and a referent to Jesus Christ

According to Ephrem, *râzâ* used for a sign or name and written with capital R, *Râzâ*, shall have only one signified, Jesus Christ Himself. It shall also refer solely

---

3   *Thesaurus Syriacus*, R. Payne Smith; The Clarendon Press; Oxford, 1957.

to Him Who for us is no longer a mystery (even less is His Economy of Salvation). He and His Economy of Salvation form rather a Redemptive Reality, an ultimate revealing phenomenon that we will call *râzâtic*. Jesus Christ is the revelation of His Father.

According to the Ephremian literature, the *Râz* will replace all the bloody holocausts of the Old Testament mentioned by Saint Paul. Having considered Nature another sacred book alongside the Bible, the Ephremian sign, the *Râz*, will also replace water, fire, air, light, wine, bread and all that is beneficial to ensure the Redemption of humans.

Within the *râzâtic* reality based on the Ephremian *râzâtic* theory, the *Râz* is a substitute above all for Sonhood and Fatherhood, but not Motherhood. Our Ephremian thesis supposes that from the beginning, before the Genesis, the three divine Persons in perfect agreement reserved by kenosis the Eternal Motherhood, to bestow It on Mary in the fullness of time when the Promise of God to Adam and Eve found fulfillment, as prophesied by Isaiah. Jesus Christ, as it fully pleased the Father and the Holy Spirit, renewed on the Cross this gift of the Eternal Motherhood. This is the span that the *Râz*, with capital R, covers and in which converge all the *râzê* whose roots descend into the first *Fiat* of the Genesis. (More references and explanations await the reader in the text.)

Maškanzabnâ (the letter *š* represents the sound 'sh')

This word is a blend of two others: *maškan* (ﻣﺸﻜﻦ) habitat and *zabnâ* (ﺯﺑﻨﺎ) time. Habitat, in Aramaic, as well as in other Semitic languages, comes from the verb *šaken* (שָׁכֵן سَكَنَ) and has two meanings: to inhabit and to rest. Biblically, *maškanzabnâ* means the place of God's dwelling among humans, such as the Tent of Meeting mentioned in Exodus 33:7-11 or the Dome which designates the blue firmament called "the Sky", etc.

A typical use of this concept, which emphasizes the state of 'resting in', can be found in Psalm 91, which begins, "Whoever dwells in the shelter of the Most High (His *maškanzabnâ*) will rest in the shadow of the Almighty." This use suggests peace and protection from all uncertainty and unpredictability. In short, it inspires repose and serenity.

Under the auspices of Einstein's space-time, we have deconstructed *maškanzabnâ* and shifted its meaning into 'habitat of time'. By this, we respect both relativities between space and time, and help bridge between the two major Creation theories, the Biblical and the Big Bang. It supports the theory of a hidden Natural Law that directs everything toward the positive fullness of Creation. We noticed, during our stay in Canada, a physical model of this shifted *maškanzabnâ* in the geodesic dome conceived and built by the engineer R. Buckminster Fuller on the Expo grounds of Montreal. This connection will be more fully explored in Chapter III.

Yât (ܝܬ)

These two Syriac letters *yud* and *tao*, the same as 'Y' and 'T', with the diacritic "â" for "alpha" that we pronounce as previously explained, basically form a definite article that determines the substantiality of a 'Being' and its specificity. In this sense, it is comparable to the definite article "the" when it designates a unique person as in the expression "the philosopher" to mean Aristotle, "the Savior" to say Jesus Christ. It is also comparable to the Latin *'se'* or the English 'self', with a specific difference. This difference is the hidden power, force, or creative energy contained in the bosom of the innovative and revivificating concept *râzâ*.

This Syriac syllable is synonymous with the Hebrew *Ït* and *Ët* (אח). It plays a crucial role in the uniqueness of the principle, substance, and essence that are visible, invisible, and meta-invisible.

Ityâ (ܐܝܬܝܐ), *itutâ* (ܐܝܬܘܬܐ)

Both nouns come from the article *yât* and indicate the substance itself, *e.g.,* *Ityâ ghnizâ* (ܐܝܬܝܐ ܓܢܝܙܐ) which means the hidden substance. Ephrem, indeed, stresses this since the Bible says b-riš-ït, (בְּרֵאשׁ ית) where *riš-it* indicates the beginning of all substance (*Ityâ, itutâ*).

"DifferAnce"

This is a philosophical concept coined by the philosopher Jacques Derrida.

Dictionaries of Philosophy and other relevant encyclopedias thoroughly explain this concept of universal scale. We found it perfectly suitable for the foundation of our *râzâtic* theory. It helped us to distinguish between realities and to shed light on a new Christological one, just "differAnt" – the *râzâtic* one. We based it on the 'chiasmus' of the Incarnation, the "crossing" between the Divine and the Human.

"DifferAnce" also affects the whole domain of signifiers, signified, and referents. It puts into practice a positive deconstructive method, also used by Derrida, which proves that opposite entities considered incompatible by traditional logical methods, are but "differAnt" and always mutually necessary for the exhaustive perfection of the Universe, *e.g.,* Light and Darkness in the book of Genesis. The negation "non-" has nothing to do with the absolute "not" or *nihil.* "Non-", according to the theory of "differAnce", indicates a different way of being and not a total absence of being. A non-being can also mean a potential being, a being in process, while the absolute "not" or *nihil* means irrevocably nonexistent, and that is absurd. We found it very inspiring to review the kerygma of our Christian Faith from this angle.

To *synesserate:*

The Cartesian proposition *"Cogito ergo sum"* (I think therefore I am) and the depth of Heidegger's *"Mitsein"* directed us to the Latin form of the verb "to be", *esse,* to generate this new verb out of its matrix. We needed it to express what is meant by "withing" when the "withing", hidden in the Name *Emmanu-El,* reaches the stage of being essentially together at the risk of losing the total existence, even the total being, once separated or split, or missing the "other", whoever the other is. To be together or not to be at all becomes the challenge.

Indeed, while according to the cosmic theory of Multiverse or the Observer Observed Universe theory, existence depends on the other's observation, being, as proposed by Descartes, depends on one's own "thinking", one's own consciousness. Hence, not to be means practically the absence of all ways and kinds of thinking, *i.e.,* the Chaos, the absurdity.

By the verb "to synesserate", we try to make it clear that the cogitating capacity is a whole 'unity' that joins the thinking humans (*Mitsein*), the other and me, to the thinking Principle, the "Wholly Other". No cogitating person can exist individually without any "other" or any "Other". It is the strength of the Triune conception of the first Thinking Being.

"To synesserate" represents a category of verbs with a comprehensive dimension that covers, without embarrassment, different realities and different plausible environments.

Kasyâ and *galyâ*

Often used by Saint Ephrem, these two concepts are usually translated by "veiled" and "unveiled" or, in other cases, by the "hidden" and the "apparent". Christologically, they belong to the domain of the Revelation, especially that of the Incarnation but, we underline that, philosophically, the two can reach the height of the Platonic-Kantian dichotomy of 'noumenon/phenomenon', or the Aristotelian-Thomistic 'Potency/Act'.

Using these contrasting terms ensures a more excellent understanding for interested persons. It is new fodder for a new, well-balanced, uncompromising reconciliation between Science and Religion so that neither discipline preaches the anathematization of the other one and no longer sees the other domain's ideas as an unnecessary hypothesis. We also hope that it will cause a shift in the traditional methods of evangelization.

Milieu, milieu, 'milieu'

This word is composed of the syllables originating in Latin '*mi-, medius*' (half, center) and '*lieu - locus*' (place, locality, environment, geographical or mathematical point at the center of distances or times, time measurement, moment like 'middle of the day – noon'= *mi-du jour, midi*). It has a specific impression when used as a signifier for a signified as Teilhard de Chardin used it. We have kept using it in our text on purpose, for its philosophical and linguistic impact as sign. That is why it can be found written Milieu and 'Milieu', milieu and 'milieu' etc. When figurative, we have added single quotation marks, being a signifier for a divine Time or divine Space or Environment ('Milieu' of the divine Trinity, 'milieu' of Rublev's Trinity icon, 'Milieu' of the Eternity, or 'milieu' of the eternal time). It is capitalized at the beginning of a sentence and is found capitalized for divine respect. When it is not playing the role of a signifier for a signified and a referent, we have used its English parallels like "center", "context" or "environment". Our main purpose is to reduce confusion in the understanding of our ideas.

TABLE OF CONTENTS

PART I

Saint Ephrem:
From "Light in the Eye" to "Fire in the Eye"

PART III

Quantum Mechanics:

From the Birth of the Photon to the Tomb of the Light

PART IV
Creation out of Love:
Eternal Motherhood and Childhood's Tent

GENERAL INTRODUCTION

*"Imagination is more important than knowledge.
For knowledge is limited, whereas imagination
embraces the entire world, stimulating progress,
giving birth to evolution."*

Albert Einstein

Humans are not to blame for their lack of faith in God, although the Bible alludes to this idea. Since the Church teaches that Faith is indeed a free and unconditional gift from God, the Father of Jesus Christ, why then blame those who did not receive it?

The saying goes, "Where there is a will, there is a way." However, in the case of Faith, the "way" exceeds human capabilities. Yet, there is nothing that prevents these capabilities from seeking the Truth, especially when it comes to "ascetic researchers" from disparate disciplines such as Science and Religion. The same can be said for Faith and Reason as Saint John Paul II underlined in his encyclical *Fides et Ratio*. Indeed, nothing stands in the way of finding solutions for a good number of scientific puzzles.

What are we aiming at by undertaking this project of reconciliation between Science and Religion?

The problematic relationship between these two disciplines has existed since ancient times. It started with the Greek philosophers and will continue its journey as long as collective morality and ethics depend on it. Descartes' *Discourse on Method*, one could say, severely accentuated a total divorce between them. Under the influence of the Cartesian dichotomy of 'clarity' and 'distinction', the European collective mind began to sharply separate what is "of God" from what is "of Man". It even managed to move the Lordship of reality from God to Reason. The era of philosophers and of the sages (those who supposedly knew everything) started losing ground to specialists in experimental sciences.

Hence, according to our vision, humans are not to blame for their lack of faith in God. On the contrary, they are to blame for any reductionism and lack of sincerity that leads to the denial of the "other", any "other", especially of the "Wholly Other", to use the terms of Rudolf Otto.

Following the publication of the encyclical *Laudato Si', On Care for Our Common Home*, we felt obliged to dedicate this book, environmentalist *par excellence*, to His Holiness Pope Francis. Moreover, the famous Encyclical in question helped extend the dedication to include scholars of consciousness in both Science and Religion. We hope these esteemed scholars will not neglect the present monograph, pretending to only find proselytism in it, aiming – they would contend – at indirectly convincing them to believe in an "unknown god".

Knowing that the Christian Faith is a free and unconditional gift, this book does not aim to convince readers into believing in a "God", not even in the specific God that Christ has called Father, my Father, our Father, the Triune God. This book is an effort to answer the question posed by scientists to the Catholic Church, concerning the God she proposes to them as the moral and ethical foundation of the evaluation of their contributions and of the progress they are making at all levels. What is the nature of God that the Church is proclaiming?

On the one hand, scientists who are already believers will find in this book an invitation to better know themselves as well as to improve their knowledge of the God in Whom they believe and Whom they most probably love *a priori*. Non-believers, on the other hand, will become aware of how to welcome the challenge posed by the "Out of Love" and by the "Luminous Bridge", half of which is already at their disposal and which stands behind almost all their discoveries.

Therefore, if God does not play dice, as Einstein said, it is up to us to do so. In the Levant countries, and especially in Lebanon, the 'dice' commonly refers to the social game of backgammon. To play it requires two opponents. 'Dice', here, is renowned for the individual and egocentric satisfaction it offers to the winner. However, it would be much nobler if both players agree, just like in a charitable tournament, that the dice roll only for common satisfaction in hopes of finding the Truth that leads to the bettering of humanity. This insight leads to the deduction that sharing efforts and inspiration is an indispensable condition to establish, through Love, a bridge of reconciliation between two opponents. At the same time, even though I personally meditate from my own perspective on the 'mysteries' of existential realities, I still admit that Childhood, in all that it symbolizes, is the common denominator worthy of respect by all humans. Childhood is our common cradle to which nostalgia takes us back daily.

By using the title *Creation out of Love vs Creation out of Nihil: A Luminous Bridge Between Science and Religion* and the additional subtitle *An Essay in Râzâtic Theology*, we aim to use the same Fire that illuminated the 'eye' of Saint Ephrem, the Syriac *Doctor Ecclesiae*, and before him the one of Saint Paul, the Architect of the Christian faith, to satisfy the expectations of both the Scientific and Religious horizons. We desire to help the Fire of the Book (Bible), together with that of Nature, light a "differAnt" candle in the darkness of our present world, instead of cursing it. To be able to do so, we found it appropriate to distribute over four parts our inspirations and the information that our imagination has been able to accumulate:

Saint Ephrem: From "Light in the Eye" to "Fire in the Eye".

Interpretation and the *Râzâtic* Reality

Quantum Mechanics: From the Birth of the Photon to the Tomb of the Light

Creation out of Love: Eternal Motherhood and Childhood's Tent

We will reserve the first part for the theological and spiritual contributions of our primary reference, Saint Ephrem. We will divide it into two chapters.

The first, entitled "The Light in the Eye", will set the stage for an accurate understanding of an 'eye' supposed to see the "Light of the World" of the prologue of Saint John's Gospel. This chapter will highlight the disappointment that human eyes can cause. It will highlight the importance of the eye-ear correlation, an essential correlation among "ascetics" who spend their lives in silence and dark shadow to hear and see, beyond the philosophical norms, what belongs to meta-metaphysics.

In keeping with the sublime poetic art of Ephrem, this chapter will emphasize the importance of linguistic genius in the rules that Ephrem imposes on any interpretation of the 'mysteries' of Salvation in both Old and New Testaments. Aiming at the problem of interpretation and linguistic limits, we will insist in this chapter on clarifying the concept *râzâ* (ܪܐܙܐ), a keyword in Syriac theology. We will highlight the importance of the interdisciplinary synergy evoked in our mind by Rublev's Holy Trinity icon.

We will affirm that the same synergy should apply between Religion and Science, as between different realities, so that together they can better express what a discipline, a language, or a reality is insufficient to do on its own, especially when it comes to making clear and distinct what is ontologically veiled or confused.

The second chapter, "The Fire in the 'Eye'", will deal with the problematic of Fire, its position *vis-à-vis* the deity, and whether or not it is created. It is specifically in this chapter that we will develop an ardent synergy between Ephrem of Nisibis and Gaston Bachelard through his work *The Psychoanalysis of Fire*. We will do so aiming to explore the delicate triune relationship between Father, Son, and Holy Spirit, and its repercussions on the intra-trinitarian acts of Generation and Procession. The issue of the *filioque* will consequently surface with some probable solution. We will close this chapter on a specific theological issue based on the symbolism of the 'Ember' that touched the lips of the Prophet Isaiah. It plays a significant role in the words of Transubstantiation of the Maronite Mass. Through the deconstruction of Fire, we will reach the symbolism of 'embers', a symbol of the Incarnation. It is also the foundation of the *râzâtic* reality, the central core of the third chapter.

We end this chapter asking if that 'ember' (*gmurtâ* ܓܡܘܪܬܐ) can serve as a unit of measurement of the *râzâtic* reality, a unit that will be compared to the quantum unit, Max Planck's constant.

In Chapters III and IV of Part II, entitled "Interpretation and the *Râzâtic* Reality", based on Plato's Myth of the Cave, we will shed light on the problematic of the interpretation and the intermediation that stands between humans and the two sources of Revelation, the Bible and Nature.

In Chapter III, entitled "The *Râzâtic* Reality", we will develop an epistemological, linguistic, and philosophical critique of the anti-Babel phenomenon, recognized under the movement of translation and interpretation. It will allow us to re-elaborate the concept '*râzâ*' and to emphasize its importance as a sign and an innovative symbol of a Christology more open to the dialogue between Science and Religion. The trick was to read Ephrem in the light of a few modern philosophers like Gadamer, Guitton, and Derrida. We will end this third chapter by introducing the soteriology of the Hymen and the Umbilicus.

In this context, we will find Ephrem replacing the Greek term "mystery" used by Saint Paul with the concept *râzâ* he picked up from the Book of Daniel, from the bosom of the Aramaic–Persian genius, because it better fits the logic of the Gospel. From the logic of the *râz* as a concept will be born the *râzâtic* reality, which is the Christological Reality as seen by Saint Ephrem. It will have an amazing role in the reconciliation between Science and Religion from the Christian perspective.

Furthermore, we matched Ephrem's capacities with those of some prodigious scientists of the twentieth century, without losing sight of the fact that it is the Substance of the Light of the World, the *Yât*, Who joins them all. It will require undertaking in the fourth chapter an introduction to the Syriac concept *Yât*, (*Ît* or *Ët* in Hebrew) which represents the primordial substance from which all has been made. The *yât* (ܝܬ) we found hidden in the *b-riš-ît* (בְּ‎רֵאשִׁ‎ית), the first word of Genesis, will be presented as the biblical replica of the quantum of energy (Planck's constant). Its relation to the primordial Fire (uncreated) is equivalent to the relation of the quantum of light (the photon) to the created fire detectable in embers under the ashes, as well as in the sun with its dazzling presence.

Chapter IV, entitled "The Substance *yât* (ܝܬ), from Ember to Sun", will extend to its maximum the interpretive extrapolation to set out the structure of the '*râzâtic* reality' with its necessary components. It will prepare us for its comparison with the quantum reality, born from the sources of quantum mechanics, to use it as a perfect element to approach the scientific domain.

These two chapters cover a wide range, including noetic, linguistic, philosophical,

Christological, and Trinitarian issues. The 'Mystery of the Trinity', for Whose nature Ephrem poetically gives irrefutable evidence, will dominate in this approach between Religion and quantum physics, in collaboration with the input of the theologian Jürgen Moltmann. They will address, among other things, the art of the dichotomy of which Ephrem makes wonderful use, especially the two types "Womb" and "Tomb" whose unveiling will reach its peak in the fourth chapter.

In the latter, we will point to the transmutation that the paradigm relating to each of the two types has undergone, thanks to the Light of the Incarnate Word that affected them. We will reveal in the same way the dimensions that emanated from the dichotomy "full" and "empty" of the tomb itself, due to the impact of that transmutation and the exigencies that emerged with it. We will close the second part of our work by opening the door wide, from an inspirational angle, to the corresponding contributions of quantum mechanics, especially those related to optics and atomics.

In general, these two parts will establish the necessary foundations for a "space" of meeting and exchange of information between Science and Religion. We will compare it with the Tent of Meeting in the Book of Exodus (Ex 33:7,11), called in Syriac *Maškanzabnâ* (ܡܫܟܢܙܒܢܐ). The latter means the "Habitat of Time" and ultimately refers to the dwelling of God among men. It will prepare an opening of Religion to the treasures of today's Science. It will establish a basis of mutual appreciation between the "ascetics" of the two disciplines and underline an existential difference in the western mentality represented by Descartes with the eastern one represented by Ephrem, underlining the difference between the rational and the exhaustive ways of thinking.

Part III, composed of a single chapter, will be reserved for the contribution of quantum mechanics. We have given it the title "Quantum Mechanics: From the Birth of the Photon to the Tomb of the Light". Its data will be condensed in Chapter V so that, as in the phenomenon of the Big Bang, there follows in the final two chapters the inflation of all that preceded it, to form, according to the *râzâtic* theory, the image of a Creation made out of Love in the "Womb" of Love. In these chapters, we will consider the unique environment of a Sacred Childhood capable of playing the role of the Meeting Tent itself (*Maškanzabnâ*), where Science and Religion, Faith and Reason, can easily relate with their relative truths, to enjoy seeking the Unique Truth together.

The fifth chapter, entitled "Planck and the Quanta", will introduce to us the founders of quantum mechanics. It will highlight the data that has come from it, which form the environment of this science and which mark henceforth its reality, to help us perfect the crux of our problematic.

The newly discovered depths of matter will confirm that the physical reality of creation, in the proper sense of the word, should be conceived differently than the

Newtonians thought. They will evoke Thomas Young's experiment of the Double Slit (1801), which revealed the wave nature of light and exposed it to the art of an expressive scientific symbolism. This experiment, key to all optical physics, deals now with the wave and the particle, highlighting the value of the anti-light, the dance of the waves, and the secret that stands behind the fringes caused by a light beam intercepted in a specific way.

Thanks to Heisenberg's uncertainty principle and its application to "Schrödinger's Cat", we have grasped that the depth of quantum mechanics reveals common denominators with the phenomenon of the Resurrection of Jesus Christ as described in the Gospels. They can cause shivers in the skin of the hardest reductionists of both disciplines. A new approach to the tomb of Christ and the Light that accompanied the scene of the Resurrection will allow us to join together Moses, Laplace, Planck, Einstein, Schrödinger, Guitton, and Ephrem in front of the same tomb, at the end of the fifth chapter.

The meeting of these seven "characters" and their deliberation in front of the wall (in reference to the wall of Planck) about the Reality that reigns behind the Wall, while projecting it on their respective realities, will reveal a "differAnt" vision of the problematic of 'Being' and of the relation between the dichotomies Reality/realities, death/life, full/empty. It will support the importance of refusing all reductionism, in order to know oneself better, as well as to know the 'other' better. The purpose is for scientists and theologians to get along with each other more respectfully on one hand, and with the "Wholly Other" (God) on the other.

Following this intuitive encounter, a new vision opens upon the universal reality with its entities. This reality, *i.e.*, the environment in which humanity has been since its genesis, will no longer be sought out by observation of appearances, contingencies, or by aprioristic ideas. It will instead be analyzed for the first time based on its objective and comprehensible substance, which until now has been invisible, such as the quantum of energy in the scientific sense and the *yât* (ܐ) in the Ephremian sense.

This vision and the recourse to quantum and *yât* to reinterpret what has already been established since Genesis under a "differAnt" angle of view, will project us into the adventure of Part IV, entitled "Creation out of Love: Eternal Motherhood and Childhood's Tent". This adventure consists in specifying the coordinates of the famous Tent of all agreements, certainties, and harmonies *vis-à-vis* the future of humanity and in revealing the nature of the 'Luminous Bridge' that leads to it.

This section will make traditional parallels converge. It will put in question the causative cause of creation to find out if it was created from *nihil*, by the pure Will of an absolute God and placed outside Him, at His exterior (*extra Deum*), or if it was created by the divine Trinitarian Love, from the same Love, by Love, and placed in Womb of this Love, *i.e.*, His central Point. This section will cover the last two chapters. They are of a very delicate composition and a somehow daring one. They push the imagination to the extremes of acceptable extrapolation, yet without succumbing to the risk of heresy. The challenge is hard, but it is necessary to unveil a facet of what we consider "new items in Evangelization" that have for a long time remained hidden.

Chapter VI, entitled "Are Space and Time Created? The Creation out of Love", will deal with the two cases of 'space' and 'time' and their so-called created materiality. We say "so-called" due to the transmutation that these two components of the universe have undergone due to Einstein's space-time discovery (*Raumzeit*). Science itself demands from Theology a clarification concerning their nature, to decide whether they belong to created or uncreated beings. This clarity will lead us, once for all, to describe almost mathematically the location of the universe relative to God's 'Space'. Could it be possible?

To make this statement possible, we will submit the concepts time, space, *Raumzeit*, as well as the concept of God, to the *râzâtic* loupe of the "Luminous Eye". We will invite people to have a better awareness of what humans say, think, understand, write, and do. The issues concerning linguistics and nominalism will surface again. The two creation theories, that of "*zimzum*" (צמצום) of the Kabbalah School of Isaac Louria and the Trinitarian one of Jürgen Moltmann, the views of the late Stephen Hawking and several scientists, most of whom are Nobel Prize winners, will be discussed to arrive at the proposal of a *maškanzabnâ* satisfying all parties. This proposal, which without fault can be called audacious, comes from a deduction that Creation took place out of Love and not out of Nothing (*Ex Amore* and not *Ex Nihilo*) and that everything is in the Womb of God Who is nothing other than Love Himself. It is because all *nihil* is absurd, and according to the *râzâtic* perspective, Love is not *nihil* as it is not at all foreign to the concept of quantum energy. This proposal can constitute, from a very original point of view, an excellent "Tent of Meeting" for God and humans, for Faith and Reason.

With the final chapter, entitled "Spin and the Trinity: the relation between cosmic gravity and Love", the consequences of our deductions will be put in a state of gestation at both the scientific and religious levels, to join, with the *Petit Prince* of Antoine de Saint-Exupéry, the greatest astonishment at the birth of "Childhood". We will propose it as the irrefutable medium for all taming, all understanding, all dialogue and progress towards a Kingdom of predictability and certainty where every sane human being would be able to say that it is indeed the same 'Reality', whether in Heaven or on Earth.

With the Spirit of Childhood as described by the French author Antoine de Saint-Exupéry in his famous book *Le Petit Prince*, we move to a different adventure that will make the universe share the Trinitarian perichoresis. We will try to embark on the orbit of Cosmic Gravity, at the rhythm of the Spin of the Trinity, to comprehensively grasp what can be considered the "Grand Unified Reality" (GUR), the *Râzâtic* One. We will confirm the *Triunity* of the Creation as well as of the Universe. Welcoming and invigorating Love will be the only means of mutual familiarization. Childhood will be its platform. This kind of Love replaces any umbilical cord and, as in the case of Tobias of the Bible, provides the gall bladder of the fish capable of splitting any membrane which blinds, which deafens, and which prevents a father from seeing his beloved son (Tob 11:4). Love by its spark inflames the 'eye', makes it luminous, and allows everyone to see, listen and understand as a 'child' what happens in the generating Womb of the Trinitarian Creator Who includes us all.

The illustrations we will use, especially in the last two chapters, are very useful in transmitting the unspeakable. Our interdisciplinary contribution, dealing with the visible, the invisible, and the meta-invisible, needs these kinds of graphic representations.

So, what allows scientists and religious to see and understand each other, as 'waves' or as 'particles', separated then united, always in phase, at the edge of each slot of the two eyes that each has, remains *par excellence* the *yât* of Childhood. Childhood is indeed the best regulator of the relation between 'eye', 'ear', 'light', and 'angles of view' according to the best synergistic efficiency. Would we succeed in providing to all eyes a 'gall bladder', even if *râzâtic*, similar to that which Tobias' son was able to provide to his father under the angel Raphael's direction?

We in turn call on divine Providence to come to our aid because, according to the reality we are living in, perfection is only a Trinitarian phenomenon in the making. It needs a full synergy between divine Fatherhood and Marian Eternal Motherhood within the perpetual birth of Childhood, between the Source of inspiration, the sending and receiving antennas of imagination, and the critical, objective mind of the readers.

Michel Rouhana Ph.D., O.A.M.

PART I

Saint Ephrem

From "Light in the Eye" to "Fire in the Eye"

*

Chapter I

The Light in the 'Eye'

> *We believe in one God, the Father Who controls everything, Creator of heaven and earth, of all that is, seen and unseen; and in one Lord, Jesus Christ, the only begotten Son of God, born of the Father before all ages ... Light from Light. True God from true God.*[4]
>
> Nicene Creed

Jesus once said: "I am the light of the world. Whoever follows me will not walk in darkness, but will have the light of life."[5] With this statement, the Nazarene, the son of Joseph the carpenter, well known to the Pharisees and whose miracles reversed all natural and scientific laws until declaring with authority, "For judgment, I have come into this world so that the blind will see and those who see will become blind,"[6] raises the perpetual challenge to the religious conscience as well as to that of the scientist. While this challenge to the religious conscience represents a theo-anthropological problematic, to the scientific one it represents an anthropo-theological one. However, dealing with this specific problematic is not in the horizon of this book except under the angle of deductions that can serve professional ethical considerations. Scientists, specifically physicists and cosmologists on the one hand and Christian theologians on the other hand, will be invited at the same time to reconsider for their own part their position *vis-à-vis* all abuses done by them to the concept 'God' and, consequently, to all other concepts like 'abortion', 'euthanasia', 'marriage ', 'cloning', etc., belonging to one or the other disciplines.

From these references of the Gospel, we have concluded that the integration between Science and Religion is a matter of light and sight. How can we approach these two paradigms to start the specific reconciliation between them, the two

4 Literal translation, made by the author, of the following Syriac text:
ܡܗܝܡܢܝܢܢ ܒܚܕ ܐܠܗܐ܀ ܐܒܐ ܐܚܝܕ ܟܠ܀ ܥܒܘܕܐ ܕܫܡܝܐ ܘܕܐܪܥܐ܀ ܘܕܟܠܗܝܢ ܐܝܠܝܢ
ܕܡܬܚܙܝܢ ܘܕܠܐ ܡܬܚܙܝܢ܀ ܘܒܚܕ ܡܪܝܐ ܝܫܘܥ ܡܫܝܚܐ܀ ܝܚܝܕܝܐ ܒܪܐ ܕܐܠܗܐ܀ ܗܘ ܕܡܢ ܐܒܐ
ܐܬܝܠܕ ܩܕܡ ܟܠܗܘܢ ܥܠܡܐ܀ ܢܘܗܪܐ ܕܡܢ ܢܘܗܪܐ، ܐܠܗܐ ܫܪܝܪܐ ܕܡܢ ܐܠܗܐ ܫܪܝܪܐ.

5 Jn 8:12.

6 Jn 9:39

wings of the unique human reality? Would the words of Jesus, Son of God, convince a scientist, as the words of Einstein, son of man, have convinced religious theists? What can make this reconciliation possible?

In this chapter, we will start to unveil the secret of the theology of Saint Ephrem of Nisibis, *Doctor Ecclesiae Universalis*, who in his life (306 - 373) faced many conflicts between Philosophy, Science, and Religion, quite similar to ours in the present time. To reach this goal, we are going to uncover the same tools that he himself used seventeen centuries ago, starting with the 'Luminous Eye' (*'aynâ šafyâ* ‎ܥܝܢܐ ܨܦܝܐ), its confrontation with the sense of 'Hearing' (*šama'* ‎ܫܡܥ), and then, the brilliant concept *râzâ* (‎ܪܐܙܐ) he developed from the book of Daniel.[7] However, the confrontation between that specific 'eye' and the sense of hearing will be based on that keyword *râzâ*, which will serve to identify, in the collective mind of the humanity of the third millennium, an environment for a specific space-time, valid for the habitat (*šekinah* שכינה) of the disciplines of both Science and Religion. We may compare this kind of space-time, called in Syriac *Maškanzabnâ*, to the Tent of Meeting of Exodus, where God and Moses used to meet, and where we invite Science and Religion to do the same and maintain their dialogue. The Tent represents space objectively. However, time, without which no space can be imagined since the discovery of general relativity by Einstein, will be represented by the continuous present of the Big Bang for which we may define the start, but of which no one knows the end.

With the hypothesis mentioned above, we wish to unveil the phenomenon of the synergy produced by the correlation between metaphysics and physics. It is inscribed in the typical Ephremian concept *râzâ* which also covers other correlations between 'word' and 'ear', 'light' and 'eye', 'fire', and 'eye', etc. In short, it is a kind of synergy between the divine *Pneuma* (*Ruaḥ* רוח) and the human one. This synergy is necessary for the liberation of humanity from its "inferiority complex" rooted in the oppressive feeling of unfair limitation, slavery, lack of perpetuity, unpredictability, and uncertainty not only toward the temporal future, but specifically toward what comes after death. This long-desired liberation is what has continuously formed, on the one hand, the subject-object of human reasoning, seeking the Truth about their state of being, and on the other hand, hoping to break their limitedness towards the unlimited, towards certitude and predictability, towards what they called the eternal state of being (divinity). That is how Faith became an enigma. Should it be in conflict with Science?

---

7   *Cf.* The Linguistic Preliminary.

The plan of this chapter will cover the following points:

1. The Eye

    1.1. The disappointing 'eye'

        1.1.1. Eye-to-ear confrontation

    1.2. The reassuring eye or the luminous One. (*'aynâ šafyâ* ܥܝܢܐ ܨܦܝܐ)

    1.3. The 'eye' of Mary

        1.3.1. The hearth in the 'eye'

        1.3.2. The foundation stones of the Ephremian bridge between the Book of Nature and the Book of Faith (*kyânâ wa ktâbâ* ܟܝܢܐ ܘ ܟܬܒܐ)

    1.4. Paul's 'eye': the Architect of the Faith (*ardiḵlâ d'haïmonutâ* ܐܪܕܝܟܠܐ ܕܗܝܡܢܘܬܐ)

2. The *râz* (ܪܙ) in the 'eye'

    2.1. *Mysterion*

        2.1.1. The definition of 'mystery'

        2.1.2. 'Mystery' according to its properties and behavior

    2.2. The *râz* (*sic*)

        2.2.1. Definition or description?

        2.2.2.The description of *râzâ* (ܪܙܐ)

3. Synergy

    3.1. Richard Buckminster Fuller: 1+2 = 4!

    3.2. Why did Ephrem use a foreign concept?

        a-Etymology and dimensions of the concept 'synergy'

            a-1: Etymology of the concept

            a-2: Dimensions of the concept

        b- Interdisciplinar synergy at both the linguistic and epistemological levels

        c- Inter-reality synergy (*sic*) and the Christological dimensions of Light and Fire

Conclusion

1. The Eye

"The eye, with which the human being is endowed, is the most essential 'globe' of creation. It is the receiver of almost eighty percent of the information acquired during a lifetime."[8]

The *Encyclopedia Universalis* describes it as follows:

> The eye is the organ of vision; a receiver of luminous phenomena, the eye focuses them in projecting the image on the retina it contains ... No need to underline the physiological importance of the eye. When the destruction of this organ is complete and bilateral, there is blindness; it is one of the heaviest infirmities, and rightly the most dreaded. [9]

Given the importance of this vital organ, the "Architect of the universe" has granted animals, including rational animals, two of them just in case they lose one. All civilizations and religions have given this eye the tribute it deserves, especially by contrasting their state of being between sighted and blind.[10]

Additionally, sight is so important that in the Bible it represented the apogee of anthropomorphism, by ascribing to God an ideal 'eye', all-seeing,[11] that helps Him to have control over everything,[12] to watch always over those who fear Him,[13] and

---

8   Akl, Said. Poet and philosopher. Lecture given at the cultural club, Aïn Saadeh, Lebanon,1973.

9   It is the case of the Blind Man of Jericho whose name is very symbolic: Dust, son of dust (*Ṭima bar Ṭima*). He represents Adam (in Hebrew *Adamah* = dust of the earth).

10   Taok, Boulos. *Annar wan-Nour fil Fikr al calami*; Collection: *Majmoucat Al Wejdaniyat wa Shakhsiyat Gibarn K.*; Dar Nobilis, Beirut, 2000, p.28. From now on we will refer to this book by *Taok*. This author shed light on the importance of the 'eye' in mythologies. He says, "As Man has an eye, the gods have also, and it is the sun: the sun is the eye of Ra of Egypt, Surya of India, Mitra and Varuna of the Veda, the eye of Ahura Mazda of Persia, and he is Helios, the eye of Zeus." Taok continues the description, "... and this divine light is the past and he knows the future ..." In parallel, Jesus taught, saying, "The Lamp of your body is your Eye..." (Lk 11:34; Mt 6:22)

11   2Macc 7:35; 9:5; 12:22; Jb 28:24; Si 15:18.

12   The Creed as cited in Syriac or Arabic describes God not as almighty but more specifically, as the One Who controls everything: *Addabet-el Kull* (Ar); *Aḥid ḳul* (Sy). Recently, with the recognition of the monk Stephan Nehmeh as Blessed, the Catholic Church reinvigorates the Biblical consideration of God Who keeps continuously an eye on human beings: God can see me (*Allah yarani*). *Cf.* http://www.estephannehme. org/home.php?url=akhestephan_grp&lgid=0&depid=4&menu=4&grpid=42; [Accessed May 2020]

13   Ps 33:18.

to reassure His beloved ones.[14] Christ actually praised the 'eye', first by considering Himself as being the 'Light', a medium without which eyes cannot manage to function and reach their ontological fullness, then by referring repeatedly in His sayings to the 'eye', sometimes as a subject,[15] sometimes as an object,[16] as we will explain later.

Saint Ephrem excelled in describing the indescribable dimensions of this 'globe', especially those the 'eye' of the Man of Faith enjoys. He considered the 'eye' to be energized by an exceptional 'power' (*haylâ* ܚܝܠܐ)[17] which makes it self-sufficient in light,[18] to see day and night, as well as to see the visible and the invisible: that is to say, an 'eye' almost the same as the One ascribed to God. For him, this 'globe' and its corollaries, 'light' and 'fire', become the instruments of Science and Knowledge, as well as of Religion and Faith.

Endowed with this 'eye', Saint Ephrem was capable of bringing out of the 'pearl' (*margânitâ* ܡܪܓܢܝܬܐ),[19] used and scrutinized daily by thousands of eyes, many more symbols, information, and dimensions than any other person. By his hymns, he showed a vast, joyful imagination fertilized by the same 'power' that energized his 'eye'. He even managed to use it as a means of relationship with 'the other' (*das Andere*), whoever the 'other' is,[20] as well as with the "Wholly Other" (*das Ganz Andere*), to use the terminology of Rudolf Otto, beyond physics, beyond life.[21] One could say, Ephrem had the "Wholly Other" in his 'eye'.

However, since Christ Himself, while speaking about the "Lamp of the body", which is the 'eye', affirmed the existence of a distinction between 'eyes' ("good eye... whole body full of light..., bad eye ... body full of darkness"),[22] how can this organ, whose reliability is so much discussed, be trusted, given its vulnerability? Can it be valid as a starting point for building a bridge between the intuitions of the Syriac Fathers, represented by Saint Ephrem and Saint Jacob of Sarug,

---

14   Zech 2:12.

15   Lk 11:34.

16   Mk 9:47.

17   *TLE* p. 41.

18   The same analogy as in the Gospel of the Samaritan woman, the living water. (Jn 4:10)

19   Ephrem talks about *margânitâ* (ܡܪܓܢܝܬܐ) which is properly the coral gemstone. However, according to the description he gives of it in his hymns, an obvious deduction made by specialists (Ephremians) is that by this word, Ephrem points to the pearl.

20   *Actes du Colloque XI-Alep 2006*; CERO, 2007, Brock Sebastian, "Saint Éphrem on Women in the Old Testament", p. 35. From now on, we will refer to this book by *Actes du Colloque*.

21   Eliade, Mircea. *Le Sacré et le Profane*, folio essais # 82, Gallimard, France 1965-2008, "Introduction". From now on, we will refer to this book by *Le Sacré et le Profane*.

22   Lk 11:34; Mt 6:22-23.

and those of the Fathers of quantum physics, represented mainly by Max Planck and Albert Einstein? Will this 'eye', objectively speaking, be able to lead us to an effective Tent in which they can meet?

Under the following subtitles, we will continue to discuss, with Ephrem, the 'eye' that, despite its effectiveness, can disappoint and, despite its 'entropy', can reassure.

1.1. The disappointing 'eye'

According to the Bible, specifically the description of the six days of Creation, it turns out that human beings were not aware, from the first day of their existence, of the reality that surrounded them. Consequently, they did not start to archive its information on the shelves of the libraries of Religion, Science, and Philosophy. Humans had first to be aware of their eyes. In the beginning, the only feasible way to reach this self-awareness was to see and to be conscious of seeing. Is it not true that when Adam and Eve had their eyes opened and began to see, the story of God started with them, since before that, like 'happy people',[23] they had no story? It all began when Eve, after having been convinced by the serpent, seduced Adam and brought him to eat. The text says: "Their eyes were opened..."[24] At that moment, our two primitive ancestors had to become aware of the role of their eyes. It was a surprising experience for them as "their eyes became opened to them both, and they knew that they were naked."[25] Ephrem, in his teaching, accentuates the surprise of Adam and Eve against the primordial deception caused by their sight, by saying:

> Thus, their eyes were open and closed:
> Open because they saw everything,
> And closed because they did not see either the tree of life or
> their nakedness.[26]

We join Ephrem in affirming that the receptive eyes of our two ancestors

---

23 This is the stage of the *pre-rational* human being, the embryonic one, who sees nothing, hears nothing and says nothing...

24 Gen 3:7. *Cf.* Feghali, Paul. *Les Origines du Monde et de l'Homme dans l'Œuvre de Saint Éphrem;* Collection Antioche Chrétienne; Cariscript; Paris 1997; p. 179. From now on, we will refer to this work by *Les Origines*.

25 *Ibid.* It would be of great and remarkable importance to emphasize, right away, the parallelism between the opening of the eyes of Adam and Eve and the opening of the eyes of Paul, the apostle of the nations.

26 *Les Origines;* p. 179. *Cf. C Gen.* 2, 22, *CSCO*, vol. 153, p. 29 and vol. 152, pp. 38-39:

ܦܬܝܚܝܢ ܗܘܘ, ܘܚܦܝܠ ܚܙܝܢ ܗܘܘ ܟܠܡܕܡ ܘܚܦܝܢ ܗܘܘ,
ܦܬܝܚܝܢ ܗܘܘ, ܕܠܚܕ ܡܕܡ ܚܙܝܢ ܗܘܘ,
ܘܚܦܝܢ ܗܘܘ, ܕܠܟܠ ܡܕܡ ܚܙܝܢ ܗܘܘ ܟܠ ܚܙܝܢ ܗܘܘ.

were opened not to see that they had become like God, according to the words of the 'serpent', but to face their nakedness, according to the expectation of their opponent. Their eyes were opened to disappointment[27] and proved to be deceiving. Are both the natural and moral eyes misleading or just one of them? This is where the question lies.

Sebastian Brock, one of the pillars of Ephremian studies, reveals in his book *The Luminous Eye* what the real role of the eyes, or at least the role of one of them, is supposed to be:

> It is through the 'eye',
> That the body, with its members,
> Is light in its different parts,
> Is fair in all its conduct,
> Is adorned in all its senses,
> Is glorious in its various limbs.[28]

Here, it is either an optimistic view of the role of the 'eye' or an independent role of one of the two 'eyes', since it turns out from the same sacred text, that this was not the case for the 'eye' of Eve.

Which 'eye' is the one that Brock described, and why did Christ insist on reminding us of a certain ontological characteristic of the eye, emphasizing in Matthew 6:22 its role as a "lamp of the body"?

Rev. Tanios Bou Mansour OML, an expert in Syriac studies and Ephremian scholar by nature, highlighted, according to Ephrem's understanding, the following quality of 'sight'. He wrote that, "At the level of knowledge, the sight is gifted with a corrective role in the face of the interpretations provoked by hearing."[29] It would be entirely true in the case where the 'eye' is the efficient lamp of the body, but the case of Eve with the 'serpent' leaves one in doubt. It can be deduced that the emphasis on the relationship between seeing and hearing, or solely between eye and ear, comes from the brilliant parallelism that Ephrem uses to prove that the human senses alone, even in correlation, are not enough to shed light on the divine truth of salvation (*šrârâ* ܫܪܪܐ) and its certainties (*tuklâne* ܬܘܟܠܢܐ). It is the case

---

27   Disappointment caused by their nakedness is nothing but seeing as it is, at the height of its materiality, the reality in which they found themselves immersed.

28   *TLE*, pp. 71-72..

29   Bou Mansour, Tanios. *La Pensée Symbolique de Saint Ephrem le Syrien* [Bibliothèque de l'Université Saint-Esprit XVI], Kaslik/Liban, 1988; p. 112. From now on, we will refer to this work by *La Pensée*.
     The original French text reads: « Au niveau de la connaissance, le regard se trouve ainsi doté d'un rôle correctif face aux interprétations provoquées par l'écoute. » Translated into English by the author.

with Eve as well as with the doubting apostle Thomas. How did the relationship between seeing and hearing work in their case?

1.1.1. Eye-to-ear confrontation

The human-divine relationality would be the only way to avoid the deception mentioned above.[30] If human sight were sufficient to correct the interpretations provoked by listening, neither Eve nor Thomas would have been mistaken. In this sense, the case of the apostle Thomas is the more significant since he even required touching, thus becoming the prototype of the positivist. However, even before Jesus invited him to do so, something had shaken Thomas. A specific 'eye' darkened by doubt, now illuminated, pushed him to prostrate himself and to transmit to humanity the first creed based on a confident objective certainty: "My Lord and my God."[31] We say that the Grace of the Resurrection must have touched the 'eye' of his heart.

Nevertheless, this objective confession remains debatable for positivists since Thomas had admitted the invisible (*lâ methaziân* ܠܐ ܡܬܚܙܝܢ) into the visible (*methaziân* ܡܬܚܙܝܢ), the non-calculable and the unintelligible (*lâ husbân* ܠܐ ܚܘܫܒܢ) into the calculable and the intelligible (*husbânâ* ܚܘܫܒܢܐ), before even touching Jesus. This is, at least, what can be deduced from the text of the Gospel. By his confession, Thomas proves that the relationship mentioned above is indispensable to grasping the 'Mystery of the Resurrection' and to allow the necessary transmutation at the level of an 'eye'. This "differAnt" (with capital A) 'eye' seems to be the synthesis of all his senses, particularly his sight, his hearing, and his touch, united. Ephrem calls it the 'eye of comprehension' (*caynâ drecyânâ* ܥܝܢܐ ܕܪܥܝܢܐ), an 'eye' endowed with an 'ear', which is believed to be or to become the reassuring "lamp of the body".[32] Under which conditions does this happen?

---

30 'Relationship' highlights more precisely the phenomenon of exchange with correlation. One type of this relationality between human and divine is found in Ephrem's comparison of the 'eye' of Eve with the 'eye' of Mary. *Cf.* (*H Eccl 37, 5*), *CSCO* vol. 198, p. 93; *TLE*, p. 72.

ܗܘ ܥܠܡܐ ܬܪܬܝܢ ܥܝܢܝܢ ܩܒܝܥܢ ܒܗ
Ho colmâ tartên caïnaïn qobcine beh

ܗܘܐ ܗܘܬ ܥܝܢܐ ܣܡܝܬܐ ܕܣܡܐܠܐ
hawâ hwât caïnâ smîtâ dsémâlâ

ܥܝܢܗ ܕܝܡܝܢܐ ܢܗܝܪܬܐ ܡܪܝܡ
caïneh dyaminâ nahirtâ Mariam

Which means, "The world has two eyes fixed in it; Eve was its left 'eye', blind, while the right 'eye', bright, is Mary." (Brock's translation)

31 Jn 20:28.

32 *H Par 1, 4*; *Cf. La Pensée*, p. 115. *Cf. CSCO* vol 174, p. 2.

1.2. The reassuring 'eye' or the Luminous one (*ʿaynâ šafyâ* ܥܝܢܐ ܫܦܝܐ)

Human sight would, therefore, need a different eye, known in Buddhism as the "Third Eye". It should be able to 'see' and 'hear' what is beyond human senses, and to scrutinize what stands beyond Aristotelian physics, beyond the sun and its fire... the depth from which all certainty comes.[33] Ephrem reassures us that God is not like the fire that harms humans when they approach it nor like the sun's light that damages the eyes.[34] That is why, despite the inaccessibility of the 'mystery' of the sun and fire, God is accessible, thanks to the Incarnation. Furthermore, once humans have become conscious of possessing that "differAnt" visual power that puts at their disposition abilities beyond the limits of their profane eyes and their profane visual capacity,[35] as if it were a sacred third eye, they cannot avoid investigating its provenance. Jesus Himself indicated the source of this hidden power by saying to Thomas, "Because you see me, you believe. Happy are those who have not seen and yet believed."[36] Therefore, it is Faith, says Saint Ephrem, that sees the 'keys' of doctrine:

> The keys of doctrine, which unlock all Scripture,
> Have opened up before my eyes the book of Creation,
> The treasure house of the Ark, the crown of Law.
> It is Scripture in its narrative which, above all its companions,
> Has perceived the Creator and transmitted His works.
> Beholding all His handiwork, it has made manifest the objects
> of His craftsmanship.[37]

---

33 *Cf. Encyclopedia Universalis,* Jean Varenne. *s.v.* Karma; Epiphyse; Yoga: "It is during the *dhyana* that the rise of the *Kundalini* and the opening of the *chakras* take place, because these entities, belonging to the subtle body, can be "seen" only by the *buddhi* (or, as they say, by " the "eye of knowledge", the famous "third eye" that one places on the forehead, but which in reality is none other than "the eye of the heart", since the heart is the seat of the *buddhi*." (Translated by Google) From now on, we will refer to this Encyclopedia by *E.U. Cf.* Taok, p. 147.

34 *H Fid* 72, 21-22 ; *La Pensée*, p. 219 ; *Cf. CSCO* vol. 154, p. 222.

35 *Le Sacré et le Profane,* p. 56 (The eye of the Dome).

36 Jn 20:29.

37 *H Par 6,1. Cf. CSCO* vol. 174, p. 19. *Cf. TLE,* p. 45-46.

ܐܩܠܝ̈ܕܐ ܕܝܘܠܦܢܐ	ܦܬܚ̈ܝ ܠܟܠ ܣܦܪ̈ܐ
ܦܬܚܘܗܝ ܩܕܡ ܥܝܢ̈ܝ	ܠܣܦܪܐ ܕܒܪ̈ܝܬܐ
ܓܙܗ ܕܐܪܘܢܐ	ܟܠܝܠܗ ܕܢܡܘܣܐ
ܟܬܒܐ ܕܡܢ ܚܒܪ̈ܘܗܝ	ܒܡܐܡܪܗ
ܐܦ ܚܙܐ ܒܒܪܘܝܐ	ܘܐܫܠܡ ܠܬܘܩ̈ܢܘܗܝ
ܘܚܙܐ ܠܟܠ ܥܒ̈ܕܘܗܝ	ܘܚܘܝ ܠܬܘ̈ܡܢܘܗܝ

Following the logic of our analysis, Ephrem confirms through these verses that, thanks to the unique light of an orthodox Faith indicated by "the keys of doctrine", the human person can acquire the reassuring eye, the luminous one that is the instrument for the correct interpretation of Scripture as well as of Nature.[38] Ephrem emphasizes this power of Faith and qualifies it as an 'invisible eye' as opposed to the visible corporeal eyes.[39] He teaches that joining the divine Light with human eyes helps to identify the Faith with an 'eye' that is capable of seeing the invisible, such as the miracles performed by Jesus, or the promise made to Abram,[40] or the angels who keep watch day and night (*'irae* ܥܝܪܐ) as shown by the following verses:

> Blessed is the person who has acquired
> A luminous eye with which he/she will see
> How much the angels
> Stand in awe of You, Lord ... [41]

Brock, as mentioned above, took this sort of 'eye' as a title for one of his most relevant works in Ephremian spiritual theology. It is an indisputable proof of the importance of this 'instrument' that scrutinizes the inscrutable things of both Scripture and Nature. Even the 'kingdom' whose signs and symbols fill the creation is not beyond the vision of this royal religious 'eye'.

Do we have prototypes of this reassuring 'eye' that highlight its applicability, or should it only be considered an ideal? To meet this challenge of applicability, let us study, in the way that scientists do, two specimens of this 'eye': that of Mary, Mother of Jesus Christ, and that of Paul the Apostle.

1.3. The 'eye' of Mary

At the level of the 'eye' mentioned above, according to Ephrem's appreciation of the divine Motherhood of Mary the relationship between eye/light undergoes a "differAnt" shift (transmutation), and Mary becomes a "differAnt" prototype. What do we mean?

It means that Mary had such a privilege to have perfectly united in her 'eye' sight, the verb "to see", its subject, its object (direct and indirect), and all that

---

38   *TLE*, p. 48-49.

39   *La Pensée* p. 80.

40   *Ibid. Cf. H Eccl* 24, 3 (Faith is the eye which can see hidden things.) and *H Eccl* 38, 2, *CSCO* vol. 198, p. 52 and p. 93; *TLE*, p. 70.

41   *TLE*, p. 73; *Cf. H Fid* 3, 5; *CSCO* vol. 154, p. 8.

ܛܘܒܘܗܝ, ܐܢܫܐ ܕܩܢܐ ܗܘ ܠܗ، ܥܝܢܐ ܕܒܗ ܢܚܙܐ

ܟܡܐ ܕܚܝܠܝܢ ܡܢܟ ܥܝܪܐ

assists its performance, specifically the One to Whom everything refers, her Son, the Light of the world mentioned above. Let us explain further.

In the case of Mary's 'luminous eye', if we want to analyze the sentence "Mary sees her Son", we discover that it has nothing to do with the rules of classical grammar, since by the fact of her Immaculate Conception on the one hand and her Virginal Conception of the divine Word on the other, as well as the predilection between her and her Son, the logical analysis of the sentence becomes quite impossible. In this case there is an interlacement between the subject and the object, *i.e.,* between two very different realities, the divine and the human, without any confusion. This "differAnt" relationality makes it grammatically illogical to distinguish between the subject, the object, and the verb that joins them. The subject, the direct object, as well as the verb, all go into a 'mysterious' union. The level of 'mystery' is the same as that of the Virginal Conception by which Mary's 'ear', her hearing, and the divine Word become united in the verb "listen" (*šamaᶜ* ܫܡܥ), as well as in the imperative answer of the young virgin, *fiat* (*hwâ* ܗܘܐ), "Let it be."

> Illumine with Your teaching
> The voice of the speaker,
> And the ear of the hearer:
> Like the pupil of the eye,
> Let the ears be illuminated,
> For the voice provides the rays of light. [42]

Consequently, it is obvious to deduce from the latter allegory that by Faith the same 'Word' illuminates the 'tympanic membrane of the ear' as He illuminates the 'pupil of the eye'. So, we wonder what would happen when we consider the Word Who became the Son of Mary, the 'pupil' of His mother's 'eye', if not of both her eyes? A mother of Aramaic origin says to her child, "You are my eyes," or even, more precisely, "You are the pupil of my eyes." It is a tradition that is still used in practice today among the inhabitants of the Middle East. Moreover, since the Son of Mary is the Light of the world, He would be *a fortiori* the Light of her eyes.

We indeed have in the Bible some proofs that this custom was in use among the Hebrews since Tobias' time, and this metaphor allows us to build on it an essential part of the symbolism toward which Ephrem had a preference. The Bible recounts that Tobias addressed his son as soon as he had his eyes open

---

42   *H Eccl 37, 1. Cf. TLE,* p. 71. *Cf. CSCO* vol. 182, p. 92.

after the prescription of the angel Raphael, saying, "I see you, my son, the light of my eyes!"[43]

Thus, since the Son of Mary is her 'eyes' and, at the same time, their Light, it becomes evident that she sees through her Son, and that it is by the Light of her Son that she sees her Son, the only Light that allows her to see the divine. This is what Ephrem teaches in one of his famous hymns dedicated to the Light:

> In Your Light, we see the light, O Jesus abounding in
> Light,
> You are the true Light that enlightens creatures.
> Enlighten us with the splendor of your radiance,
> And delight us with the dawn of your morning.[44]

Therefore, this is what made it hard to understand grammatically the sentence "Mary sees her Son." It has nothing to do with the rules of classical grammar.

Symbolically speaking, the French philosopher Derrida would say that the space and time separating the sign, the signified, and what is referred to is reduced to the maximum extent, and in some cases is abolished.[45]

It is from this approach that we have drawn the title "The Light in the Eye", while being reassured by Ephrem that this Light does not blind. However, the discernment of this transmutation at the level of Mary's 'eye' was not in the reach of Ephrem without the astonishing love that inflamed his heart for her as the Mother of God.[46] This idea of 'inflaming Love' leads us to another causative cause of transmutation, which is the very Source that generates the Light of Mary's eyes, namely the divine Fire. We will later come back to this metaphor as being an indispensable instrument for Ephremian theology. For now, let us consider that

---

43   Tob 11:14. Note that it is the only time in the Bible that the same verse joins son, light, and eyes together.

44   Very popular Maronite Syriac hymn traditionally attributed to Saint Ephrem.

> *Bnuhrâk ḥozenan nuhrâ Yeshuᶜ mlê nuhrâ*
> *Dat-hu Nuhrâ sharirâ dmanhar beryâtâ*
> *Anhar lan bnuhrâk gayâ*
> *w ḥadâ lan bdenḥeh dsafrâk .*

45   An example would be that of "Love" (to believe the hero of the *Song of Songs*). "Set me as a seal upon your heart." *Cf.* Song 8:6. This gives pause to reflect on the distance between the seal and the heart.

46   "My bones shout from the grave that Mary is the *generatrix* of God."
Cf. http://www.aramaic-dem.org/English/History/Mor Ephrem is an inspiration of our time.pdf; [Accessed Jun 2019] *Cf.* also, Feghali, *Mariam El cAdra wal Ephḳaristia*, where he mentions this saying in Arabic.

since, from Ephrem's time, there can be no Light without Fire, the latter should be of the same nature as the generated Light. Where can such Fire be found? What is its hearth?

1.3.1. The hearth in the 'eye'

Ephrem, as a 'mystic' whose 'luminous eye' scrutinizes the inscrutable, actually saw the 'reality' he was preaching as an indivisible whole. It did not take him long to solve this enigma, since for him it is from the womb of Mary that the Light is born. Therefore, it is in her womb that the Fire must be found: "The Fire entered Mary's womb, put on a body, and came forth," he wrote.[47] Moreover, it is in the 'womb' of the 'eye' of Mary that Ephrem discovered this truth and sang:

> The world you see
> Has two eyes fixed in it:
> Eve was its left eye, blind,
> While the right eye, bright [the Luminous one], Mary.[48]

Here, we notice that Ephrem no longer speaks of the 'eye' of Mary, but of Mary herself, who has become the sacred 'eye' opposed to the profane one of Eve, an eye simultaneously luminous and a supplier of light like an incandescent lamp, like a hearth or a bread-oven (*Beth-lehem*).[49] Vatican II, based on Pope Saint John XXIII's encyclical *Mater et Magistra,* ascribed to that eye every kind of 'eye' needed for the salvation of souls and of the universe: the 'eye' of a *Mother and of a Magistra,* crowned with her 'eye' of *Caritas,* just to mention these three 'eyes'. That is why Ephrem introduced into his lyrical vocabulary all the 'eyes' that he needed for his catechetical and pastoral purposes. However, he never hid that he wanted to announce just one 'eye' that he wished every believer to receive, the 'eye' of Mary.

---

47   *La Pensée,* p. 81; *TLE,* p. 38; *Cf. H Fid* 4, 2- end; *CSCO* vol. 154, p. 9: The author of *La Pensée* analyzes the symbol 'womb' and informs us that, "In relation to this polysemy of light, centered on divinity and faith, the polysemy of the 'womb' (*ᶜubâ*) is of paramount importance. It is so because of its capacity to signify and link together different stages of Salvation history." Furthermore, he adds, "The same goes for the 'womb' of Baptism which, in the *Hymns on Epiphany* … represents the place where one receives the garment of the only-begotten Son, King of the heavenly Kingdom." (*H Epiph* 9, 2; 13, 14) Concerning our input, the 'womb' (*ᶜubâ*) is the only space where every vital chiasm takes place and from which all Christological globalizing synergy sets off to form the *râzâtic* reality. In this lies its primordial importance.

48   *TLE,* p. 72; *Cf. H Eccl* 37, 5, *CSCO* vol. 198, p. 93.

49   **Beth-lehem**, written in Syriac ܒܝܬ ܠܚܡ (*Bêt lḥêm*) means literally the "house of bread", that is the bake house .

It is through this 'eye' that everything must be seen, and nothing must be seen apart from it. Otherwise, any *hierophany* would be absurd: "The holy sacraments (*qudšê* ܩܘܕ̈ܫܐ) are given to holy persons (*qadišê* ܩܕ̈ܝܫܐ)," sings the Maronite Liturgy just before the Communion.[50]

Furthermore, according to the reference made to "the Lamp of the body" mentioned above, the one who possesses the 'eye' also possesses the corresponding thought. Saint Paul confirms this truth by writing, "And we have it, the thought of Christ."[51] Finally, thanks to the following verses of the *Hymn to the Light*, we grasp the span of Ephrem's adherence to the Mother of Christ, the indescribable Hearth of the divine Fire, and to the Son of Mary who shines from His Mother's midst as He shines from the divine midst of the Trinity:

> Here is the Light leaving His 'milieu' and coming toward ours;
> Humans will never know the location of His 'milieu':
> To Him be glory and to the One Who sends Him,
> For it is in the Light (that) His glory shines. [52]

It is a hymn of which Ephrem and his parishioners have been proud in the face of the surrounding civilizations which worshiped the sun and fire or any other creature of this kind. However, the fundamental enigma raised by this hymn is the location of that divine 'milieu'. We should resolve this enigma to facilitate the meeting of the 'eyes' of scientists with those of theologians. Did Ephrem not mention that humans would never be able to know where it is situated, *"waïnao atreh bnaïnâšâ memtum lâ yâd^cin"* (line 2 above)? Therefore, a scientific effort, more precisely a certain *cosmography* or *'ophthalmography'*, based in both cases on a specific optical physics, would be necessary to try to locate that ontological source of Light. We will also have to figure out its path and its point of interception,

---

50 *Cf. Maronite Missal (*Arabic), Bkerke, Lebanon, 2005, p. 742.

51 1 Cor 2:16.

52 *Šḥimtâ Morunâytâ* (Maronite Breviary), Imprimerie Catholique, Beyrouth, 1981, Monday morning prayer, 3rd *Mimrâ*, p. 124. (It may be that this hymn is not written by Ephrem, but tradition does attribute it to him.)

ܗܐ ܢܘܗܪܐ ܢܦܩ ܡܢ ܐܬܪܗ ܘܐܬܐ ܠܐܬܪܢ
Ho nuhrâ nfaq men atreh wetâ latran

ܘܐܝܢܘ ܐܬܪܗ ܒܒܢܝ̈ܢܫܐ ܡܡܬܘܡ ܠܐ ܝܕ̈ܥܝܢ
Waïnao atreh bnaïnâšâ memtum lâ yâd^cin

ܫܘܒܚܐ ܠܗ ܘܠܫܠܘܚܗ, ܕܒܢܘܗܪܐ ܫܪܐ ܐܝܩܪܗ
Šubḥâ leh wal šâluḥeh, dabnuhrâ šrê Iqâreh

hoping to be able to describe its behavior in our world. More tellingly, we will have to imagine its relative 'location' in the space-time of our cosmos, as well as in the corresponding 'wombs' of the human eyes. It would be a similar experience to the eyes of the Baptist's disciples looking for Jesus' dwelling place.[53] After all, if it is not at the very Source of Light that Science and Religion shall have their rendezvous, where will it be?[54]

1.3.2. The foundation stones of the Ephremian bridge between the Book of Nature and the Book of Faith (*kyânâ wa ktâbâ* ܟܬܒܐ ܘ ܟܝܢܐ)

The heretics who surrounded Ephrem and his parishioners worshiped creatures, precisely the sun and fire that were in reach of their senses. They refused the Gospel that would awaken them to the "differAnt" reality described by these various signs and symbols. They had to recognize that these beings should have sacred value only if they referred to a Creator of all that is visible and invisible, and therefore are pure and straightforward hierophanies.

Ephrem, as a man of Faith and Reason, was given charge of the theological and catechetical formation of the seminarians as well as other faithful in his diocese by his bishop and was obliged to defend the orthodoxy of his faith against the heresies that compromised it.[55] He was in charge of refuting the different false truths of other religious currents, especially those of the Gnostics, which are echoed in today's postmodern religious currents. At the same time, after St. Paul's model, Ephrem felt the duty to evangelize and win everyone over to his Lord and God. He wanted to convince every person that Christ, this God and Savior, this Light of the world that he preached, resides within everyone's eye, even if it is asleep. Therefore, to make all persons discover their own 'luminous eye' in the reflection of their own corresponding 'mirror', was for him to bring those persons out of their slumber or their laziness and to initiate them in the discovery of the proposed 'pearl' which is the one of *râzâtic* certainty and

---

53 The Gospel is clear on this point. Once His mission started, Jesus had a strange address for His lodging. According to Jn 1:38-39, the two disciples of the Baptist, at the invitation of Jesus, "came and saw where He dwelt, and abode with him that day." The Gospel does not say where. Further on, Jesus will give a strange indication of His dwelling place: "The foxes have dens and the birds of the sky have nests; the Son of man has no place to rest his head." (Mt 8:19-22) According to these indications, we too ask ourselves the question, "What kind of GPS could indicate to us the place where the 'Light' of the world dwells?"

54 Let us imagine a sea or a continent of light rather than water and earth: what would be its reef or shore?

55 *Les Origines,* pp. 18-19.

predictability.[56]

Whatever the advantage or the disadvantage to humanity of enjoying the reassuring eye or of suffering from the disappointing one, it remains that for Ephrem any spiritual awakening requires effort. Likewise, it needs more self-consciousness and in-depth introspection regarding what is visible as well as what is invisible, especially in what relates to the absolute Truth of the religious dimensions of the soul. For him, any laziness in this area is fatal.

Consequently, to be able to join what Mary has seen, or to be satisfied with just what the eyes of the disciples John or Paul or even Ephrem's eyes have seen, and feel ready "to sell everything to buy the field where the treasure is hidden",[57] one needs the 'luminous eye', the 'eye' of faith and love:

> You are entirely a source of amazement,
> From whatever side we may seek you:
> You are close at hand, yet distant –
> Who shall reach You?
>
> Searching is quite unable to extend its reach to You:
> When it is fully extended trying to attain to You,
> Then it is cut off and stops short,
> Being too short to reach Your mountain.
> But Faith gets there and so does love, with prayer.[58]

These poetic words of Ephrem send us back to the scene of Creation to excuse the fall of Eve who had not yet known either love or faith, or non-faith because her 'eyes' had not yet been 'open'. In short, she was not yet ready for the adventure of self-consciousness, let alone to be conscious of the other, and less of the "Wholly Other".

Indeed, returning to the disappointment of Adam and Eve who found themselves naked, the Bible highlights the driving force behind Eve's decision to eat the forbidden fruit: "When the woman saw that the fruit of the tree was good to eat and pleasant to the eye, also desirable to gain wisdom, she took some and ate it."[59] At that moment, Eve could not yet discern between the self and the 'other' (*das Andere*), or even the "Wholly Other" (*das Ganz Andere*), and consequently,

---

56   It is a relational expression from the domain of quantum physics, to which we will return later.

57   Mt 13:44.

58   *TLE*, p. 70; *H Fid 4, 11*; *Cf. CSCO* vol. 154, p. 13: "Faith is therefore the eye able to see all that is hidden and it must be, in turn, accompanied by love and prayer."

59   Gen 3:6.

for her there was no difference between love and non-love. How would she feel loved if she did not discern the difference between the self and the other? The interpretation inspired by the 'Ephremian lyre' says that Eve ate the fruit of the Tree with her eyes first, before swallowing the 'hook'.[60] Saint Ephrem underlines this, saying:

> It is clear that Mary
> Is the 'land' that receives the Source of light;
> Through her, it has illumined
> The whole world, with its inhabitants,
> Which had grown dark through Eve,
> The source of all evils.
> Mary and Eve in their symbols
> Resemble a body, one of whose eyes
> Is blind and darkened,
> While the other
> Is clear and bright,
> Providing light for the whole.[61]

Why then blame Eve if she yearned for Mary's 'eye'?

Let us compare her situation with that of Moses who decided, pathetically, to save his people from slavery, before being ordered by God to do so; consequently, he killed the Egyptian soldier. Eve fell in the same mis-evaluation of the divine plan and coveted knowledge long before its timing in the Economy of Salvation, without any invitation from the Lord of all Times (*Chronos*) and all 'eyes'.[62] For this misbehavior, she can be blamed.

Then, by his astute parallelism, similar to Platonic maieutics, Ephrem urges his interlocutors to deduce, by themselves, what he wants them to understand from the Truth. He indeed supports the biblical idea that Eve is the cause of all evils; however, at the same time, he excuses her, considering her as "the blind and darkened eye", eager for sight and light. The problematic lies in this issue.

Moreover, what did the serpent suggest to her presumptuously? Is it not the

---

60   In his Arabic article entitled "Eucharist and the Presence of Christ", Feghali recalls an ancient Maronite tradition that symbolizes the Holy Communion through the eyes before it is taken by mouth. Feghali writes what this means: "Communion was taken in the hands ... The faithful took the Host and made it touch his/her eyes before eating it." (Translated by the author). *Cf.* https://boulosfeghali.org/2017/frontend/web/index.php?r=site/text&TextID=3325&CatID=368&SectionID=41 [Accessed June 2019]

61   *TLE*, pp. 71- 72; *H Eccl* 37, 3-4; *Cf. CSCO* vol. 182, p. 92.

62   Ex 2:11.

'light' of the knowledge of good and evil and the capacity to grasp it? Why, then, blame her if she thought she was 'eating' the eagerly desired 'light' that would allow her to 'see'?[63]

By that, Ephrem refutes all those who have accused him of misogyny, because if that were true, he could never have gathered around him the Daughters of the Covenant (*Bnât Qiâmâ* ܒܢܳܬ ܩܝܳܡܳܐ) in confraternities and choirs. In the most modern sense, this activity triggered what we call the movement of women's emancipation. Such confidence helped them to regain their true human dignity. Besides, far from an attack to dishonor the human Mother or even the women of the Old Testament, Ephrem intended to make it clear to female parishioners that the 'eye' could also be a woe-bearer. It also has this possibility since, for the knowledge of oneself and of the 'other', it is free to resort to different sources of 'light' other than that of Jesus Christ.[64] By this contrast, he brings into relief as an excellent pedagogue what the Church teaches, and intimates at the same time to young girls and women, who are by nature the most sensitive to the attacks of heresies, to be wise and not to follow the 'inclinations' of their eyes.[65]

By lifting the two female prototypes in front of the *Bnât Qiâmâ*, Ephrem lays the first two foundation stones of the 'bridge' between Reason and Faith, *i.e.,* between the two 'books' mentioned above, that of Nature and that of the Scriptures. He invites the woman to make good use of her freedom and to assume her role in the Church just as the man does, her 'brother' in Baptism.[66] He invites her to acquire the 'eye' of Mary, her 'Mother' in Salvation and her 'Sister' in Baptism, to fully assume her responsibility in the divine Salvation process (*Mdabrânûtâ* ܡܕܒܪܢܘܬܐ). It was sufficient, for Ephrem, while rehabilitating the women of the Old Testament in the eyes of the female parishioners, to remind the latter of the words of Christ: "And if

---

63   This case is worth comparing with that of Moses killing the Egyptian controller. We say that Ephrem should have accused Eve of only what Moses was accused of, and that is to have started a "mission" regardless of God's will. The symbolism here teaches that the human initiative, no matter how right and good it may appear, remains without roots in the divine Economy and, consequently, fails. Being a disordered activity, an untempered intervention in the affairs of the "Wholly Other" (God), it constitutes a sin for its author. *Cf.* Paul Féghali, *Al Muḥit al Jameᶜ fil kitab al mukaddas wa-shark al kadim, Al Maktabat Al Bulussiah wa Jamciat Al Kitab Al Mukaddas,* First edition, 2003, s.v. "Exodus". From now on, we will refer to this dictionary by *Al Muḥit al Jameᶜ.*

64   *Actes du Colloque*, Introduction, p. 2.

65   The story of the challenge between Ephrem's eyes and those of the woman of Edessa that Brock mentioned at the beginning of his article, "Saint Ephrem on Women in the Old Testament", is in this context very significant in determining the inclinations of the eyes and the equality of error between man and woman.

66   *Actes du Colloque*, p.43.

your eye is for you an occasion of sin, take it off and throw it away, far from you...".[67] This advice became more acceptable, more applicable, especially as these girls and women made sure that the substitute proposed to their eyes is the 'eye' of the Mother of God, a substitute so honorable and precious compared to their profane eyes.

So, needless to say, as soon as Christ introduced Himself to us through the 'womb' of the 'eye' of his Mother as the Light of the world, the themes of the 'eye' and the 'sight' have become two primordial foundation stones for all that is self-consciousness, relationality, and freedom, for humanity and the Church alike.[68] Consequently, to become capable of saying to Jesus, like Mary, "You are my eyes" or "my eyes and their light at the same time", is the answer to the challenge raised above. The challenge is not insurmountable.

What was Saint Paul's experience with these two foundation stones and what did he transmit to those who considered that Mary, Blessed Among all Women, was an exception?

1.4. Paul's 'Eye', the Architect of the Faith (*ardiḵlâ d'haïmonutâ* ܐܪܕܝܟܠܐ ܕܗܝܡܢܘܬܐ)

Once we consider Mary, as Immaculate, an exception, no one is better able than the Apostle of the Nations, Paul, to witness to the applicability of these two foundation stones.[69] We must ask his eyes, which were baptized by the Fire and the Spirit on his way to Damascus. His eyes, which by force of the reductionist behavior inflicted on his mind by the rigidity of the Law and the Pharisaic teachings, needed an *ophthalmo-theological* intervention to acquire the discernment Jesus spoke of after His healing of the man born blind:

> Jesus heard that they had driven him out, and when he
> found him, he said, "Do you believe in the Son of Man?"
> He answered, "And who is he, sir? Tell me, so that I may
> believe in him." Jesus said to him, "You have seen him,

---

67   Mt 18:9 ; Mk 9:47.

68   Ephrem distinguishs between what God has created by Sign *"remzâ"*, by Word *"meltâ"*, and by Command *"fuqdonâ"*. (*Cf. Les Origines*. p. 55). If the first "being" (*Itutâ*) that God created is Light, this means that God had foreseen eyes for the human being He was planning to create. It would have been on purpose to underline the relation of eye – light (ܚܝܢܐ ܕܢܗܝܪܐ). This third verse of the Book of Genesis leads to such symbolism. We would say that the synthesis of all the religious dimensions with which all other religions were inspired resides in the relation between eye – light, the physical cognitive visibility. Moses would have learned all this as one of Pharaoh's sons.

69   Is that not why he begged Jesus to free him from his bodily discomfort? (Rom 7:24) Only Mary, thanks to her predilection and her umbilical union with her divine Son, could overcome this impediment.

and the one speaking with you is he." He said, "Lord, I
believe." And he worshiped him. Jesus said, "I came into
this world for judgment so that those who do not see may
see, and those who do see may become blind." [70]

What a parallel between the two miracles. It highlights the two foundation
stones of the bridge that exist in the same 'eye', the darkened 'eye' and the luminous
one! Paul was no longer concerned about protecting his life, his gender, or even
his Jewish religion, *i.e.,* his way of seeing and interpreting God. His concern has
become only to preserve the phlogistic Light[71] that "violently charmed" him and
pushed him to step beyond traditional expectations. We say "violent", because of
the Light that blinded him, and at the same time "charming", because the same
Light entitled him to the divine vision and allowed him to access the third heaven.
However, the chiasmus[72] between 'light' and 'eyes', considered by Rudolf Otto
as *"Mysterium Tremendum"*, is but a description of the falling into the hands
of the living 'God'.[73] Otto also qualifies it as a fascinating mystery, *"Mysterium
Fascinans"*, and all this is at the level of the *Numen*.[74] It seems that the phenomenon
of "charming violence" or "violent charm" goes hand in hand with the 'Mystery
of Salvation' as this 'Mystery' deals with divine-human fertility and vice versa.[75]
Otto similarly states that the same chiasmus establishes the foundations of what is
known as the 'Pauline paradox': "... Where suffering abounds, glory overflows..."[76]
Ephrem affirms this after his meditation of the famous 'pearl' pierced by humans
to become a gem of higher value:

> In your beauty, the Son's beauty is depicted:
> The Son who clothed Himself in Suffering,
> Nails went through Him.
> Through you the awl passed; you too did they pierce,
> As they did His hands. But because He suffered, He reigns,

---

70   Jn 9:35-39.

71   From the Greek masculine name *Phlox* (*Φλοξ*) which properly means "flame". We use
this term, although it is considered archaic, to designate the nature of divine fire or of
the primordial one. This term, more than being strong and incomparable, is symbolic,
par excellence. For more resources about phlogistic fluid, read about Nephthar
(*νεφθαρ*) or Naphtha (*νεφθαι*) in 2Macc 1:36.

72   Chiasmus, here, as throughout the text, is taken in the symbolic sense. *Cf. E.U., s.v.*
"Chiasme, symbolisme".

73   2 Cor 12:2-10.

74   *Cf. Le Sacré et le Profane,* Introduction.

75   *Cf.* René Girard, *La Violence et le Sacré*, Pluriel NO 897; Hachette, July 2011:p. 54..

76   2 Cor 4:15-16.

Just as your beauty increased through your suffering.[77]

Furthermore, Ephrem accentuates this paradox by accusing Peter of not being aware of this truth when he tried to save Christ from suffering by defending him with the sword. Peter did not realize that the greatness of glory should be at the level of suffering. From this point of view, the Cross remains for Ephrem an instrument of victory and glory, and this is where lies, in spite of the annihilation and the infamous death on the Cross, its primary and primordial significance.

We even find mentioned in the New Testament witnesses who have experienced this observation which links suffering with glory. Below is a brief list:

- Zechariah's reduction to silence; Elizabeth's old-age pregnancy → birth of the Heir;
- Mary's virgin maternity and accusation of adultery → divine Maternity;
- John the Baptist's predilection from the womb, harshness of the desert and martyrdom → the Precursor;
- Jesus Christ Himself: kissing the Cross and dying → kissing a Mother, Resurrecting from death, Savior, etc.

The case of Mary touched Ephrem the most and opened his eyes to the unspeakable value of the Incarnation. [78]

Then, a new 'life' was fertilized in Paul's eyes, and a new mission was entrusted

77 *H Fid* 82, 12; *Cf.* Brock, Sebastian. *The Harp of The Spirit, Poems of Saint Ephrem the Syrian,* 3rd enlarged edition, Aquila Books UK, 2013, p. 45*; Cf. CSCO* vol. 154, pp. 253-254. *Cf. La Pensée*, p. 280.

78 A stanza of *Bo^cuthâ dmor Ya^cqoub, Morning Prayer, Sunday of the Dead, Book of Tešmeštâ* (manuscript), describes, according to Jacob of Sarug, the moment of the divine decision to create the world as follows:

> The Father has taken in His holy hands the dust prepared for the creation of Adam. He called the Son and said to Him, "Behold, he whoever on the cross will lift you up, and will mock You; behold, he who, in the tomb, will lead You and by him, You will be humiliated and despised. If You admit it, I create him. If You do not, I do not create him." The Son said to Him, "Create him, for I will carry him through Mary, and will endure sufferings, and will save the world."

Although this hymn is found in the Sarug heritage, we are adamant in linking it to the Ephremian Marian spirituality. We reproduce this paragraph as it is in the manuscript to be at the service of whoever is interested in the Syriac language.

ܫܩܠ ܐܒܐ ܥܦܪܗ ܕܐܕܡ ܒܠ ܐܝܕܘܗܝ ܩܕܝܫܬܐ ✤ ܘܩܪܐ ܗܘܐ ܠܒܪܗ ܘܗܟܢ ܐܡܪ ܠܗ ✤ ܗܐ ܗܢܐ
ܠܨܠܝܒܐ ܡܪܝܡ ܠܟ ܘܡܒܙܚ ܒܟ ܡܬܒܙܚܢܘܬܐ ✤ ܗܢܐ ܠܩܒܪܐ ܡܚܬ ܠܟ ܘܡܒܙܚ ܒܟ ܡܨܛܥܪܢܘܬܐ ✤ ܐܢ ܨܒܝܬ
ܒܪܐ ܐܢܐ ܠܗ ✤ ܘܐܢ ܠܐ ܨܒܝܬ ܠܐ ܒܪܐ ܐܢܐ ܠܗ ✤ ܘܐܡܪ ܠܗ ܒܪܐ: ܒܪܝܗܝ ܡܛܠ ܕܡܢ ܡܪܝܡ ܛܥܢ
ܐܢܐ ܠܗ ✤ ܘܣܒܠ ܐܢܐ ܚܫܐ ܘܦܪܩ ܐܢܐ ܠܥܠܡܐ ✤

to him. In an instant, his 'eyes' became luminous (*šafyâ* ܨ݉ܦܝܐ) similar to Mary's, a 'hearth' for all the signs, symbols and properties that shifted under the same Power that transmuted Saul's 'eye' into Paul's. Therefore, all the interpretations undergo the same conversion and, by this kind of 'hearth-eye', – let us suppose it nuclear, thanks to the catalyzing Fire (Energy) which resides in it, – all the signs and the symbols, His name included, will be purified, baptized in the Eternal Spirit, and transformed into One and Only 'Mystery', the One named Christ. Thereby, all Paul's assertions describing a certain duality in his person become plausible.[79] However, what did this cost him?

Is it at the cost of a numinous experience that leads to a psychological paradox in the one who assumes the responsibility of a 'revelation' that presents itself, at the same time, in the form of a "frightful attraction" (*fascinans*) and a "terrifying impulse" (*tremendum*)?[80] Or is it at the cost of losing all self-control, to the degree of saying, "I no longer live, but it is Christ who lives in me"?[81] Thus, we have joined, with Paul, the point where the applicability of the 'Luminous Eye' culminates, and where the *tremendum* balances the *fascinans*, just as it occurred with Mary, between the "how" (*Quomodo*) and the "let it be" (*Fiat*). It is necessary to elucidate next the enigma of the concept 'mystery' which Ephrem calls *râzâ* (ܪܐܙܐ) in Syriac, a concept which will play the indispensable role of the third foundation stone for the 'bridge' between the Scriptures and Nature. The 'eye', 'light' and *râzâ*, whose plural is *râzê* (ܪܐܙܐ), will ensure our adventure with the correlation between the divine and the human, with the synergy that stands behind it, as well as with the rapprochement between the two conceptions of creation, the *Ex Nihilo* one of the Bible and the Big Bang of quantum physics.

2. The *râz* (ܪܐܙ) in the 'eye'

Is having the 'mystery' in the eye not the same as having the *râz* in the eye? The term *râzâ* appears for the first time in the Bible in the Book of Daniel (2:18), a verse from a book of conflicting canonicity between the Catholic Church and Hebrew authority. This part of the Book of Daniel was initially written in Aramaic rather than in Hebrew, and the word *râzâ* found there is of Persian origin. The original text as it appears in the Syriac *Pešita*:

ܡܢܒ ܪܐܙܐ ܕܢܝܐܠ ܠܒܝܬܗ ܐܙܠ.
Hoyden Dâniel lbayteh ézal.

---

79 2 Cor 4:16.

80 Heb 10:31. N.B. Paul's words join two realities not of the same realm: one superior and one inferior.

81 Gal 2:20.

ܘܠܚܢܢܝܐ ܘܠܡܝܫܐܝܠ ܘܠܥܙܪܝܐ ܚܒܪܘܗܝ.

wal Ḥananyâ wal Mišâyel wal ᶜAzaryâ ḥabroy

ܡܠܬܐ ܗܕܐ ܓܠܐ: ܕܢܒܥܘܢ ܪܚܡܐ ܡܢ ܩܕܡ ܐܠܗܐ ܕܫܡܝܐ

meltâ hodé glâ: dnebᶜoun raḥmé men qdom Alâhâ dašmayâ

ܥܠ ܪܐܙܐ ܗܢܐ ܕܠܐ ܢܐܒܕܘܢ ܕܢܝܐܝܠ ܘܚܒܪܘܗܝ

ᶜal râzâ hono dlâ nébdun Dâniel wa ḥabrao

ܥܡ ܫܪܟܐ ܕܚܟܝܡܐ ܕܒܒܠ.

ᶜam šarko dḥakimê d-Bâbêl.

ܗܝܕܝܢ ܠܕܢܝܐܝܠ ܒܚܙܘܐ ܕܠܠܝܐ

Hoyden l-Dâniêl bḥêzwâ dlilyâ

ܪܐܙܐ ܐܬܓܠܝ. ܘܕܢܝܐܝܠ ܒܪܟ ܠܐܠܗܐ ܕܫܡܝܐ.[82]

râzâ etgli. W Dâniêl barêk l'Alâhâ dašmayâ.[82]

Which is translated:

> Then Daniel went to his home and informed his
> companions, Hananiah, Mišael, and Azariah, and told
> them to seek mercy from the God of heaven concerning
> this 'mystery', so that Daniel and his companions with
> the rest of the wise men of Babylon might not perish.
> Then the 'mystery' [emphasis mine] was revealed to
> Daniel in a vision of the night, and Daniel blessed the
> God of heaven.[83]

To answer the comparative question raised above, we will try to unveil the variations between the Greek *mysterion* and the Persian–Aramaic *râzâ*. It is significant for our subject. We will do this first by discussing the term 'mystery', then *râzâ*, and finally the particular synergy that accompanies them and allows no reductionism.

Before we start, we insist on saying that reproducing the text of Daniel as it is in its original Aramaic draws us closer to the 'Force' (*Ḥaylâ* ܚܝܠܐ) which inspired it and which stays hidden at all times behind its letters. It is a biblical matter of language and conviction.

Moreover, this scene helps us to see the words drawing a road map that leads

---

82   Dan 2:17-19; *Syriac Bible 63DC; United Bible Societies* 1979; UBSEBF 1987-3M; ISBN 564 03212 3; p.692.

83   We find some English translations using "mystery", such as NRSVCE, and others using "secret", such as KJV and ASV. This is but a sign of diverse understandings of the term "*râzâ*".

to a specific destination beyond the horizons of the sages, magi, and the seers of Babel. The text underlines it clearly through the lips of the chief of the Babylonian religious ranks who apologized to King Nebuchadnezzar saying:

> What you demand, O king, is too difficult; there is no one who can tell it to the king except the gods who do not dwell among men.[84]

The biblical text, as it stands, points to the abode of the gods. It is a three-dimensional 'milieu' vital to the development of our research comparing 'mystery' and *râzâ*, because the latter also resides in that 'milieu'. Let us start to unveil the meaning of 'mystery' according to its origin from the Greek *mysterion*.

2.1. *Mysterion*

We will consider this point under two headings:
    1. The definition of 'mystery'
    2. 'Mystery' according to its properties and behavior

2.1.1. The definition of 'mystery'

Even before the Incarnation takes place, a mystery in the universal sacred sense, and not in a particular one, can be deduced from the same scene that took place in 'Babel': it is a kind of information beyond the intellectual capacity of a human being, something difficult or impossible to understand or explain. It turns out to be a concept that contains its verb, subject, and both direct and indirect objects that form together a specific manifestation of what is considered sacred, *i.e.*, a hierophany.

Before moving forward in the 'dissection' of this concept, let us look at its encyclopedic coordinates.

Two facets can be discovered here. One is from the mythological world, and the other from the Christian one, where the Greek word *mysterion* as used by Paul of Tarsus seems to have reached its apogee.

On the one hand, regarding the mythological facets, *mysterion* originates in the esoteric religions of antiquity. We will not focus for the moment on the origin of the term in those religions, particularly the mythological religions in the lands of the Bible, but we will refer to an encyclopedic sentence which is always valid: "To speak of a mystery is to profane it, in other words, to destroy it".[85] Therefore, the fundamental rule of the ancient esoteric religions that flourished in the

---

84   *Ibid.* Dan 2:11.
85   *E.U. s.v.* "Mystère". Article written by Édouard Jauneau.

Mediterranean world was "silence". The initiated ones kept it well.

However, that sacred silence would have been of minimal value if it had not contrasted with the event of the Word of God who became Man to unveil what was for a long time kept hidden (*kasyâ* ܟܤܝܐ) in silence. The distinction between a 'mystery' understood as something secret and a 'mystery' understood as a *râzâ* begins to take shape without ever reaching a total and definitive divergence. The first 'mystery' invites one to keep silent, while the other, invites one to speak. This present paradox reaches its climax in the case of the annunciation to Zechariah: to stay silent so that silence speaks. Not just any silence, but the silence of the sterile uterus that becomes pregnant.[86]

Consequently, if a unique, clear and precise definition of mystery cannot be deduced, it is because of the wide range of uses that it assumed, especially inside the Scriptures, because of the continual change of places, gods, and situations endured by the Jewish people. Similarly, we have to consider the phenomenon of 'Time' and 'Process' through which the Economy of the Redemption of Humans unfolds following successive prophecies and hierophanies. By that phenomenon, the concept of 'mystery' (*sôd* צד) grew in dimensions and depth. It seems that the symbols too were deaf-mute and sought their meaning in the imagination of the Jewish people. Therefore, we would say that *mysterion* was a concept in progress until the Annunciation to Mary, Zechariah being the last of the confused ones. It was not the symbol, John the Baptist, who gave meaning to Jesus. It was instead Jesus, the "*Râz*", according to Saint Ephrem, Who gave all meaning to the existence of the Baptist three months before his birth, and that through the "luminous eye" given to Elizabeth, his mother.[87]

Ephrem seems more than satisfied to have grasped the confusion that reigned in Nisibis between hierophanic mystery, gnostic mystery, and other similar ones. The problem was essentially etymological, directly related to the range of verb, subject, direct and indirect objects, and precisely to the syntax that should govern their relationship and their behavior. In the fullness of time, the Incarnate Word that baptized the Alphabet and Language *in se* will henceforth require a specific partnership with this range and its syntax. The words, terms, concepts, and signs in general, with the different syntax that governs them, will have to become 'luminous' in order to serve *Christic* hierophanies.[88] It is no longer just any word or grammar that

---

86   Lk 1:39-45.

87   Mk 1:7 and Lk 7:27, and especially Mt 3:14-15.

88   *E.U. s.v.* "Mystère". Article written by Édouard Jauneau. "The term mystery was, in a way, baptized by Saint Paul, and starting from this fact, it imposes itself on Christian

can serve the cause of Ephrem, whatever language they belong to or how resounding or poetic their genre is, whether in Homer's Greek or Bardaisan's Persian poetry, if not illuminated by Christ's light according to Christian faith.

Hence, 'mystery' seems to be a concept that, from the secular point of view, has gone too far from the sacred realm over time, whereas according to the profane meaning, it continued to hide something static and relies on the occult for its sacredness. According to the Christian point of view, its sacredness has nothing to do with occultism. It resides, instead, in the main subject-object it used to hide dynamically (Jesus Christ) before the Incarnation and dynamically revealed itself afterwards, to become capable of being revealed in the continuous process of proclamation without any fear of profanation.[89]

Ephrem assures us that the symbols, whether they come from the Scriptures *ktâbâ* (ܟܬܒܐ) or from Nature *kyânâ* (ܟܝܢܐ), have been filled and energized by the incarnated *Râz*. Consequently, the old power of esoteric religions based on temporal mysteries, like that of the Gnostic movements, is no longer reliable. We will reach the peak of Ephrem's apology for the *Râz* later, under the point that will deal with the symbol of the 'sun', a tri-unitary symbol, Ephremian *par excellence*.

In short, 'mystery' should have more than one definition due to its etymology. It all depends on the context in which that word (or better, that concept) is used. Accordingly, we will try to get out of this difficulty by focusing on the development of its meaning, but this time, from its properties and behavior.

2.1.2. 'Mystery' according to its properties and behavior

Is it possible to desecrate a 'mystery' or, in other words, to destroy it? How would it behave in such cases?

If 'mystery', taken in the profane sense, cannot avoid being profaned, given its vulnerability *vis-à-vis* the human inclination to treason, in the Christian sense it seems indestructible, knowing that Christ even encouraged Judas Iscariot to do what he had plotted against Him.

The *Dictionnaire de Théologie Catholique* enlightens us on the three essential properties the Church attributes to mysteries and that have been subject-objects of a wide range of controversies parallel to the same range of interpretations of the Messianic event.

---

authors." We should not be surprised by the idea of having terms, concepts, signs etc., baptized.

89   Mk 16:15.

- Truths proportionate to divine intelligence, infinitely superior to all created intelligence, human and even angelic,[90]
- Truths whose knowledge can only reach us by revelation,
- Truths which, even though known by divine revelation, are still covered with the sacred veil of Faith and shrouded in a dark 'cloud'.[91]

The challenge of these properties added to our preceding analysis seem to have intrigued Ephrem in the Nisibean context, especially as director of catechetical teaching and defender of the orthodoxy of the Christian faith. Since it is in such kinds of ' dark clouds' that the usurpers of religion, the traitors, will hunt to try to unleash their interpretations and gain followers, Ephrem, as a good shepherd and pedagogue, saw the need to eliminate this opportunity.[92] At a certain point, for him 'clouds' are a significant inconvenience. 'Clouds' as veils, mentioned in the second letter to the Corinthians,[93] should be waved away between God (Who has incarnated by Love) and His next of kin, especially those who have loved Him and from whom He has taken mother, flesh, weakness, even their physical death, with the exception of their sin. But even this sin He assumed, and by his martyrdom, He shifted it to become an opening to His mercy which allowed Saint Augustine to describe it as a "happy sin" (*felix culpa*).

In spite of this opening, the concept of 'mystery' remains very mystic and mythic, sophisticated, and "Greek" *vis-à-vis* Ephrem's good, simple, needy, anxious Nisibean parishioners, Syriacs and Persians alike. Is it possible that the 'Most High', Who by kenosis 'stooped down' and made Himself 'Very Low',[94] out of love for the poor people He is in charge of, leaves between Him and them a 'cloud' that obliterates their communication? Something must change.

Mysterion, according to Ephrem, with its 'dark cloud' property, is no longer suitable for Christian spirituality, as proved by the intervention of Jesus in Saint Paul's sight and life on the road to Damascus. Darkness does not allow the faces of

---

90 How can divine and angelic intelligence be defined by humans?

91 French meaning of *Mystères.Cf.* https://www.universalis.fr/encyclopedie/mystere/2-les-theologiens-modernes. [Accessed Jan 2021]
Also *Cf.* MYSTERY (IN THEOLOGY): https://www.encyclopedia.com/religion/encyclopedias-almanacs-transcripts-and-maps/mystery-theology [Accessed Jan 2021]

92 The 'cloud' of the Transfiguration scene as well as the one of the Ascension did not obstruct the disciples' vision as long as Christ-Light was with them. It was just that after the ascension the disciples had to wait for the Holy Spirit to fill them with the necessary power to break through the 'cloud' again.

93 2Cor3:15-16.

94 Phil 2:7-8.

the sons and daughters of the Covenant to shine. How then, under this condition, can He ask them to be perfect as their heavenly Father is? According to Ephrem, the concept of *mysterion* must be tamed, and taming a "fox", as the Little Prince of Antoine de Saint-Exupéry will later assert, begins by giving it a familiar name.[95] It will help, little by little, to soften its properties, as well as its behavior, in a way that suits the Children of the Kingdom. Ephrem made his search in the Book of Daniel, borrowed the term *râzâ* (ܪܐܙܐ) and made it a substitutive concept suitable to the Nisibean context. How is *râzâ* to be understood?

2.2. The râz (ܪܐܙ)[96]

We will also consider this point under two subtitles:

1. Definition or description?
2. The description of *râzâ* (ܪܐܙܐ)

2.2.1. Definition or description?

Sebastian Brock informs us of the origin and the philological dimensions of this key word. In his article, "In Search of Saint Ephrem", Brock similarly states that the choice of *râzâ* by Ephrem was to avoid any definition of the sacred.

Brock said, "Saint Ephrem was suspicious of the concept 'definition' which, from the linguistic point of view, intrinsically contains limits (*fines*, in Latin) and inspires the idea of limitation. He hated to limit the unlimited." He added, "The essential term he uses in Syriac is *râzâ*."[97] Accordingly, would it not be pretentious, on our part, to give a proper definition to *râzâ* (ܪܐܙܐ) which, in essence, is averse to any limitation? What, then, are we to do? Can there be a reality, or a science, without definitions?

To limit the unlimited was a scruple for Ephrem. He needed a concept that simultaneously 'limits' and 'does not limit'. The *râz*, originating from the time of the Babylonian exile and from the Persian language,[98] is in itself symbolic, as

---

95   *Cf.* http://gutenberg.net.au/ebooks03/0300771h.html#ppchap1; [Accessed June 2019].

96   *Cf.* "Linguistic preliminary", p. 12.

97   *Actes du Colloque*, p. 22.

98   Dan 2:24-47. Indeed, this part of the Book of Daniel was originally written in Aramaic. (*Cf. Al Muḥit al Jameᶜ*) The Persian word is used because the course of events was happening in Babylon, and Daniel was addressing the King in Persian. The late Dr. Victor El Kik, a specialist in the Persian language, confirmed in a personal interview that this guttural accent "Ah" which has been muted (*mbatlânâ*), comes from a Shibboletic tradition that still exists in Persia. *Cf.* the following Persian dictionary: *Mobin Culture: Arabic – Persian, Qasim Boostani*, Faqih Publisher, Tehran 1373 AH.

فرهنك مبين، عربي - فارسي، كامل المنجد ابجدي، مترجم قاسم بوستاني، إنتشارات فقيه، تهران 3731 هـ. : (سرّ)

Brock liked to point out. It is liberating, just as the "Cup of Joseph", son of Jacob, and the "Staff of Moses" of the Exodus. The *râz* symbolizes the liberation of the people of Israel. Thus, it may well assume the role of a soft revealer in the unveiling of all that is indefinable. This comparison with the "Cup" and the "Staff" – and its link with "liberation" – is meaningful, but not sufficient to delimit a 'milieu' to *râz*, or to delimit the *râz per se*.

Given the impact this term has on the scene that unfolds in the Aramaic text, *i.e.,* the relation between Daniel and the Source of his intuitions, Brock adds a cause or supplementary reason that ratifies Ephrem's choice of *râzâ*. It is the aforementioned relationality which, thanks to the Incarnation, has become mutual. Indeed, mutuality has always existed between God and Man. It is part of the divine Justice (*zadiqutâ* ܙܕܝܩܘܬܐ) without which neither the *râz* nor the Incarnation could ever have taken place.[99]

It is with conviction that we support Ephrem's sensitivity to definitions in a religious culture, especially a Christian one. Once qualities such as relationality, mutuality and sometimes 'relativity' intervene, definitions capitulate. For example, in a description based on common sense, the meaning, from a behavioral perspective, can move from conceptual to moral. This implies that to make known a dynamic value, we must move beyond rigid, dogmatic definitions. It is preferable to describe a concept through its properties and behavior. This is the case with the Ephremian *râz*. Let us proceed.

2.2.2. The description of *râzâ* (ܪܐܙܐ)

To have the Ephremian *râz* in the eye is to have the eye empowered with relationality and mutuality, *i.e.*, a kind of 'mirror'.[100] What more would we need to succeed in taming it? What is still lacking is a place of encounter. Would it not ultimately be the 'eye' itself? However, the 'eye', especially if it is like that of Mary or Paul, unveils rather than conceals. Do we not, then, risk the desecration and destruction of the *râz* itself?

The answer is negative because, as mentioned above, humans can desecrate only secular mysteries, whereas confidential information coming from the 'phenomenon' Jesus Christ, His Incarnation and His interaction with His Mother's kind is holy and unassailable. Jesus, indeed, challenged His disciples by ordering them, at the moment of His ascent to His Father, to go and spread His

---

99   At the same time that God reveals the *râz* to Daniel, the latter reveals it to Nabuchad-nezzar. Relationality, here, is Trinitarian.

100   This is the type of philosophical and psychological instrument that came into use from the Platonic era.

Good News throughout the world, that is to preach Him, His proper Self, as the substitute for the Godhead-Father, as being His *Râz*, rather than His symbol or a sign that refers to Him.

It was evident for Him that the more they expose Him to profanation and the more they make Him public, the more His *Râz* becomes sacred, immune, and convincing. Therefore, to silence this 'truth' and to be ashamed of it is what will profane it and risk its loss. This risk caused Jesus to warn them that He would treat them the same in front of His Father.[101] Verbs like go, eat, take, evangelize, baptize, heal, forgive, reconcile, serve, watch, pray, etc., and above all love, once accomplished on His behalf, in His memory, become *râzified*'. He intends that their subject, object, and complements – all complements without exception – are Himself, the *Râz* of His Father Who lives and dwells in humans. He is the commander, the commandment, the commanded, and at the same time, the One in Whom, and by Whom, is fulfilled the Will of the Almighty Who "controls it all".[102] He is the Father's Son by Whose Name all that we ask the Father is to be considered already granted, but nothing apart from Him.[103]

All this benefits our 'ego', our self-salvation. As for what comes back to His own 'ego', His self-satisfaction, He assured us, through the Good Samaritan, that it is to the "other" that it should come back, and that all we do to one of the least of his brethren, we do to Him.[104] So He is the "all in all humans, as well as in all action".[105] He is All in all.

He is the shared one who is given and revealed. He is Love, Truth, Charity, Mercy, Agape, Eros, Justice and Prayer etc. He has irrevocably affirmed, "By loving one another, they recognize that you are My disciples,"[106] "Where two or three meet in my name, I will be among them,"[107] and, "I truly tell you, you will not have gone through the cities of Israel before the Son of Man (the *Râz*) will have come."[108] Therefore, given these properties and this behavior, and trusting in the words of the Incarnate Word, we can say with certainty that *râzâ* stands on the

---

101 Mt 10:33.

102 It is only in the Arabic version of the Creed that we read "Who controls it all" (*Abon dabit el kol*). The Syriac *Aḥid kul* suggests much more the English "the Almighty", which leaves a certain margin for human freedom.

103 Jn 16:23-24.

104 Mt 25:40.

105 1 Cor 15:28.

106 Jn 13:34-35.

107 Mt 18:20.

108 Mt 10:23.

opposite side of *mysterion*.

We can only show our respect to this Ephremian contribution. Any mystery-secret breaks-down once betrayed and thus shatters, without any hope of resurrection. Only the *Râz*, the theophany of the second Person of the Trinity, the immortal *Verbum Dei*, dynamic, alive and stimulating, never dies. He is the Light of the world that no bushel can cover. He is the Salt of the earth that revives the taste of all kinds of spilled salt and gives it hope and reason for being. [109]

To reach a clear idea of what we mean by *râzâ* without either defining or delimiting it, we will summarize what we have planned as a description. Then we can move forward with certainty.

The *Râz* is not a static term like 'sign' or 'symbol'. It is He Who endows signs and symbols with their meaning and their reason for being. He is a living concept and in continual becoming, mutually relational between two sacred realities, one by Essence, the other by Incarnation. He has nothing to do with the profane or the occult. He is the intermediary of mutual taming clearly described in the account of the Rich Young Man. At the same time, He is the force of attraction as the parable of the Prodigal Son unfolds. His instruments are the eye and the kenosis as emphasized by the same two parables. From this, we can understand why Brock highlighted this relationality that would make out of any symbol, any 'mystery' in the sense of enigma, secrecy, or incomprehensibility *vis-à-vis* the behavior of the personalities of the Old Testament, a *râz* in process, awaiting the Incarnation to become a *râz* in act.

Brock wrote: "Inherent in the *râzâ* is what Ephrem calls the hidden Power (*Ḥaylâ kasyâ*)."[110] It is this same Power that illuminated everything that prepared the coming of Christ and continues to do so with everything that follows Him, from *Olaf* (ܐ), the pre-Genesis, to *Tao* (ܬ), the post-Final Judgment, and which will fill all of existence with meaning.

Moreover, if we cannot attribute this 'power' to *râzâ* (with small r), or even to *râzê* (the Syriac plural of *râzâ*), as the Maronite Eucharistic prayer implies,[111] we can dare to do it at least to the extent that these symbols contain the primordial *Ḥaylâ*. However, if this 'graft' were not there from the beginning, all symbols would have remained within the frame of profanity, and the world would have been quite different.

---

109   *H Nat* 15, 10. *Cf. CSCO* vol. 186, p. 83.

110   *Actes du Colloque*, p.18.

111   '*gmurtâ mḥasionitâ w malyat râzê*' (ܪ̈ܐܙܐ ܡܠܝܬ ܡܚܣܝܢܝܬܐ ܓܡܘܪܬܐ), *Cf. Manuel de la Messe Maronite*, Bkerke, Liban, 2005, p.738.

Additionally, if it happens that someone accuses us saying, "But you are equating *Râz* with Christ Himself," we will gladly affirm, prostrated with Ephrem before the Eucharist, "It is well said; He is indeed the *Râz*."[112] He is at the same time the *Râz* of his Father,[113] and the *Râz* of Man,[114] of this Man who, in turn, once baptized, becomes the *Râz* of Christ. This is what it means to see far, to see exhaustively, to see multidimensionally.

Starting from this illuminating vision, we can well say, with Ephrem, that thanks to the Incarnation, the symbolism has undergone a radical shift. As a result, Ephrem succeeded where the famous Origen failed. Ephrem, with his *râzâ* avoided the risk which Origen involuntarily exposed himself to, *i.e.*, the problematic of the inequality between the Father and the Son as expressed by the verse Jn 14:28 in which Jesus says, "For my Father is greater than I."[115]

This hypothesis of the shift that the symbolism has undergone at the level of its *ontos* sends us straight back to the heart of the cognitive theory, which determines by its paradigms, from Plato till now, the foundation stones of general epistemology. We will develop this point in the following chapter, which will deal specifically with what we shall call "*râzâtic* reality" and its applications.

For the time being, and according to the way Brock read Ephrem, we can deduce that any latent symbol, whether from the Scriptures or from Nature, was activated by the Incarnate Word. More precisely, we say, it was ignited, made active as a verb in itself, made a living 'being' that acts and reacts like the Star of the Magi at Christmas, to indicate the 'Bethlehem' of the Universe.[116]

This is what we mean by relational. Even humans who are 'images' (*surâtâ* ܨܘܪܬܐ) and 'likenesses' (*salmê* ܨܠܡܐ) of God before baptism become after it *râzê* of God. In this case, the *salmâ* in the following stanza, having undergone the unification (*ḥlat* ܚܠܛ) and the mixture (*mzag* ܡܙܓ) can only be taken in the sense of *râzâ*.

---

112 Jn 18:37.

113 Jn 14:6-14 and Col 2:2.

114 Lk 11:30.

115 It is about the problematic of the ontological superiority of the Father in relation to the Son.

116 *Les Origines*, Note 4, evokes the philosophic axiom: "God alone is Being, everything else is creature." We have no objection to this, but after the Incarnation, in the case of the union of *Ḥaylâ Kasyâ* (ܚܝܠܐ ܟܣܝܐ) with the created, what has become of this rule of Greek logic? Is not Jesus Christ the man who is the *Râz* of God the Father, a Being (*Ityâ*)? Is this not what Ephrem anticipated by *râzâ*?

Glorious is the wise One Who allied and joined
divinity with humanity,
One from the heights and the other from the depths.
He mingled the natures like pigments in an image and
ṣalmâ came into being: the God-man.
Refrain: Glory to Your Dawn, divine and human.[117]

Having become the respective and personal *râzê* of their heavenly Father, the words and ideas of the baptized should no longer, after baptism, be the same as before.[118] The 'graft' of the sanctifying Fire must have taken possession of them all, both male & female. Moreover, with a bit of Ephremian poetics, we would say that the latent 'symbol' of the past is baptized in the Fire and the Spirit at the Jordan, and becomes a *râz*, *i.e.,* a permanent living symbol. It is no longer a dead 'sign' like the ones that indicate the road directions toward a particular city. It instead becomes alive; it speaks like the "Ass of Balaam".[119] It shines like the bush of Horeb. It accompanies like the Christmas Star. It sings and glorifies, as all beings of nature, to point and direct to the One Who vivified it and will always do so — to the truth (*šrârâ* ܫܪܝܪܐ) that stands beyond the reality of the same symbol. *Râzâ* is a bit of all of that.

After this description, so delicate and so exhausting, what or who could absolutely be the "Hidden Power" (*Haylâ Kasyâ* ܚܝܠܐ ܟܣܝܐ) who penetrated Creation through the Incarnation other than the Word of God as Isaiah described in the oracles:

"By myself I swear, truth goes out from my mouth,
a word that will not be revoked."[120]

"So is my word that goes out from my mouth: It will not
return to me empty, but will accomplish what I desire
and achieve the purpose for which I sent it."[121]

Do we not feel behind these flaming words a certain synergy? Let us find out.

---

[117] *H Nat 8,2. Cf.* Kathleen E. McVey, *Ephrem the Syrian Hymns,* The Classics of Western Spirituality, Paulist Press, NY. Mahwah, 1989, p. 119. *Cf. La Pensée,* pp. 48-49. Note 38, included in Bou Mansour's text, reads: "The constant in *Pešitta* is rather to translate by *Ṣalmâ* the man created in the likeness of God ..." *Cf. CSCO* vol. 186, p. 59.

[118] 1Pet 2:1-3; Phil 2:5.

[119] Num 22:2 – 25:9.

[120] Is 45:23.

[121] Is 55:11; *Cf.* also Heb 4:12-13.

3. Synergy

Although Ephrem wrote in Syriac, he used, as previously mentioned, the Persian term *râzâ* to designate what was usually considered, in his traditional Christian education, mysterious, enigmatic, and secret. Why would an author do this?

This question indeed arouses in our mind the concept of synergy and the important role it could have on the horizon of our contribution. If we fail to convince our readers of its importance, our position *vis-à-vis* Ephrem and his *râzâtic* phenomenon will remain lame, and the 'Luminous bridge' we plan to build between Religion and Science will be of little significance. It is thanks to the book *Mind's Eye of Richard Buckminster Fuller,*[122] which was the "womb" that gave birth to an *Energetic and Synergetic Geometry,* that we will try to succeed in our task.[123] Who is Buckminster Fuller and how does his scientific contribution help us discover the value of 'synergy' and satisfy our question?

3.1. Richard Buckminster Fuller: 1+2 = 4!

Donald Robertson, the author of the aforementioned book, was Fuller's lawyer who accompanied him intellectually and professionally throughout his career. In his volume, he introduces Fuller to us in two ways: first in prose and then in verse.

It begins by describing the qualities of the person in question and highlights his ability "to be born to test every preconceived notion, and to reject every 'can't do' of man".[124] Robertson then sees in Fuller the possessor of a rare ability to subjugate fragments of the truth within ancient knowledge while gaining a broader perspective through the contemplation of what he called "the totality of a problem". He says of Fuller that he was, "...the tutor and mentor of the excited imaginations of students, scholars and intellectuals of all races, putting at their disposition the fresh and surging power of patterns of unhindered thought...",[125] among others the famous equation $1 + 2 = 4$ which will be key to our subject.

The poetic introduction highlights the intuition that Fuller enjoyed, which allowed him "to make a difference" and to introduce his students to "the reality of

---

122 Robertson, Donald W., *Mind's Eye of Richard Buckminster Fuller*, St Martin's Press, New York, 1974. From now on, we will refer to this work by *Mind's Eye,* a title that sounds very Ephremian.

123 *Ibid.* Chapter 3.

124 *Ibid.* p. 13. Can this be applied to Saint Ephrem?

125 *Ibid.* p. 13. Can this also be applied to Saint Ephrem?

the unseen". We read:

> They asked, "Why houses in the round? "
> Why make them square? Said he.
> But more, why tie your thoughts at all,
> To round, or square, or old geometry,
> That's dead and strange to all reality.
> For Universe is life and motion.
> There's more of form and energy
> Than of material things we see.
> We must think comprehensively.[126]

Would that not have pleased Saint Ephrem?

To find the answer to the question that we asked regarding Saint Ephrem, "Why would an author do this?", let us find out what stands behind the original behavior of Buckminster Fuller who, thanks to "his mind's eye", was able to distinguish himself. Is it something mysterious or is it our *râzâ* that made him capable of discerning 'the unseen reality'?[127]

Fuller invited everybody to think in an exhaustive way. He helped his students understand what this means. Robertson tells us the following story:

> Beginning his explanation of the new geometry, he would
> say, holding before his class a simple model consisting of
> three triangles hinged together in a chain:
> One equilateral triangle...
> Hinged to two others...
> Can be folded into a three-sided 'tent' whose base is a
> fourth triangle:
> Having suited action to the words, he tilts the tent
> backwards to show the base triangle.
> Now, Bucky continues with mounting excitement:
> "The *inadvertent* appearance of this *fourth* triangle is
> a demonstration of synergy: which is the behavior of a
> system unpredicted by its parts: 1+2=4."[128]

---

126 *Ibid.* Epigraph of the book. Robertson left a note to imply that it was Fuller himself who wrote these verses. Poetry enables new forms of expression and thinking.

127 *Ibid.* p. 15. Now comes the problem of how best to explain Fuller's discoveries when we understand that the most fundamental of them probe so deeply into the unseen dynamics of force and precession.

128 *Ibid.* p. 25.

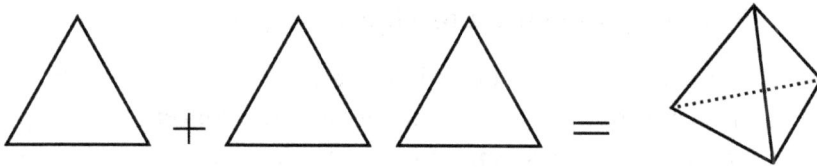

Figure 1: Vision of Buckminster Fuller

We are sure that any theologian who reads this paragraph receives a strange shock that illuminates him and awakens in him a desire to be 'more', not knowingly, but in a capacity to surpass oneself to understand exhaustively how it applies to Ephrem's concept '*râzâ*' and, specifically, to the *Râz* of the Holy Trinity. Which of us, for example, has tried once to understand comprehensively Andrei Rublev's Holy Trinity Icon and to deduce from it that 1 + 2 = 4: Father + (Son + Holy Spirit) = Father + Son + Holy Spirit + Creation, specifically the human being, perfect image and likeness?

Figure 2: A little exercise in artistic criticism: did Rublev think exhaustively?

We will answer these hypotheses in our syntheses, after comparing Fuller's fourth dimension with Einstein's. Next we will relate the whole to the dimensions Ephrem used to prove to the heretics, worshipers of fire or of the sun, that the two latter are also signs that refer to their Creator; indeed he used the sun to prove His Triunity.[129]

For the moment, we will try to answer the question: why would an author do that?

---

[129] *La Métaphysique Primalitaire* de Tommaso Campanella.
Cf. Michel-Pierre Lerner, "Campanella Tommaso - (1568 - 1639)", *Encyclopædia Universalis.* http://www.universalis.fr/encyclopedie/tommaso-campanella ; [Accessed June 2019]

3.2. Why did Ephrem use a foreign concept?

All that has just been said about Fuller and his genius, plus the different uses made of his comprehensive design theory and the concept of synergy in the twentieth century,[130] lead us to suppose that the phenomenon of synergy was indeed a target for Ephrem and, therefore, must be of crucial importance for our essay. Let us explain this point briefly under the following three titles:

> Etymology and dimensions of the concept 'synergy'
>
> Interdisciplinar synergy at the linguistic and epistemological levels
>
> Inter-reality synergy (*sic*) and Christological dimensions

a - Etymology and dimensions of the concept "synergy"

We subdivide this title into its two components:

> a-1: Etymology of the concept
>
> a-2: Dimensions of the concept

a-1 : Etymology of the concept

Synergy, according to the various dictionaries and encyclopedias, comes from the scholastic Greek feminine noun *sunergía* (συνεργία). The word already existed in the Greek literature of the time of Ephrem. St. Thomas Aquinas used it to deal with the intelligible form that causes trustworthy knowledge and unveils its dual origin.[131]

In a synthesis of the various explanations given for this term, we can say, "Synergy means cooperation or association of several different elements, belonging to various disciplines, for the accomplishment of a function." It also generally has the meaning of "Pooling of several actions contributing to a unique effect with an economy of means."[132] According to the French philosopher André Lalande,

---

130   *Cf.* Homily of His Holiness Benedict XVI, Saint John Lateran, Thursday, 15 June 2006. https://www.vatican.va/content/benedict-xvi/en/homilies/2006/documents/hf_ben-xvi_hom_20060615_corpus-christi.html ; [Accessed June 2019].
"...The bread is fruit of heaven and earth together. It implies the synergy of the forces of earth and the gifts from above, that is, of the sun and the rain. And water too, which we need to prepare the bread, cannot be produced by us..."

131   *Cf. E.U.*, Édouard-Henri Wéber, "Intellect & Intelligibles",: http://www.universalis.fr/encyclopedie/intellect-et-intelligibles ; [Accessed June 2019]. *Cf.* Thomas Aquinas, *De veritate*, qu. 8, art. 6.

132   According to *Larousse Dictionary,* the 'economy of means' can be explained in the

there is a distinction to be made between "free, active, independent and living synergy" and "mechanical and passive interdependence of co-action".[133]

According to Lalande's compatriot Alfred Fouillée, synergy also marks the theory of "main ideas" (*les Idées-Forces*) that is to say, of the mind as an efficient cause of the propensity of ideas to be realized by a conscious action.[134]

a-2 : Dimensions of the concept

All of these explanations of the concept 'synergy' will shed more light on how Ephrem uses it. We will call this the Ecclesial dimension.

Ecclesiastically, we are somewhat accustomed to the term 'synod', which is also rooted in the Greek word *sunodos* (συνοδος) and which means "get together":

> It is a meeting of specialists following a certain decline aiming to renew the deliberation of efforts, capacities and skills, to restart and advance again together, with an adequate synchronization.[135]

Joined to *chronos*, the prefix '*sun*' is also found in synchronization, from the Greek *sunchronos*, to indicate the necessity to also regulate the time, the rhythm, and the calendar of each member, so that there never be setbacks, and, therefore, any halt to the process. Accordingly, any concept that has the prefix 'syn' (*sun*) is incompatible with any form of anachronism, monolithic thinking, reductionism or isolation, all of which are daunting enemies of *koinônia*, the 'being together' (Heidegerian *Dasein*), the essence of the *Corpus Christi*.[136]

---

mechanical sense, based on the economy in "fuel" in the relation weight, speed, energy consumption. This can be translated, economically, by the consumption of personal or national capital. Analogously speaking, 'the economy of means' in the divine Economy of Salvation (*Mdabrânûtâ*) is also based on the relationship weight, speed, consumption, where weight is that of the sins mentioned in the liturgical chant *Abâ d'quštâ,* the speed is mentioned in the hymn that begins with "Be careful, beware, you desperate ones..." (*Ettᶜir, Ettᶜir duwyâtâ*), and what is consumed is Christ's Love for the salvation of the human person, this *Minima Naturalia* of the ecclesial community, the *Corpus Christi*. The sinner is the one who disturbs this synergy or even behaves against its means.

133 Lalande, André. *Vocabulaire Technique et Critique de la Philosophie*, PUF, 1988.

134 *Ibid.* Fouillée, *Morale des Idées-forces*, Conclusion p. 352.

135 *Ibid.* From Greek *sunkhronos*, of *sun* "with" and *khronos* "time".

136 *Cf.* Mt 18:20; Jn 13:34-35.

Ephrem brilliantly applied these various "*sun*" to maintain the solidarity and unity of his Nisibean Church subjected to all kinds of atrocities: "Enemies from the outside and enemies from the inside", he wrote, until the fall of the city into the hands of the Persians.

Ephrem did not limit himself to what is traditional in the initiation of catechumens and in the vitalization of the parish. After having experienced, among the enemies of orthodoxy, the synergistic weight of the poetic dimension that inflames the spirits, he went so far as to introduce it in his method of teaching. Thanks to this poetic dimension, it was possible for him to go beyond the 'finite' reality of others and to unveil (*galyâ* ܓܠܝܐ) what was left hidden (*kasyâ* ܟܤܝܐ) of the *râz* of the continual virginity of Mary, "Virgin before, during, and after the birth of her Son", without fear of desecrating it.[137]

The advantage that the poetics of Ephrem had over that of his opponents was the element of his love for Mary, which he nourished in his heart. She is, for him, the eternal Virgin-Generatrix of the God-Man (*yâldat alâhâ* ܝܠܕܬ ܐܠܗܐ), by Christ, her Child, as by His Church, His Body, called the 'Mystical Body' while not mystic at all.

His own poems testify that this remarkable element of love has synergized his life and his theology. Ephrem, as a pedagogue, did not limit himself to writing about and teaching his love for Mary and her Son. He put it into co-action, entanglement, with the synergistic force of the choir-chanting factor, described by the following saying: to chant is to pray twice. To this end, he formed the group of the Daughters of the Covenant (*Bnât Qiâmâ* ܒܢܬ ܩܝܡܐ), the Covenant of Chastity, and encouraged them to compare themselves to their brothers in baptism in the service of prayer and liturgy. Likewise, to help the emancipation of the baptized women within his parish, he activated, as master of dichotomies, the synergistic element of challenge and complementarity between groups of men and women. He felt, by the Force (*haylâ*) of the Holy Spirit, that once the educational poems (*madrošê* ܡܕܪܫܐ and *Qolê* ܩܠܐ) composed from inspired words were sung by male voices mixed (*hlat* ܚܠܛ) with female ones, they would bring forth an Angelic Choir. Therefore, according to Fuller's equation (1 + 2 = 4) we deduce that:

> Chant of inspired words + (male voices + female voices) = the
> first three elements + an unexpected Angelic factor.

---

[137] Lk 8:10.

Ephrem's *râzâtic* feeling was right. In our ecclesial reality, we have the empirical proof of it, since his hymns and melodies are still alive and frequently used until this day. The Maronite Liturgy accentuates this synergistic element by recalling that it is in union with the Seraphim, the Cherubim, and the inhabitants of Heaven, that the faithful sing the glory of God.[138]

Saint Ephrem proved to be right with regard to the synergization (sic) by the râz of the finite reality and its transmutation, through *râzâtic* union with the Holy Spirit, into an infinite reality. Since, sacramentally, the two realities, the finite and the infinite, unite in the Eucharist and give birth to an unexpected "differAnt reality", the Body of Christ, the Church, that comes qualified *râzâtic*, the equation 1 + 2 = 4 finds itself well applied:

> *râz* + [finite reality + infinite reality] = the three of them + the *râzâtic* "differAnt" reality.

We will later, in chapter III, return to this hypothesis of the '*râzâtic*' reality and compare it with the quantum reality of which Science now speaks.

In what preceded, we approached some dimensions of the concept 'synergy' from the angle of our theological contribution. The outcome of the activity of the Prophet Daniel's friends who chanted all together in the furnace represents its true type (*ṭûpsâ* ܛܘܦܣܐ): the appearance of the unexpected fourth element. Although other types and figures of synergy punctuate the Bible, the Book of Daniel was the key text for Ephrem. This type of "synod – synergy" will be supported by the divine Master when He confirms much later that what took place in the furnace with Daniel's friends will happen, synergistically, every time two or three come together in His name. That is because the fourth unexpected 'Element', which caused the astonishment of Nebuchadnezzar when he saw 'Him' with the three Hebrews, will be there, without fail.[139] What would then be the case if two or more disciplines, or languages, unite in His name?

---

[138] *Maronite Missal* (Arabic), Bkerke, Lebanon, 2005; *Ephramiat; Ya saliḥan abda lil wujud*, 2005, p. 963.

[139] Dan 3:24 -25, or else 3:91-93 (Jerusalem Bible)

b - Interdisciplinar synergy at both the linguistic and the epistemological levels

The equation 1 + 2 = 4 allowed Buckminster Fuller to build the following geodesic dome in Montreal, Canada.

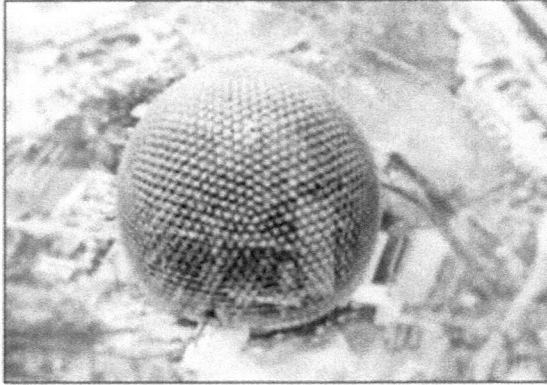

Figure 3: Comprehensive design [140]

For us, according to the architectural sciences, this dome is the perfect theoretical type of spherical form, symmetry, coordination, closure and openness. From the outside, we can see it as a universe, a sun, a moon, the earth-globe, the eye-globe, a 'pearl' (Ephrem's pearl), a Host, a Word (whether verb or noun), dot, digital bit, atom, even quantum (to see as small as that). Any architect would like to achieve such a perfect entity, even Saint Paul, considered by our Syriac patristic literature the 'architect' of the Christian faith. He is, indeed, the architect who stands behind the 'Dome' of the Church, the *râzâtic* Body of Christ, that should lack nothing in playing the role of *maškanzabnâ* for the whole of humanity, even in the present day.

To this type, or prototype of perfect form, the unseen, the 'inadvertent appearance' of which Fuller speaks in his pyramidal experience, does not apply. From any angle we observe this geodesic form, it is the same. Where then would its unexpected element be?

We find the answer in the same Icon of Rublev that we have just submitted to a specific artistic critique, part of which is an architectural one. The answer is in the center, at the meeting point of the heights, the point where Rublev placed the Cup of Incarnation and of Redemption.

---

140 Expo 67 Center, international World Fair held in 1967 in Montréal, Québec, Canada. Dome conceived and executed by Richard Fuller in 1965.

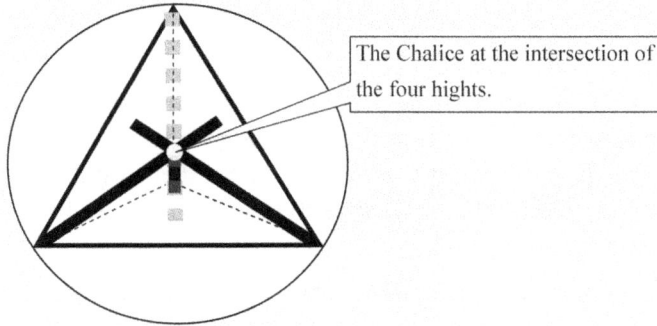

The Chalice at the intersection of the four hights.

Figure 4: The Center

The problem that arises here is the following: how to see this center or to enable others to see it. There, the "Bucky-ism" does not apply. Only a kind of 'endoscopy' like the one made by Ephrem's 'luminous eye' towards the 'pearl' allows it:

> One day, my brethren, I took a pearl into my hands;
> In it I beheld (I saw ܚܙܝܬ) symbols (*râzê*) which told of
> the   Kingdom, Images and figures of God's majesty.
>
> It became a fountain from which I drank the mysteries
> (*râzê*) of the Son.[141]

This instrument, or rather this prototype instrument, which Ephrem used to see the invisible, should be, in our opinion, applicable to both epistemological and linguistic domains. It is quite enough to replace the concept 'sphere' or 'pearl' with 'language' (*lešânâ* ܠܫܢܐ) or 'word' (*meltâ* ܡܠܬܐ) to grasp its importance. Every language or discipline of human intelligence considered upstream (that is to say, from the outside) to be pure, perfect and self-sufficient, becomes downstream (considered from the inside) in need of co-action with elements of other languages and disciplines, as well as contributions from them, so that the 'Heart' of the 'Sphere of Truth' becomes knowable.

In the case of the Gnostics and sun-worshipers, Ephrem had to probe the 'womb' of the concept 'sun' to prove to those pretentious people that their truth was insufficient. He then undertook to explore the 'womb' of the words and names to teach them how to read what these words were veiling, not what they were revealing. Ephrem urged them to acquire the divine power (*ḥaylâ*), that additive value which makes their eye luminous and, therefore, more penetrating

---

141   *H Fide* 81, 1; *TLE*, p. 106. *Cf*. *CSCO* Vol 154, p. 248. Ephrem says "I saw in it" (*ḥzit beh* ܗܘ ܚܙܝܬ) because, for him, the goal of all meditation is the 'Vision'. We also notice that Ephrem inspires certainty when he says, "I saw", "I drank". Thanks to his *ʿaynâ šafyâ'* he leaves no room for doubt.

of the unfathomable and capable of foreseeing every unpredictable element that contradicts the Truth of the Incarnation and the Redemption.

In the effort to polish the 'luminous eye' like the face of a mirror that allows introspection and, at the same time, wins everyone over to Christ, nothing should be excluded from the use of the interpreter.[142] Any 'unit' that synergizes another 'unit' to bring forth a third 'unit' is welcome,[143] especially if it is a 'unit' of a language other than Hebrew to synergize the 'unit' of the Hebrew of the Book, and vice versa.

Through Christ, as St. Paul taught, there is no longer any difference between one language and another. Any language that arose from the so-called 'curse of Babel' and which might be a necessary return to the primordial language that allows dialogue, understanding, predictability, certainty, peace and serenity between God and humans, as well as between humans themselves, is most welcome.

So, to the question, "Why would an author borrow a word from a foreign language?", the answer could be twofold: either his lexicon lacks a concept that indicates exactly what he means in the sense appropriate to his context, or he finds that the relevant concept that exists in his language and that comes from an already outdated reality is now too polysemic. It is also possible that these concepts have lost their relative denotations or at least suffer from remnants of a mythological or alchemist background,[144] to the point that they lost their effectiveness compared to the new reality. Henceforth they lead to a lot of confusion.[145] Their use would first require their demystification and regeneration in order to rehabilitate and transmute them to the desired purpose. This ends in a very complicated operation that runs against simplicity. This difficulty generates either one of the following two needs: to forge a new concept, a new word or verb, or to borrow one from any other language, especially from a neighboring culture.[146]

In the first case, it would take a long time to integrate this new word, publicly

---

142 1 Cor 9:19-23

143 The duty of interpreting a soteriological reality obliges us to force the English grammatical rules and to derive from the name "synergy" the verb "to synergize."

144 *Ibid.* p. 17; "C'est sur quelque trace du mythe archaïque que sont greffées les significations les plus prophétiques du sacré" (Ricœur). This means, " It is onto some trace of the archaic myth that the most prophetic meanings of the sacred are grafted".

145 *Ibid.* p. 131

146 A culture of peoples who share, in spite of differences of language or accent, the same climate, the same manners and the same mentality, especially in our case, the religious one. This case recurs frequently even today, whenever words from the ancient Greek, Persian or Latin language are used in a text of a different language. It remains to emphasize that few biblical, patristic or even exegetical texts make an exception to this synergy. Words like *Abba, Tabitha, Maranatha, Eli Eli Imâ šbaktan*, etc., will always resound differently in the listener's ear and energize any text that uses them.

and pastorally, into the widespread collective understanding. In the second case, borrowing a term free of all that is proper to one's own language, a term of the same orbit that one wishes to join, and then introducing it into a new linguistic context, it seems more practicable since this case allows the enjoyment of a triple veil of mystical strangeness:

> 1 - The veil of a term that sounds familiar and unfamiliar at the same time, logical in context, readable, but remains incomprehensible without initiation because it is just a sign of itself and refers only to the idea that the author makes it adopt. (We will return to this point by quoting the philosopher Jacques Derrida.)

> 2 - The veil of an unknown author, who becomes recognizable only through the interpretation of the causes that required the use of this strange concept.

> 3 - The veil of a remaining incomprehensibility, even after initiation, especially in the case of sacraments (so-called 'mysteries'), like the Christian ones.[147]

It is the second case to which Ephrem resorted by adopting the term *râz* (ܪܐܙ),[148] and by this, he lets us feel, very deeply and strongly, that he has ventured into new 'territory'.

Indeed, he ventured since what he was seeking was a foundation stone supposed to help him get out of a particular uncertainty in which he found himself. Unfortunately, that was not the case because the term he borrowed was Persian, the language of the enemies of Christianity and the residents of his own town, Nisibis. It would have been understandable if he had synergized his Syriac language with a Greek word since both languages were already Christianized, but to have done it with a Persian concept was very delicate, mostly from the sociopolitical point of view.

First of all, Persian disciplines were based on a dual ontological principle (two opposing gods, a good one and an evil one); therefore, they were foreign to monotheism and considered heretical.

Secondly, Persia was also socio-politically unfriendly. The Persian Empire was the

---

147 Not to be accused of contradicting ourselves, we underline that Ephrem who refuses any 'cloud' between the Godhead, Father, and His "children", just as Jesus did not want it between Himself and his disciples (Jn 15:15), admits with holy submission a certain insurmountable distance between the human condition and the divine one, as established according to Mt 24:36 by the Son in His human condition.

148 This contribution of Ephrem by which he far outstripped his contemporaries is very close to Carl Gustav Jung's position. Jung, proposing an alternative definition to the concept 'symbol', distinguishes it from the concept 'sign'. *Cf.* C. G. Jung, *Psychological Types*, 1921, p. 601. H. Godwyn Baynes translation, 1923).

awful enemy attacking their town. So, although the concept '*râzâ*' (ܪܐܙܐ) was taken from the biblical Book of Daniel, it did not compensate for the fact that Ephrem borrowed it from a foreign reality hostile to that of the Gospel. Moreover, this word had belonged, and still belonged, to a completely different philological family.

Ephrem, quite sensitive to all these obstacles, decided to venture anyway, as did before him the mythical god, Prometheus. He went beyond the dimensions given to this term in the Book of Daniel because he saw it from a different angle. In fact, between the *râz* of Daniel and his own, the Incarnation took place and became, from then on, the filter through which Ephrem saw, read, and analyzed everything. The God in whom Daniel put his trust to solve the enigmas of the message passed to Nebuchadnezzar made Himself 'flesh' to deliver to all humans what He delivered to Daniel,[149] and before him, in Egypt, to Joseph son of Jacob.[150] The Incarnation became, from the moment of its fulfillment, the key to the unveiling of all the puzzles that obstruct the correct interpretation of Creation and its relationship with its Triune Creator, God the Father, Son, and Holy Spirit.[151]

Thanks to the three foundations above, the Eye, the Light and the *Râz*, Ephrem went beyond the dimensions of which St. Paul spoke,[152] and, henceforth, pointed to eight dimensions in the post-incarnational reality, six of which were already familiar. There are, first of all, the four dimensions coming from the contemporary consideration of the world as being a plane with borders included between four corners,[153] then the two dimensions ardently evoked by Jesus, the 'above' and the 'below'.[154] Ephrem, seeing things according to another scale, added the 'inside' and the 'outside'.[155]

At this level, the interdisciplinar synergy at both linguistic and epistemological levels typically begins to exceed the norms of reality limited to temporal lexicons as well as to interpretations circumscribed in horizons not revivified by the divine. It exceeds them, toward the norms of the "differAnt" reality mentioned above, synthesized by the sages of Nebuchadnezzar. In that reality, the secular and the sacred match, and the Power (*Haylâ*) that the angel Gabriel spoke of to Mary seems to have made pregnant not only Mary, but further, due to her *Fiat*, all names, words, verbs, and syllables known to humans.

---

149   Dan 2:27-28.

150   Gen 41:15-16.

151   Mt 5:48; *Cf.* also 1Chr 17:13. According to Jung, based on the Words of Jesus Christ inviting humans to be as perfect as their Heavenly Father is, the text would replace 'Father' with 'Archetype'.

152   Eph 4:9-10.

153   *Les Origines,* p.133; *La Pensée,* pp. 23, 66, 88 (pluridimentionality), 308; and footnote 25, p. 542.

154   Jn 8:23.

155   *Les Origines,* pp.131-133; 224.

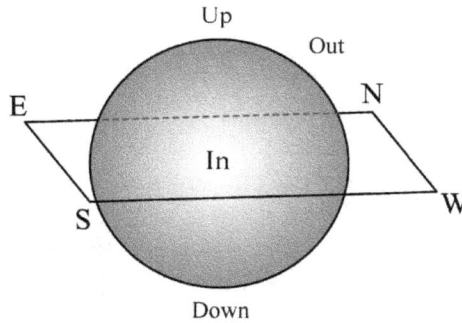

Figure 5: When it turns, everything turns around the Center

It is in this particular chiasmus of the Incarnation that Ephrem saw the *Râz par excellence*, considering the *Râz* to be the Truth veiled behind a symbol, a type, an image or hidden in a paradigm.[156] Therefore, *Râzâ* becomes, linguistically, the center of convergence, or rather, the entirely transparent prism through which any secular name becomes transmuted into a sacred one, regardless of its provenance, direction, or angle of diffraction. *Râzâ* becomes the catalyst of all relationality between the two realities, the so-called secular and sacred, and no longer allows impediment to communication between the two.

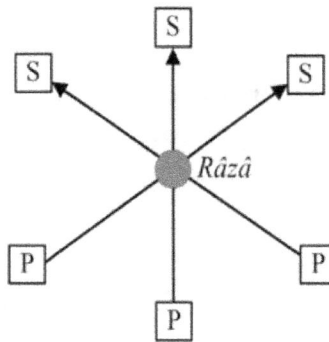

Figure 6: From secular (profane) to sacred,
"Through Him, all things were made"! (The Creed)

Most likely, without having fully opened the door to this interdisciplinar synergy, there would never have been the progress that we know today in Biblical exegesis at both the linguistic and epistemological levels. The long-sought key, with its small 'k', was humility, or even the kenosis of the Incarnation, which the Master of the two Testaments freely gave to His one unique Church and, through

---

156   *Cf. TLE*, pp. 40-41.

it, to humanity. Without this kenosis, no progress would have been possible and never would be.

We conclude by inviting linguists and epistemologists to consider, as Ephrem has done, in their research as well as in their analysis of phenomena and causalities, the foundation stones mentioned above derived from the chiasmus of the Incarnation, as they do for the chiasmus of quantum mechanics and that of Einstein's relativity. We also invite them to keep in mind, during the dissection they apply to words and names, that He who shifted the name Abram into Abraham and Jacob into Isra-El (*sic*) has left everywhere letters and syllables, visible and invisible, which would help to synergize disciplines, languages, and names,[157] including ours and His own, YHWH.[158]

Linguists, biblical scholars, theologians, and architects of exhaustive knowledge should, through kenosis, *râzify* themselves to intercept these fragments of the *râzâtic* reality. They would also do well to allow these fragments to penetrate their soul and spirit so that they become *synergized* in turn. These letters and syllables are alive and bear sounds from the Voice that synergized the Baptist's. These *râzâtic* fragments constitute a linguistic and epistemological discipline of their own that lacks neither music nor poetics. Humans who are interested in Science, as well as in Conscience (Religion), would benefit from adopting this kenotic – synergistic attitude to allow humanity and the universe to benefit with them.[159]

Let us turn to the third point, to the study of the phenomenon of synergy between realities (inter-realities) and the Christological dimensions of Light and Fire.

c - Inter-reality synergy (*sic*) and the Christological dimensions of Light and Fire.

The divine Word, Son of God, by inflaming the immaculate womb of the "New Eve", takes flesh (*sarx* σάρξ) and assumes, by this fact, all the *malum* of this flesh, the *malum culpae* as well as the *malum poenae*, and leaves that womb as Son of Man.

Relatively speaking – with a well-understood and well-administered relativity – Christ, by assuming all of the human condition with the exception of sin as a deliberate and executed act also transmits similarly to humans the condition of His nature, as it was known in Him after His kenosis. Was it not He who affirmed that all the works He did can be accomplished by those who believe in Him, and that

---

157   *Cf.* Derrida in his commentary on his own name Jacques. (pp. 103-106; 169; 182)

158   Yahweh; *YhWh* יהוה; has its roots in *"haya"*. It is the proper noun of the Eternal God. It means "Who is the Being". This would be the pronunciation of the Hebrew Tetragram. *YHWH*; *Cf.* https://www.penn.museum/sites/journal/75/ [Accessed Sep 2020]

159   We recall here the French saying, "Science sans conscience n'est que ruine de l'âme".

they can do even more?[160] They will undoubtedly be able to do so on the condition that they keep the synergetic boost among themselves, as with Him and His Father.

The Maronite liturgy has the honor of having in its Eucharistic text a reminder of the basis of this exchange of nature, starting from substance. It also supports the synergy that comes into effect at the very moment of the transmutation of one substance into another, what is called transubstantiation:

> O Lord, You have united Your divinity to our humanity and
> our humanity to Your divinity, Your life to our death and our
> death to Your life; You have assumed what is ours and granted
> us what is Yours...[161]

The perfect correlation between the divine and the human natures, which Ephrem recognizes within the womb of Mary, is the source of this inter-reality synergy: "Fire entered Mary's womb, put on a body and came forth."[162]

If this is the case between divine reality and human reality, what would be the case of the correlation and the resulting synergy between the divine reality and the cosmic one?

It is evident that the Word of God who inflamed Mary's womb, also inflamed the 'womb' of Nature. He did it through the baptismal water of the Jordan. By this means, He assumed the water's entropy, transmitted to it His Providence, and came out of it the Son of God. The Holy Spirit was there actively present as in the time of Genesis, and as in the time of Mary's conception. Consequently, based on traditional kerygma, we can say that the 'curse' of the 'break' between the creation and the Creator caused by the sin of Adam and Eve did not mean that the two realities, the divine and the secular, no longer cohabited the cosmos. It merely meant that the synergy that existed between the two broke down. Both realities, with their relative energies, have always been 'there', each on one side, because God can never contradict Himself. He proved this truth in His Covenant with Noah, just after the Flood. (Gen 9:17)

So, since the time of what we consider original sin, the universe has continued its course. Adam and Eve have continued their way of life. Even Cain, the first murderer, has not been exterminated, and God, behind the 'Flaming Sword' of Genesis, has always waited for the return of the Prodigal Son. The divine Fire that

---

160 Jn 14:12.

161 *Maronite Missal (Arabic)*, Bkerke, Lebanon, 2005; final Consecration words, p. 739.

162 *H Fid 4, 2-end. Cf. TLE*, p. 38 and *CSCO* vol. 154, pp. 9*ff*: The two dominant Christological currents are, universally speaking, Christology from above based on the Johannine scriptures, and Christology from below based on the synoptic Gospels. Consider a fresh Christology from chiasm, whereby it is the encounter of the 'descendant' with the 'ascendant', with all the dynamics and synergy that result. It is just the moment of the fertilization of the divine in the human.

settled the challenge between the prophet Elijah and the priests of Baal is a great witness to the difference between the absence of this synergy and its presence. Consequently, if the Incarnation of the Word of God made the human being fit for deification, the so-called material reality – the cosmos, nature, the human body among others – thanks to the Baptism of Christ in the Jordan has been made worthy of it too:

> For that cause for which He entered the womb,
> For the same cause, He went down into the river.
> For that cause for which He entered the grave,
> For the same cause, He makes us enter His 'Chamber':
> Perfecting humankind with every cause.[163]

The statement "I am making everything new!"[164] in the Book of Revelation does not mean, as Jesus proved, to destroy and rebuild, but to trigger again this inter-nature and inter-reality correlation in such a way that the synergy between the hidden powers (*ḥaylê ksayâ* ⲕⲗⲃ ⲕⲥⲙ), and the apparent powers (*ḥaylê glayâ* ⲕⲗⲃ ⲕⲗⲅ) does not fall once again. It is, we would say, the primary role of the Holy Spirit mixed with Fire (*i.e.*, Baptism), that Jesus so much highlighted, to enlighten humans so that they hold the tonus of this synergy between divine and human, between sacred and profane, until no more profanity perpetuates either in potentiality or in act.[165] It is also that of which the Christological dimension of the Economy of Salvation consists: a Fire and a Light which, after having been potentially existing for a long time, come to be in act at a specific time, to continue being in process of actualization and realization until the end of times, and of Time.

So, again, it is the *râz* of the Incarnation, as we have seen, that Ephrem considers the *Râz* towards which converge all *râzê*, those potential ones of the Old Testament as well as all those actual, from the birth of Jesus Christ until today.

Likewise, in an exceptional symbolism, and in order to "fulfill all righteousness", this *Râz* also covers all the dimensions of cosmic salvation.[166] It is thanks to the

---

163 *H Epiph* 10,3; *Cf. CSCO* vol. 186, p. 181. *Cf.* John Gwynn, *Hymns and Homilies of St. Ephraim the Syrian*, Veritas Splendor Publications, San Bernardino, CA; USA, 2018, p. 284.

ⲥⲙ, ⲕⲗⲃ ⲕⲙⲃ ⲙⲃ ⲕⲗ ⲥ
ⲥⲙ ⲕⲗⲃ ⲥⲙ ⲕⲁⲙ ⲕ ⲥⲙ
ⲥⲙ, ⲕⲗⲃ ⲕⲙⲃ ⲕⲁⲙ ⲕⲗ ⲙⲃ
ⲥⲙ ⲕⲗⲃ ⲙⲃ ⲕⲗ ⲥⲁⲙ
ⲕⲗ ⲙⲃ ⲕⲁⲙⲃ ⲙ

164 Rev 21:1 and 5.

165 The relation will be, as Ephrem grasped, between Sacred and sacred, between *Râzâ* and *râzâ*.

166 Mt 3:15. *Cf.* François Cassingena, o.s.b; *Hymne sur l'Épiphanie; Hymnes Baptismales de l'Orient syrien, Spiritualité Orientale,* n° 70, Abbaye de Bellefontaine.

baptism in the Jordan that the same *Râz* of the Annunciation, the Fire of the Holy Spirit, and the Light of the World are realized in 'act', and the 'God-Son', the *Verbum Dei*, becomes naturalized as 'cosmic'.

The Fire/Verb dichotomy will enter the 'womb' of water and inoculate it with the power of the transmutation of men and women into 'gods', into divine particles, and of the cosmos into Eden.[167]

It is, properly speaking, the sanctification of the universe.[168] Jesus, yielding to the water the *phlogistic* power of His Fire and rendering it a Uterus for His *râzâtic* body, *i.e.,* the Church, aims to transmute into *râzâtic* all that is visible and invisible in the universe and to make out of it a *râzâtic* reality. What kind of synergy must have resulted? Is it not what Peter mentioned in his letter, that what used to need a thousand years to happen, will now happen in the blink of an eye (*rfâf ᶜaynâ* ܪܦܦ ܥܝܢܐ)?[169] Ephrem especially liked this 'blink of an eye'.

One word remains to be said with regard to the above-mentioned kenotic synergy. This time, it is at the practical level and serves as an answer to the question "What must be done?" asked by the rich young man of the Gospel.[170] As a matter of fact, this question, raised by a person so well educated and so wealthy, does not seem to be a supplication but instead falls under the category of a rational resistance to kenosis and, by the same token, to the taming proposed by the Good Master who loved him.

The peak of this kenotic synergy, produced by the concourse of the two gazes, is the conjunction of the divine kenosis with the human one. This conjunction implies the success of a unique effect with an economy of means. In fact, the 'Young Man' did not approach the Good Master empty-handed. He had done a lot in his life and was applying himself to be a righteous man. However, Jesus shocked him by asking of him total kenosis to merit his discipleship. He went away sad, but Jesus too went away sad because He was just at the point of achieving 'here below', between Himself and the Man, the miracle of the synergy that He enjoys 'up above' with his Father. Unfortunately, the 'Young Man', whose soul and mind Jesus tried to penetrate through the eyes, disappointed Him. Jesus tried to make him understand that He was looking for someone like him to realize a precedent for that synergistic experience, but the youth's wealth and worldly self-esteem prevailed.

Kenotic synergy is the phenomenon that will from now on govern the ebb and flow of humans by having a role in the Salvation Economy. It is the synergy between fatherhood as well as motherhood and the child, between the Creator

---

167 *Ibid.*

168 *Ibid.* By His descent in the Jordan, Christ, as it can be deduced, inoculated the waters with the sanctifying virtue of his divinity. *(H Nat, 5)*

169 *Les Origines,* p. 56; 2Pet 3:8. *Cf.* Ps 90:4.

170 Mk 10:17-22.

and the creation, between the divine laws and the natural ones, specifically the scientific ones that derive from them and which helped the *Verbum Dei* to become human, to have a mother, to live and die in order "to make the universe new".

Conclusion

It is for all these reasons that Ephrem used the Persian root *râzâ* (ܪܐܙܐ). Expert in human languages as well as in the Language of the Holy Spirit, as he proved to be, he had to recognize in the DNA of this lexicon what the women of the Old Testament recognized in the DNA of Abraham, Jacob, David and their entire lineage from which the Messiah should come.[171] Ephrem had to admit that *râzâ* is the only expression that is a sign and a symbol of a particular 'mystery' that does not allow confusion. He had to recognize in this "keyword", foreign and non-foreign at the same time, a unifying element of what is veiled and unveiled at the same time, and that, while being 'mysterious', is meant to be visible, understandable, touchable and even edible, without fear of it ever being desecrated and 'destroyed'. It requires instead that it be made as public as possible, and it blames those who do not "preach" it.[172]

The Ephremian foundation stones elucidated above (Eye, Light, Fire, Synergy, *Râz* and Kenosis) will be of great use to us in speaking of Generation and Procession that represent the intrinsic, perfect and self-sufficient movement of the three divine Persons as painted by Rublev in his Trinity Icon. It is from this movement, as we have epistemologically supposed, that the *râzâtic* synergy emerges and illuminates the eye of the mind of those who seek Truth. This synergy between Father, Son and Holy Spirit is the source of the 'veiled power' (*ḥaylâ kasyâ*) of which Ephrem speaks and which witnesses to the Creator's presence in His creation. This divine *ḥaylâ* can only be seen by a 'luminous eye' similar to that of Mary or Paul.

Now, we have to aim ontologically at Fire, as we claim to do in the next chapter, that is, as an indispensable instrument of the Ephremian method. The Fire should be of the same nature as the generated Light. This puts us relentlessly in the face of the following challenge: "Is this 'Fire' from which 'Light' and 'Heat' come and whose main hierophany seems to be the sun created or not?" Therefore, what kind of 'Fire' is it?

We pray the Mother, who conceived by the 'Fire' and who gave birth to the 'Light', to lend us her attention for a 'blink' in order to succeed in the adventure of answering this question.

---

171   Rev 22:10-20.

172   1 Cor 9:16.

Chapter II

The Fire in the 'Eye'

> *When Pentecost day came round, they had all*
> *met together, when suddenly there came from*
> *heaven a sound as of a violent wind, which*
> *filled the entire house in which they were sit-*
> *ting; and there appeared to them tongues as*
> *of fire; these separated and came to rest on the*
> *head of each of them.*

Acts of the Apostles [173]

Before his ascension, Jesus had foreseen this famous event of Pentecost which will establish the Church, His *Râzâtic* Body, like a "Tower", parallel to that of "Babel". He had warned his disciples that it was in their interest that He left because if He did not, the Holy Spirit (Paraclete) would not come to put an end to the "Curse of Babel".[174] If in "Babel" God had descended and made the unilingual builders of the Tower no longer able to understand each other,[175] at Pentecost, He descended by His Fire to remedy the wound caused by the presumption of humans and to realize the opposite. It is now the Church that proposes herself as a meeting place of understanding and peace, as the Garden of Eden was.

We will attempt in this chapter, entitled *The Fire in the Eye*, to reveal the identity of this "fire" and to take a stand in the controversy that Ephrem waged against those who took it for divinity, adored it, and therefore considered it uncreated. How did Ephrem address this problem, and how did he use it as a means to guide the lost people to the True Light?

This issue is what we will try to find answers to, relying on some data from deconstructionist philosophy as used by Jacques Derrida. We will do the same to answer the following question: What fire is it? We will use the key, *râzâ*, to elucidate or describe in a better way the *râzâtic* reality, and find for it a unit of measurement valid to compare with the quantum. To reach our goal without burning our fingers, we will respect the following plan:

---

173   Acts 2:1-3.

174   Jn 16:7.

175   Gen 11:7.

1. The Godhead / Fire controversy

    1.1. The Divine Fire

        1.1.1. Ephrem's position

        1.1.2. The created fire

        1.1.3. The non-created Fire

        1.1.4. The God-Fire

    1.2. The Son-Fire

        1.2.1. Parallelism or kinship?

    1.3. The Spirit-Fire

        1.3.1. Epiphany unveils the essence of Divine Synergy

        1.3.2. The mortal doubt

        1.3.3. Sheol (שְׁאוֹל)

    1.4. The Father-Fire

        1.4.1. The Fire that consumes or the Fire that inflames?

2. What is Fire?

    2.1. The deconstruction of fire

    2.2. The Fire of Faith

    2.3. The friendly Fire and the hostile one

    2.4. Can 'ember' (*gmurtâ* ܓܡܘܪܬܐ) serve as a unit of measurement for the *râzâtic* reality?

Conclusion

1. The Godhead / Fire controversy

The various aspirations of civilizations to find an answer to the question "What is fire?" must have cost tens of thousands of years and millions of deaths caused by the wars of the gods and the loss of many kingdoms, to believe the different mythologies. Various mythologies, specifically the Egyptian, the Persian and the Roman, resolved to consider fire, given its etheric nature and its capacity to etherize all that it joins, "God" *par excellence* or a god or, at least, a divine instrument. That is on the one hand. On the other hand, where mystics have been able to honor the sun with its luminescence and heat as the most perfect hierophany of fire, civilizations were quick to take the sun as the god of gods, the most high, the omnipotent on whom all life depends in the reign of the day and who alone is able to set limits to the reign of the night. These mythologies, therefore, considered the conservation of fire in temples and in houses as sacred,[176] and its theft from the realm of the gods, once humans were short of it, as heroic.[177]

The Bible, having reflected the sacred matters of neighboring civilizations as in a mirror, similarly reflects, already from Genesis, the superiority of fire over the elements of water, air and earth, as classified by the Greeks. One of the signs of its superiority lies in the fact that God used fire to prevent Adam and Eve from returning to Eden: "… the flashing of the flaming Sword to keep the path of the tree of life".[178] God could have encircled Eden by an aqueduct since the term 'fire', until this period of the Genesis, did not yet exist among the terms of the Book, while the aquatic vocabulary abounded. The word "flame" comes to be used for the first time only at this point.

All these mythological and symbolic issues piqued the curiosity of the mind of the human being eager for 'clarity' and 'distinction'. A series of issues forced itself as civilizations intertwined through trade and exchange of information or as they entered economic wars, religious wars, or mere wars of satiation of the superego. The consequent controversy coming out of those issues that preoccupied intellectual and religious circles, among others the issue raised in the Book of Daniel, was whether to believe in the fire as god or to consider it just a creature, *ex nihilo*, of the unique God of Abraham, Isaac and Jacob, as suggested by Genesis.

But even the Scriptures are not definite nor precise in this respect. What does

---

176 Bachelard, Gaston., *La Psychanalyse du Feu; Le complexe de Prométhée*, Folio-essais, Gallimard; 2008; p. 68. This is what the candles of our liturgies, our vows and our heroic acts refer to. Stealing fire in the manner of Prometheus is equated with the acquisition of the Kingdom of God by 'violence'. (Mt 11:12) *Cf.* also Taok, p. 165. From now on we will refer to this book as Bachelard..

177 *Ibid.*

178 Gen 3:24.

this Flame, which protects the boundary between the reign of the Tree of Life and the reign of the tree of death, represent objectively? From where did God take it? Is it really of the domain of uncreated matters and therefore divine, or would it be an angel, or the objectification of a divine quality, such as Justice? Does it belong to the realm of creatures not mentioned in the Scriptures? In Ephrem's time, such questions nourished the sects and the different intellectual currents, as they now nourish atheistic scientific theories. They disturbed the Church and delayed the mission of spreading the Christian faith. Ephrem had to address them in order to defend the Christian faith. How did he do it?

1.1. The Divine Fire

Indeed, the first word of the phlogistic domain appears only in Genesis 11: 3, meaning "to burn, to cook" (*s:rephah* שְׂ רֵ פָ ה), where humans decided to build the Tower of Babel after controlling this fire and discovering its importance in the manufacture of bricks. The single mention suggesting the existence of something flammable before this event, as mentioned earlier, is the flame of the Sword in Genesis 3:24, and the nature of this flame is to be discovered.[179] Its nature also represents another enigma that, according to the universal equation that states "like fire, like flame", is added to the first. This is the enigma of the fire, the source of the flame. But we must always remember that, in the Bible as in civilizations, fire and flame were often instruments of destruction and annihilation, whether among the gods, to cite only Zeus, or among kings, to cite only Nebuchadnezzar. Fire and flame expressed the rage of their lords and carried out their punishment.

Biblically speaking, the most atrocious of these punishments was included in the concept 'hell'. Therefore, fire and flame, in the apex of their action recognized in lightning and thunder, were indispensable weapons which the gods used and, in circumstantial alliances, lent to kings or to their respective peoples. Besides, the crucial point for the feudal lords of this world was to preserve the sacred and exclusive right to the use of these weapons, so that they are used, neither by the gods nor by commoners, but only by themselves in the service of their own interests. This laid the basis for the conflict of interest between them and Jesus of Nazareth.

In this sense, the Christian faith was not considered politically naive by proposing to the peoples as savior a revolutionary man, crucified by his own

---

179   Even in the Arabic version of the Jerusalem Bible, we read what refers to a flaming Sword. Nothing is reminiscent of fire, unless the Cherub is like the Seraph, a spiritual being of 'fire' with four faces and six wings that controls over an angle of 360°. The name Seraph comes from the Hebrew "Saraph" or "seraph" (שָׂרָף) which means fire. It is not by coincidence that Ephrem qualifies the angels as *annariyine* which means beings made of fire, the singular being *annari* (ܢܘܼܪܝܐ).

people on a cross, surmounted by the inscription "I.N.R.I.". The authorities, then, put him to death because he defied their system and preached a kingdom of justice inconceivable to them. The worst was that this revolutionary declared himself immortal, eager to sow 'fire' and 'sword' everywhere. He claimed to hold in his hand the sources of light, of which the sun is apparently formed, as well as the sources of the water of eternal life. He also claimed to control the sources of fire and of all ethereal things, including souls, since He imposed on them baptism by Fire and the Spirit and He revived the dead.[180] Ephrem even taught everyone to sing all these qualities of this God-Man who, before dying, also made of His Body and Blood the Holocaust *par excellence* and perpetuated them in the form of bread and wine, so that His followers feed on them daily and never die.[181] His Holocaust is believed to be the substitute for all burnt blood offerings of all civilizations and the fire that consumed them.

Even though He had confirmed that His kingdom is not of this world, all that He said and taught was a sufficient alibi for the feudal lords of kingdoms and empires to defend themselves more fiercely against this intruder 'king', and to persecute his disciples and followers.

The very fact that the Bible does not mention God is the creator of fire allowed all other peoples, who were well "at home" in their own religions, to refuse this new religion and to defend the beliefs in which they had found their certainty and predictability. Indeed, this analysis allows us to ask ourselves if the Magi did not come on the scene at Christmas just to assert to the Persians that their divinity united itself, substantially, with the Divinity of the Jews to form the Exhaustive Christ. The same thing could be said of the *'Logos'* (λογος) who replaced the *'Sophia'* (Σοφια) of the Bible (*Hikmah* חָכְמָה) to suggest to the Greeks the same hypostatic union between their highest divinity and the Christ of the Jews. At this existential level, the apostasy of the emperor Julian that Ephrem had attacked becomes lawful as a defense of a 'Sun god' who was behind all the glory of his ancestors.[182] It remains that, in order to defend his faith, Ephrem also had to find a stable footing in the face of this scriptural lapse, the linguistic lapse, and the aggressiveness of other civilizations, especially that of the Jews. The latter categorically refused to recognize the Christian faith and, even worse, were striving for its annihilation. It was, therefore, indispensable for Ephrem to shed light on the origin of this Fire from which all 'light' comes at the divine and the human levels. Ephrem, having carefully weighed the pros and cons of battle, built a fortress to position himself there. His answer to the overarching question "Where does the fire come from?"

---

180   *Cf. H Fid* 1, 4-7. *CSCO* vol. 154, pp. 3-4.

181   Jn 11:25.

182   The Roman era represented by the ruins of Baalbek (Heliopolis) is proof.

will provide this stable footing in this battle. Let us, then, look at it more closely.

1.1.1. Ephrem's position

Ephrem wrote:

> Heaven, earth, fire, wind, and water were created from
> nothing as Scriptures bear witness. However, the light, which
> came to be on the first day, along with the rest of the things
> that came to be afterward, came to be from something...
> [Moses] said, God created heaven and earth. Even though it
> was not written that fire, water, and wind were created, it was
> not said either that they were made. Therefore, they came to
> be from nothing just as heaven and earth came to be from
> nothing. (*C Gen.* 1/14)[183]

Without looking at where Ephrem got his ideas and his inspirations, as well as his certitude, we find him endeavoring to distinguish between a created fire and a non-created one. He proceeds by differentiating between the created light of the beings that the myopic people worshiped as gods and the non-created Light that Christ declared to be His exclusive personification. By the same fact, he finds himself forced to point out the difference between the fire that produces the first one and the Fire from which proceeds the second. And since human reason had already learned to distinguish, thanks to Aristotelian logic, between the causative cause and the other causes, it also knew how to relate 'light' to 'fire' and to recognize in it, because of its ethereal power, a god to worship. That is why, once in Edessa, the capital of the veneration of the sun god, Ephrem had to clear up the enigma that stands behind the created fire and the non-created One and, consequently, what stands behind the nature of each relative 'light'.

1.1.2. The created fire

Regarding the created fire, Ephrem disconcerts his adversaries by his commentary on Genesis with a rare apology which may well be called Aristotelian. He will succeed in proving that fire and its light, and even the sun, are creatures (*ityê* ܐܝܬܝܐ). He cleverly unveils what has remained hidden in the narrative of the

---

183 *Les Origines*, p. 50. *Cf.* E.G. Mathews & J.P. Amar, *Saint Éphrem the Syrian, Selected Prose Works: Commentary on Genesis; Commentary on Exodus; Homily on Our Lord; Letter to Publius.* Translated by E. G. Mathews Jr. and J.P. Amar. Edited by K. McVey [The Fathers of the Church, 91], Washington DC, 1994. p. 85. *Cf.* https://books. google.com.lb/books?id=kjukkUUG3gYC&printsec=frontcover&source=gbs_ge_ summary_r&cad=0#v=onepage&q&f=false (Accessed Jan 2019)

Scriptures by evoking the underlying meaning so well concealed in the bosom of the words themselves.

Although by the statement of the aforementioned paragraph against the Gentiles (*C Gen. 1/14*), Ephrem, as a good pedagogue, insisted with some modesty on emphasizing the vital importance of the objects adored by his adversaries, by giving them a Biblical basis. He wrote to them, saying:

> Because that first light was indeed created good, it rendered
> its service by its brilliance for three days and it also served, as
> it is said, in the conception and birth of everything that the
> earth brought forth on the third day. [184]

So Ephrem does not deny that the light, the fire that stands behind it, and its warmth were of paramount importance in the created world. However, he emphasizes that they acquired this importance only in the second place, in rendering service and fulfilling their mission just for the time reserved for them.[185] In the meantime, the sun that will be the substitute for them by its emergence on the fourth day was there, hidden in the firmament.

> It is said that from light, now diffused, and from fire, which
> were both created on the first day, the sun, which was in the
> firmament, was fashioned...[186]

Ephrem concludes this crucial point of the creation *ex nihilo* and of what followed the first day of Genesis by affirming with clarity, certainty, and authority that the verb 'to create' is specific to what was described of the divine activity of God at the first day, more precisely, at verse one of the first chapters of Genesis. After that, nothing is going to be created, properly speaking, but instead made or fashioned. Even Adam's body is going to be modeled.

At this level of our analysis, a question arises: what would happen to the Light of which Jesus was the referent, and also happen to the Fire of the Holy Spirit mentioned in our epigraph? What would happen, then, to our Nicene Creed, which asserts that the Son-Father relationship is Light from Light and not Light from Fire?

---

184   *Ibid.* p. 81.

185   *Ibid.* pp. 80 & 82. Ephrem gives great importance to the succession of causes and roles of the intermediate creatures, especially the elements created *ex nihilo*, the day before the first day. They are creatures predestined for a specific purpose. Once the purpose is reached, they disappear from the text.

186   *Ibid.* pp. 81-82.

1.1.3. The non-created Fire

This beautiful apology that we have just expounded, and whose effectiveness would seem to have a specific scientific aspect timed not by the second, but rather by 'God's blink', must have sharpened the mind to get an answer to the previous question: would Ephrem, while refusing the adoration of fire and light as created beings, dare to preach their adoration as non-beings (*lâ îtyê*), that is to say, as God, with a capital 'G'?

It is evident to Ephrem that Genesis says nothing about fire. What he will now consider with regard to this controversy will have its source in the famous biblical chaos, the *tohu wa bohu* (ܬܗܘ ܘܒܗܘ ܘ ܘܗܘ, תֹהוּ וָבֹהוּ), which would be a transient environment ('milieu') and which could serve as a type of chaotic reality resulting from the *nihil* (*lâ medêm* ܠܐ ܡܕܡ) or as a source of symbols, *par excellence*, for all that is ambiguous, amorphous, unorganized, in short, imperceptible since it is not intelligible.

As a matter of fact, the '*tohu wa bohu*' does not make sense. However, for Ephrem, all its "sense" lies in the fact that it has "no sense". Having "no sense", this nothingness, this *nihil* (*lâ médêm* ܠܐ ܡܕܡ), also taken for a 'milieu', an environment, a field or a space *(atrâ* ܐܬܪܐ), becomes a representation either of the pre-rational cacophonous stage of the human psyche or of the domain of divinity, which is imperceptible and non-intelligible.

It is a 'milieu' to which we can refer all that is potentially grasped by common sense, without necessarily being so in act, as in the present case of the non-created 'fire'. This is the only hypothesis that provides an answer to the question asked previously about the source of the flame of the dazzling Sword. Furthermore, since at the end of the sixth day, and before resting, God saw that everything He created was good, even totally, and since nothing can be 'good' apart from God Himself, this would lead to the deduction that:

1- The *tohu wa bohu* is good.

2- The *nihil* (from which Genesis took place) is good because nothing good can come from a bad source.

3- If, after all, there is something good in what was created, it must be the imprint of the Good One who created it. Therefore, we are obliged to determine both the place (*atrâ*) of each of these traces and the place where this Good One placed all creation: is it outside of Him or in Him?

How does one distinguish between the created fire and the non-created One, and then designate the respective places where each is located? This is the mission that Ephrem would begin by meditating on the Mirror,[187] the Pearl, the Prologue of the Johannine Gospel etc., and murmuring to the rhythm of the beating of his heart the following words: "Behold, the Light left His 'milieu' and came into ours" or "The Light appeared to the righteous ... Jesus, our Lord, has shone upon us from the Womb of His Father ..."[188]

Only a childlike logic at the height of the symbolism of the story of Genesis, a logic consistent with the invitation of Jesus to become children anew, would allow us to say that it is indeed, in the *nihil*, the non-being and non-place, that this invisible and illegible Fire of the first day took place. This Fire that Ephrem shows to have been in existence even before the first day also belongs to non-time. What would be the nature of this Fire and that of the 'milieu' from which this same Fire came?

The answers of classical scholars are divided into two categories. Some say that all this is absurd because it goes beyond our concepts. Others assert that this Fire can only be God Himself because He alone, by concept, has always existed. What does Ephrem say?

1.1.4. The God-Fire

The same allegory of the created fire hidden in the concept "earth" can be applied to say that the divine Fire was also hidden in the concept "*ha šamayim*" (הַשָּׁמַיִם) which, according to Psalm 115, means "heaven" and "heavens" at the same time, since the Hebrew root of this word does not admit a regular singular. Based on this particularity the various translations of the Bible confront us with inevitable confusion. Some of these translations say, "our God is in heaven", others "in heavens"; the concern of the nations receiving the Gospel remains the same in both cases: how to conceive of the habitat of the Creator.

Verse sixteen of the same Psalm suggests a similar 'heaven' while singing, "The heaven, even the heavens, are the Lord's." The plural gives additional flexibility to this psalm and satisfaction to the *Pater Noster*. The importance of this point becomes more sensitive once compared to verse 4 of Psalm 148 which says: "Praise him, ye heavens of heavens, and ye waters that *be* above the heavens", which reserves, at least, some Heaven beyond the water of creation.

In short, considering the implication, the hidden meaning (*kasyâ* ܟܣܝܐ), as real and valid for the apology as the apparent one, Ephrem opens a window on the Sun which is in the "heaven of heavens", as it appears in the Syriac text, where Fire,

---

187 *TLE*, p. 39; "Ephrem was clearly fascinated by mirrors... the metal ones that had to be kept polished in order to reflect the light and the image of the beholders..."

188 *Cf.* Footnote 53.

its Light and its Heat of divine nature have always resided.

Based on this, the three qualifications, Fire, Light and Heat, must belong naturally to the three Divine Persons. They will no longer be attributed to them by analogy. They will become rather analogically applicable to humans. To say, for example, that a person is loving, radiant and warm, is to attribute to him/her qualities that are originally divine.

Ephrem no longer doubts the link between Fire and God, especially at the level of the Trinity. The palpitations of his heart assure him. He will apply his theory to bring human good sense closer to the divine Good Sense, and not farther away. He will develop through his ascetic life all his spiritual, theological and poetic qualities in order to become a Lyre of the Holy Spirit. Through poetry, he will strive to convince everyone, without distinction, that the Father, the Generator of the Man named Jesus of Nazareth, the Light of the world, is Fire. Furthermore, he shows that the Triune God is *par excellence* the Source of all life and all that is necessary for life: the sun, fire, light, heat, etc. All that is at the disposition of humans and their welfare, whether created or made, comes from Him.

What did Ephrem do to popularize the frustrating Christological enigmas that flow from the conception God-Fire contained in the following expressions: "I am the Light of the world", "Baptize with the Spirit and with Fire", and in the description of the descent of the Holy Spirit on the disciples in the form of tongues of Fire"?

1.2. The Son-Fire

While ridiculing those who worshiped creatures (among others, fire),[189] and putting an end to this polemic with his theory of *kasyâ* and *galyâ* (ܓܠܝܐ ܘ ܟܣܝܐ), Ephrem, was at ease to bind, as we have seen, to the extreme limits, the 'fire' and to the Triune God of the Bible, Father, Son and Holy Spirit.[190]

Sebastian Brock wrote under the title "The divinity as fire" in his famous book *The Luminous Eye*: "Ephrem very frequently describes the divinity as fire."[191] Brock added quoting Saint Ephrem: "Fire entered Mary's womb, put

---

189   *Les Origines*, p. 51.

190   *Cf. Les Origines,* p. 107. If Ephrem was reluctant to describe the celestial bodies in his *Hymns against the Heretics*, written in an environment close to the Chaldean influence, it is not the same for the *Hymns on the Faith* written in Edessa where he is not afraid to speak of the sun with admiration. His purpose was to prove to the pagans that this Globe they adore is merely the symbol of the Trinity and the divine Filiation.

191   *TLE,* p. 38.

on a body and came forth."[192]

Ephrem rejoices in the Spirit and interprets the Event to the faithful:

> Blessed are you, my brethren,
> For the Fire of Mercy has come down
> Utterly devouring your sins
> And purifying and sanctifying your bodies.[193]

They are blessed because, by experience, Ephrem is sure that by this Compassionate and Purifying Fire, though it is in some cases 'consuming' and even 'devouring',[194] the "wombs" (*ʿubê* ܥܘܒܐ) are reassured – whether they be wombs of mothers, ears,[195] eyes, water, even the womb of wood, – and therefore the hearts and consciences (*reʿyânê* ܪܥܝܢܐ). Even the Womb of the divine Father is reassured, because He finds predictability and satisfaction thanks to the synergy between Him and the Son, especially after the Incarnation.[196]

With the concept 'womb', so dear to Ephrem, we find ourselves right at the heart of our subject, at its core. In fact, the angel Gabriel said to Mary in the famous Annunciation: "The Holy Spirit will come upon you, and the Power of the Most High will cover you with its shadow. And so the child will be holy and will be called Son of God".[197]

---

192 *H Fid* 4, 2-end, *TLE,* p. 42. *Cf. CSCO* vol. 154, pp. 9-16. "To those who possess it, he also gave to see the visible fire' of the Son who heals Simon Peter's mother-in-law from her visible fever. (*H Virg* 25,14) The presence of Christ in the womb of the Virgin Mary and in the 'Bread' is similarly described as a 'fire' (*H Fid 10,17)." Cf. TLE, p. 94 ; CSCO* vol. 154, p. 51. The original text:

ܗܐ ܢܘܪܐ ܘܪܘܚܬ ܒܥܘܒܐ ܕܝܠܕܬܟ
See, Fire and Spirit in the womb that bore You,
ܗܐ ܢܘܪܐ ܘܪܘܚܐ ܒܢܗܪܐ ܕܒܗ ܥܡܕܬ ܗܘܝܬ
see, Fire and Spirit in the river in which You were baptized.
ܢܘܪܐ ܘܪܘܚܐ ܒܡܥܡܘܕܝܬܢ
Fire and Spirit in our Baptism,
ܒܠܚܡܐ ܘܒܟܣܐ ܢܘܪܐ ܘܪܘܚ ܩܘܕܫܐ
In the Bread and the Cup, Fire and Holy Spirit.

193 *TLE, p. 38. Cf. H Epiph* 3, 10; *Cf. CSCO* vol. 186, p. 148.

ܕܢܚܬܬ ܢܘܪܗ ܕܫܝܢܐ ܠܟܘܢ ܕܐܡ ܐܟܚܕ ܠܚܛܗܝܟܘܢ
ܘܕܟܝܬ ܘܩܕܫܬ ܦܓܪܝܟܘܢ ܐܟܠܬ ܠܚܘܒܬ ܚܘܒܝܟܘܢ

194 Deut 4:24; Heb 12:29.

195 Just as from the small womb of Eve's ear Death entered in and was poured out. So, through a new ear, that was Mary's, Life entered and was poured out. (*H Eccl* 49, 7) *Cf. TLE,* p. 33 and Rom 10:17-21.

196 Lk 12:49.

197 Lk 1:35.

Therefore, there is no doubt that, in this context, it is the Son, the Word,[198] who is considered Fire. It is the Son, the *Verbum Dei* (λογος), Who will later say of Himself that He is the Light of the world, Who entered into a human "womb", put on a body, and appeared. Ephrem, thanks to his *râzified* intellect receptive to the waves of Revelation, was able to relate the Word and the Fire in a human 'womb'. How was that possible?

1.2.1. Parallelism or kinship?

Bou Mansour, commenting on Beck, the father of the resurgence of Syriac literature, emphasizes the parallelism on which Ephrem's poetics relies. Two parallels meet, although belonging to different realities, as in the case of the divine Nature meeting the human one in the maternal womb. The poetics of Ephrem helped do it without the risk of falling into the errors of monophysitism or of dualism, both of which he categorically refuted. Bou Mansour explains saying: " Ephrem draws a parallel between God's Nature and man's, as well as between the divine Nature, described as Fire, and the physical one of the human body. Moreover, he expresses his astonishment at their meeting in Mary's womb:"

> For He is God by His entrance and Man *(barnâšâ ܒܪܢܫܐ)*
>   by His exit.
> Astonishment and confusion to our understanding,
> Fire entered Mary's womb,
> Put on a body *(lebšêt pagrâ ܠܒܫܬ ܦܓܪܐ)* and came forth.[199]

Although we praise the parallelism as we praise the challenge of the contrasts in the Ephremian poetic style, we insist on the verb "to relate" when referencing "womb", for kinship and parenthood come from it. Ephrem pointed out that God and Fire on the one hand, and Human and body on the other hand, which are "parallels" out of the "womb", become related by parenthood that will have no end *(lâ ʿâbar ܠܐ ܥܒܪ)* once they meet in the maternal womb, beyond their hypostatic nature *(qnumâ ܩܢܘܡܐ)*, to fulfill the Economy of Salvation. This is where synergy takes on its full splendor.

---

198   *On the Trinity*, XV, 20; Cf. http://www.newadvent.org/fathers/130115.htm [Accessed June 2019]
      We find in Augustine (*De Trinitate*, XL 20) a very interesting description of the relation Word – Verb. He wrote: "Accordingly, the word that sounds outwardly is the sign of the word that gives light inwardly; which latter has the greater claim to be called a word... And as our word becomes an articulate sound, yet it is not changed into one; so the Word of God became flesh, but far be it from us to say He was changed into flesh." This is perfectly applicable in case the Word, even the Verb of God, is considered Fire.

199   *H Fid* 4, 2. Cf. *La Pensée*, p. 234 ; Cf. *CSCO* vol. 154, p. 9 ; Cf. also *TLE*, p. 38

What kind of kinship would that be?

It should be of the same nature as the relationship between Nebuchadnezzar's fire and the three 'young men' in the furnace of 'Babel'.[200] It should also be of the same nature as the relationship between Jonah and the whale.[201] Here, the "hidden" element that stands behind Mary's visit to Elizabeth becomes apparent. What does that mean?

This means that, even without the visit between their mothers, Jesus and John would be relatives all the same, but with the visit, the situation becomes different. He who gave the child to Elizabeth does not recognize, for His Kingdom, the kinship of human flesh and blood. He will declare this, later, as a condition *sine qua non* for the acceptance of believers in his domain.[202] He recognizes only the kinship that comes from the Spirit and Fire because it is from it alone that the Light springs. For this reason, we would say, that Mary hastened to visit Elizabeth, driven by the same impulse that made her say *Fiat*.

It is important to emphasize, from now on, this figurative kinship, the 'maternal womb', which exceeds all expectations. Out of it, Ephrem joins several types (*ṭûpsâ* ܪܚܘܘܬ) to make possible the shift in meaning from mystery to *râzâ*. It is only in that 'womb' where parallels are able to cross, in the same plane, that of the Economy of Salvation, without having their relative substance affected or their relative nature undergoing an ontological change. All that changes is Creation, by "chiasmus". It sublimates itself into a *râzâtic* reality, radiant, luminous, phlogistic, where 'parallels' ratify and bless each other. This is what *râzification* means, since by it, the Might of God makes possible what is not possible for humans, and the Trinitarian synergy wins by its momentum.

What about the Spirit of God? Is He also Fire?

---

[200] *La Pensée*, p. 234, footnote 70: "A third intermediary is always necessary for this kinship. The angel whom Nebuchadnezzar compared to a son of the gods (Dan 3:92) and who blew in the middle of the furnace on the three young men like a coolness of breeze and dew (Dan 3:50) would represent a prophetic sign of a Messianic baptism that would make the believer immune in the sense that "one of his hairs would no longer fall independently of His heavenly Father" (Mt 10: 29 - 30). It is a sign of the hidden Power (*ḥaylâ kasyâ* ܚܝܠܐ ܟܣܝܐ) which will come later from the *râzâ* as the Maronite Liturgy intones in a Hymn sung after the reception of Holy Communion. It says: "I have eaten your Holy Body, the fire eats me no more." (*deklet paghrâk qadisâ nurâ lâ teklan* ܕܐܟܠܬ ܦܓܪܟ ܩܕܝܫܐ ܢܘܪܐ ܠܐ ܐܟܠܢ) The kinship with the divine can only be by the Spirit and the Fire, or by the Spirit Who is divine Fire and Who is in the Breath, the Water, the Bread, the Wine ...

[201] Jon 2:1-2.

[202] Mk 3:33-35. "... Who are my mother and my brothers?" ... "Whoever does the will of God is my brother and sister and mother."

1.3. The Spirit-Fire

The scene of Pentecost bears, by itself, a sensitive proof of the phlogisticity of the Holy Spirit. Brock continues to write under the same theme: "Fire is, too, the symbol of the Spirit."[203] This assertion is confirmed in other Ephremian figures:

> The Spirit is in the Bread, the Fire is in the Wine; [204]
>
> See: Fire and Spirit are in the womb of her who bore You;
>
> See: Fire and Spirit are in the river in which You were baptized.[205]

However, it is in Ephrem's *Hymns on the Epiphany* that one is in the presence of the polysemy of fire, sometimes seen as a fire fanned by the wicked,[206] other times as the fire of baptism representing the Holy Spirit.[207] What was the role of the Holy Spirit in the baptism of the Son?

Let us imagine, with Ephrem, the Fire descending into the water so that "all justice is accomplished".[208] It is clear here that Ephrem alludes to the one Baptism that Christ could receive and through which He will baptize humanity and all Creation: fire is only baptized by fire,[209] for water, scientifically and logically, would extinguish it or stifle it.[210] Even John the Baptist did not grasp what Jesus meant by, "Let all righteousness be done." If Jesus insisted on descending into the profane water of the world, it was because He saw that He was thus fulfilling the Will of His Father. The Father will approve on the spot through remarkable words that remind the Son that as He divinely generated Him from all times, He

---

203   *TLE*, p. 38; (*H Fid 40, 10*)

204   *Ibid.* (*H Fid 10, 8*); *Cf. CSCO* vol. 154, p. 131 for 40, 10 and p. 50 for 10, 8.

205   *Ibid.* p.94 & 108; (*H Fid 10, 17*); *Cf. La Pensée,* p. 234, and more precisely *La Pensée,* p. 397. *Cf. CSCO* vol. 154, p. 51.

206   Böer, Paul A., Sr. *Hymns and Homilies of St. Ephrem the Syrian*, Veritatis Splendor Publications 2012, p. 278. Later on we will refer to this book as Böer.
       We would say that 'wicked', here, means the human mind subjected to cosmic entropy, and which, having not yet received the Holy Spirit, worshiped the profane fire as god, thus aiming at all phlogistic deities probably existent ... *H Epiph* 8, 7. *Cf. CSCO* vol. 186, p. 170.

207   *Ibid. Cf. La Pensée*, footnote #70. *H Epiph* 8, 6; *Cf. CSCO* vol. 186, p. 170.

208   *Ibid.*

209   Mitchell C.W., M.A., C.F., *Against Mani*, Volume 2, p. 6/11, (1921). "And if Fire was mixed with Fire, and Water with Water and Wind with Wind, it necessarily follows that Light also (was mixed) with Light! Now that these 'natures' are akin one to the other, all reasonable beings know, apart from madmen – but perhaps even madmen apart from the Manicheans." *Cf.* http://www.earlychristianwritings.com/fathers/ephraim2_7_mani.html; [Accessed June 2019].

210   *Ibid.* p. 5/11; "...and how does Water love Fire that absorbs it or Fire love Water that quenches it?"

is also generating Him at this proper time. The Son, for His part, by His act of obedience, proves to the Father that He ratifies at once both His generation and the purpose of His mission.

Epiphany, considered the event of the appearance *(denḥâ)* of the Trinity for the first time in the created universe, becomes an act of renewal of the Pact between Father and Son for an eternal generation and obedience.

The baptismal water that will once again sanctify the nature of the whole universe has been transmuted into a phlogistic "womb" from which the purifying and sanctifying Fire could be drawn without harming anyone. The passage from the eternal to the temporal, which the Economy of Salvation requires, must not change the synergy of the Father-Son relationship. It will instead be the subject of reactivation, as said above, of the synergy between Creator and creature, 'broken' by the disobedience of a single day. We invite ourselves to dive again into the baptismal water in search of the essence of this synergy.

1.3.1. Epiphany unveils the essence of Divine Synergy

The Father, Who eternally generates the Son in a way unnoticed by humans, highlights at this specific moment the power of compassion between Father and Son. It goes beyond both eternal and temporal qualifications. Between Them, everything happens in the Holy Spirit, in a reality "differAnt" from both the divine and human. It is in the reality of the Incarnation, in which the Baptism is only a stage, that everything happens. The Holy Spirit is its witness as He is the guarantor of the resulting Flame that will be distributed on the heads of the twelve Apostles, and on Mary [supposedly][211] and the other women at Pentecost. However, words like those read in Matthew and Mark, "You are my beloved Son; You have all my favor, my affection" and "In you I found all my pleasure", are what should be heard in the same way by every baptized person who desires to be taken by the same synergy of the divine generation.[212]

The Holy Spirit Who proceeds from the Father and the Son was there, in the form of a dove whose moving wings recall the movement of the flame of Moses's

---

211 We added "supposedly" on purpose in agreement with verse 35 of the Annuciation event: "The Holy Spirit will come on you, and the power of the Most High will overshadow you. So the holy one to be born will be called the Son of God." Furthermore, the sacred text of Acts does not specify that Mary received the same Flame, nor does it mention any role for her at all in the birth of the Church, the *râzâtic* body of her Son, a role that our book fully supports.

212 Lk 3:22; Mk 1:11; Mt 3:17.

bush, to accomplish with Him, the baptized One, the baptism of the water.[213] Water should itself be baptized by the Spirit and Fire[214] and transmuted into a phlogistic "womb" to give birth to the children of God, His Father.[215]

By saying, as reported by the Gospel of Luke, "You are my beloved Son, you have all my favor – affection–", the Father confirms the words, "You are my Son; I myself today, I have begotten You."[216] This synergy, coming from the continual generation of the Son by the Father, a generation that is supposed to be without the intermediation of the Holy Spirit before the Incarnation and by the Holy Spirit at the Incarnation in space and time, has caused so many troubles and controversies to the Church. In response, Ephrem was prompted to write:

> Whoever is capable of investigating
> Becomes the container of what he investigates;
>
> A knowledge which is capable of containing the
> Omniscient is greater than Him,
> For it has proved capable of measuring the whole of Him.
>
> A person who investigates the Father and the Son is thus
>   greater than them!
> Far be it, then, and something anathema,
> That the Father and Son should be investigated,
> While dust and ashes exalts itself![217]

"You are my Son, I have today begotten You" can only be an echo of the symbolism that is oriented toward the five senses of humans, since their eyes are not sufficient to truly see God. Ephrem will indeed repeatedly go back to the eye-ear relation to prove that under the *râzâtic* angle the ear can substitute for the eye to see the divine realities. Thus, the Oracle of the Father that crowned the Epiphany did not bear witness only to the Son but fulfilled His begetting. The Epiphany for Ephrem, as for the Eastern Church, was the great feast after Easter,

---

213 Augustine talks about a 'globe of Fire' (*De Trinitate*, III, preface and Chapter I, 4.*Cf.* http://www.newadvent.org/fathers/130103.htm [Accessed June 2019]. *Cf.* also Taok, p. 164. Referring to J. Fräzer, Taok evokes a mythological bird, creator of fire: "... The creator of the fire is a small beautiful bird that carries on its tail a red sign that is only a stain of fire."

214 *H Fid* 10:17. *Cf. TLE,* p. 108. (See, Fire and Spirit in the river in which you were baptized; fire and Spirit are in our baptismal font.)

215 1 Cor 15:51-58. Paul offers a wonderful application of this transmutation. We wonder why not say transubstantiation? We prefer to reserve this substantive for the Eucharist, as the term hypostasis is reserved for the divine Persons.

216 Heb 1:5; 5:5; *Cf.* 1Jn 5:1; Ps 2:7.

217 *H Fid* 9,16. *Cf. TLE,* pp. 26-27.

not Christmas, notwithstanding that some Eastern Churches continue to celebrate Christmas with the Epiphany to this day. For them, each of these festivities implies an Epiphany different in form, but identical in substance. The words of the Godhead, the Father, can be understood only in the light of an affirmation of His perpetual generation of the Son and, this time, with its actualization in the space-time of our universe. As with the Incarnation, it is through the Holy Spirit that this generation is perpetuated.[218] The emphasis on "today" in the temporal sense is for us humans, since the calendar does not exist in eternity. It is nothing but the expression of a continuous progressive present tense which is supposed to be the unique tense to be conjugated in eternity and which confirms that the generation of the Son by His Father, as with the procession of the Holy Spirit from the Father 'through' the Son, form the unique and eternal Synergy at the Womb of the Trinity.

The 'Son-Fire' in the profane water was not alone for sanctifying the water of the abyss.[219] The Holy Spirit also descended on this water and the power of the Most High covered it with His Shadow, and the Holy Spirit with the abyss, for this abyss, generated the Messiah.[220] A transmutation, similar to that which will take place in Cana, took place in this baptism, but this time, it is the profane water that has become sacred water. It is, we would say, from this water that Jesus will later bring out the wine.

Did Mary, in asking her son to provide for the hosts, know that the water in the jars was "differAnt", as she already knew that John the Baptist himself was "baptized" by her Son at their meeting with Elizabeth, her six-month-pregnant relative? The preparation of the water by the Fire for the birth of the 'Wine' is similar to the preparation of Mary by the "immaculation" in view of her acceptance to conceive in her Uterus the *Râz par excellence* and to clothe Him from her maternal flesh.[221]

This *Râz* will replace water, fire, light, and wine, as well as bread and all that is

---

218 *H Fid* 10-17. *Cf. TLE,* p. 108. (See, fire and Spirit are in our baptismal font; in the Bread and the Cup are Fire and Holy Spirit.)

219 *Les Origines,* pp. 45-46; 57; 68: *brîthun da-šmayô* (the creation of heaven) or *brîthun da thumô* (the creation of the Abyss); *Cf. La Pensée,* p. 155.

220 A typological nuance is to be noted between the Son's generation and the Messiah's generation.

221 Clothing the adjective 'maternal' with a transitive verb makes the style more direct and therefore, the meaning stronger. The verb that we have allowed ourselves to forge for the occasion suggests an act and an agent behind the act and highlights the direct object of this act which is Mary and not her Son.

good and is pleasing to humans.[222] Moreover, He will replace everything that was offered in oblation to redeem all life, according to the Jewish law, namely lambs, bulls, ewes, in short, all kinds of sacrifices, especially the bloody ones.[223] In the case of the elements that He does not replace, gold for example, they would at least be taken by the synergy of the Incarnation and would become fit to be 'burnt offerings' of the New Testament. This is the scale that the *Râz*, with a capital R, covers and in which converge all *râzê* whose roots reach back to the first *'fiat'* of Genesis.[224]

The enigmatic word *râzâ* that has been grafted onto humanity by the Incarnation and is now enjoying the qualities of the graft submits in its own way to the same continuous present tense, to the progressive transmutation toward the fullness of its meaning, as the scholar Bou Mansour mentioned for this case. It is a transmutation in the making as long as Creation is in the making, and as long as the Economy of Salvation is as well.

What stands behind the modification of the meaning of *râz* is, on the one hand, the dialectic of freedom and duty between Creator and creature, and on the other hand, the hidden Power (*ḥaylâ kasyâ*). Both energize the *râz* and make it capable of revealing itself through Science and Grace, as the human intellect becomes empowered to better distinguish between the dimensions and between the realities that influence its unveiling.

If 'the *Râz*' covers the meaning of "generating" in the absolute from the Womb of the Father, it does not mean the same when it covers generation in the concrete through a 'womb' oriented toward a specific mission. It is, after all, the art of understanding verbs and words. To say, for example, "I love", in the absolute, is not the same as saying it to a particular person who represents a center of affection or passion. The same goes for "donating". The Son, generated from the Womb of the Father before the beginning of time, was donated as a gift to humanity through the womb of Mary and to the universe, wounded by the sin of Adam, through the 'womb' of Nature, the water of the Jordan.

That the divine Voice repeats each time, "You are my Son; I, today, have begotten you", implies that we must understand it in coordination with the *Sitz im Leben* of the event, as well as with the audience to whom the message is addressed. "Those who have eyes and see, ears and hear"[225] would believe and confirm that

---

222 *Les Origines*, p. 126.

223 The Maronite Liturgy insists on the non-sanguinity of the offerings. *Cf. Maronite Missal* (Arabic), Bkerke, Lebanon, 2005, p. 728.

224 The 'pearl' to which Ephrem dedicated a number of hymns is a symbol of the aforementioned *Râz*. This is what Ephrem says: "This 'pearl' satisfied me, replacing all the books, all their interpretations and all their readings." (*H Fid* 81, 8) *Cf. TLE*, p.106.

225 Mt 13:15; Act 28:27.

all the so-called "acts of generation", in spite of their multiplicity, accomplish one unique act and also represent one unique movement between Father and Son, which in turn, and through the procession of the Holy Spirit, causes the unique synergy between the three divine Persons.[226]

Generation, procession, and hypostatic movement, which were not on the horizon of humans, have entered, for once, into this horizon of *râzâtic* reality that we consider spatio-temporal reality, and it has become possible for humans to describe this happening with their imperfect instruments, submitted to the different tenses, and providing only relative and limited meanings. The Generation in the created universe was, is, and will always be by the Holy Spirit, that Fire Which ceaselessly proceeds from the Son and the Father, and with Which the Son will baptize all, first by the water of His baptism, then, after the crucifixion, by the 'Fire' of His blood,[227] to make everything new.[228] It seems that after consultation among the three Persons of the Trinity nothing is done in this world of relativity except by the Holy Spirit. He is the comforter; He is the illuminator.[229]

He is the one who stays with us and teaches us the things of the Kingdom. It is He, and not the Son, Who is the guarantor of umbilical synergy. Is this not why the Son emphasized to His disciples the necessity of His leaving because if He did not, the Paraclete would not come to them? Has He not assured powerful protection to this Spirit? Do we have the right to doubt the phlogisticity of the Holy Spirit?

1.3.2. The mortal doubt

To doubt that the Spirit is fire will lead to a crisis in the fire-water relationship, that is to say the relationship between the two formulas: to baptize with the Holy Spirit and fire, and to baptize with the Holy Spirit and water.[230] Let us compare these two declarations:

---

226   *Cf.* Roublev's Icon.

227   Lk 12:50; "I have a baptism with which to be baptized, and what stress I am under until it is completed!"

228   Rev 21:5; "And the one who was seated on the throne said, 'See, I am making all things new.' Also He said, 'Write this, for these words are trustworthy and true.'"

229   According to the consecration words of the Maronite Mass, the Person of the Holy Spirit is considered ܪܫܝܬܐ ܘܫܘܡܠܝܐ ܘܚܪܬܐ, which is translated in the *Book of Offering* (2012, p. 766) by "the beginning, the end, and the perfection of all that was and will be in heaven and on earth..." For more precision, according to our linguistic deconstructive method we understand it as: the initiator, the fulfiller and the perfecter of...

230   Jn 3:4. Why was Jesus not satisfied with saying "of the Spirit" as is written in Lk 3:16, and how do we satisfy what John himself wrote in Jn 1:33?

The Baptist said, "Jesus will baptize with the Holy Spirit and fire."

Jesus said, "Unless you are born of water and Holy Spirit..."

Can we deduce from these two baptismal formulas that water is fire, or merely the recipient of fire? What fire? In the case where this water is the container, what kind of water would it be? Would it be the kind of water that flowed from Jesus' side on the Cross, knowing that the Maronite Liturgy mentions this water, once with the wine in the preparation of the Oblations that will become the Blood of Jesus Christ, and another time, as a womb in its baptismal rite?[231] Or would it be of the kind of *nephthar* (νεφθαρ) or *naphtha* (νεφθα), the phlogistic water whose symbol existed during the time of Nehemiah and the Maccabees?[232]

The question that imposes itself here as for all *râzê* reads: what makes a type or symbol like water become the substance of a sacrament, a *râz*, and not a 'mystery', in the ecclesial meaning?

According to Ephrem, the answer is unique and definitive. It is the Holy Spirit, just as happened at the Creation, at the Incarnation, at the Epiphany, as well as at Pentecost, the foundation of the Church, called the Mystical Body of Jesus Christ. We propose, henceforth, to qualify it as His '*Râzâtic* Body'.

Here, the legend of the birth of the 'pearl' (*margânitâ* ܡܪܓܢܝܬܐ)[233] that Ephrem highlights in his hymns of the Epiphany satisfies any doubt or questioning. Sebastian Brock wrote:

> It was widely thought in antiquity that the pearl came into existence when lightning struck the mussel in the sea. This birth through the conjunction of two disparate elements is seen by Ephrem as a symbol of Christ's birth in the flesh from the Holy Spirit, the Fire, and from Mary, 'the watery flesh'. Christ is thus both the Eucharistic Pearl and the Pearl born of the Holy Spirit and Mary. It is to the parallelism between Incarnation and Eucharist that we now turn.[234]

It is then by the flame, the spark, the lightning (*šalhêbitâ* ܫܠܗܒܝܬܐ), that the divine reality joins the human reality while respecting the order of signs and

---

231 *Maronite Manual of the Sacrament of Baptism*, Bkerke, 2003, p. 48.

232 2Macc 1:36. This term is found only twice in the Bible. The second time is in Dan 3:46 where the liquid will be used to increase the flames over the furnace.

233 Pearl and coral are both products of the warm seas and both are used in jewelry and may have the same whitish color. It would be very interesting to know the most popular ornament worn by girls and women of Ephrem's area and whether it was related to some superstition, such as the horseshoe, for example.

234 *TLE*, pp. 107-108.

symbols. How to ignite a fire in our profane water if not by the same spark that ignited the Fire in the uterus of Mary?[235] The water which Jesus mentioned to Nicodemus signifies fire, and so it comes by the Holy Spirit. Indeed, according to the concept '*râzâ*', the sign water refers here to fire without ever ceasing to signify natural water, a real and non-figurative fire, a fire of a "*differAnt*" reality without which there can be no synergy, transmutation or transubstantiation.[236] Water becomes a non-water and fire a non-fire.[237]

To doubt that the Holy Spirit is Fire breaks the correlation between the two ends of the reality of the Incarnation, between the Alpha and the Omega, as this lack of belief makes all 'inter-realitarian' synergy fall. To preserve 'Fire' is to preserve baptism and protect the door back to Eden, *i.e.*, the access to the Kingdom of God. Are there other points of support to remove any doubt from the phlogisticity of the Holy Spirit?

1.3.3. Sheol *(שְׁאוֹל)*

Indeed, we find this support in Sheol, the first place visited by Christ to restore to the righteous dead the rights which they had been expecting impatiently for a long time. Here, the vibrant symbolism of the 'fiery furnace', mentioned in the Book of Daniel, takes all its dimensions.[238] As the angel of God descended into the furnace and saved the three young men from the profane fire, Jesus descended into the 'womb' of Sheol to "make everything new" for the righteous who had been

---

235 What does the name Mary mean? The *Encyclopedic Dictionary of the Bible* says: "In Greek Maria or Mariam is from Hebrew *Miryām*; the scientific etymology of the name is not clear, despite many attempts at explanation. It is very likely that the name is like that of Moses, of Egyptian origin, the first element probably being *mri*, "to love".
The popular etymology is illuminating. This name derives from the Hebrew *ra'ah*, "to see", from which comes "seer" or "one who makes see", *i.e.*, "Prophetess" (Ex 15: 20. *Cf.* Gen 22:2-14, where Jerome translates *moriyya* by *terra visionis*), or from '*wr*' (glow), hence *mē'ry'yām*, "illuminator of the sea", or '*wr*' (awake) from which comes *mē'îryām* "arouser of the sea"; Jerome includes *mar yām* (drop of sea), in Latin *stella maris*, that becomes "sea star ". But *mar yām* can also mean "bitterness of the sea".

236 Only quantum physics can explain how.

237 According to the Book of Hosea, Yahweh, offended by the betrayal of His people, uses this unheard-of metaphor, kept as it is by all translators of the Bible. God commanded Hosea to call his daughter, "*lo-ruhâmâh*" ["no more my beloved"], for henceforth I shall have no mercy on the house of Israel, and then to call his son, "*lo-ammi*" ["no more my people"], because you are no more my people, and I do not exist for you. But, it is obvious that through the last verses of the first and second chapters Yahweh rebuilds hope of reconciliation, and He says to Hosea, "I will sow her in the land, I will have mercy on '*lo-ruhâmâh*', I will say to '*lo-ammi*' you are my people and he will say, my God!" (Hos 2:23-25)

238 Dan 3.

waiting for Him all along, in accordance with the prophecies that they received.[239] Let's imagine the divine Fire entering Sheol to burst the locks of the closed doors and release the righteous who were waiting for the Parousia. The "heaven of heavens" above the Cross was already open. The Spirit had taken a stand in this new scene and accompanied the Son moment by moment. The power of the Father took Sheol under His shadow and everything changed. Sheol was *râzified*. Instead of gobbling up into its 'bosom' the divine Fire penetrating it,[240] Sheol was caught in the momentum of the same synergy of the generation, was transmuted by the Verb-God, and Christ was generated there for the umpteenth time by the Father, was born by the Holy Spirit, and resurrected from death.

This time Christ was not generated in eternity, nor in secular space and time, nor in quantum space-time, but in a "differAnt" reality, the *râzâtic* one, where the dead are not dead but await the arrival of Christ. This is where Sheol exists now. A dogma would be necessary to envelop it. What should we call this Sheol after its transmutation? What is going on, or could go on, in it? Before answering these questions, let us continue to see if the Father's Person, who is often recognized by the speaking Voice, is also phlogistic in nature.

1.4. The Father-Fire

It turns out, according to the biblical account, that fire was also the symbol of God in the Old Testament, long before the Trinity was unveiled, and long before God was recognized as Father,[241] or perceived as Unique.[242]

First, during the Mosaic era, the era of the establishment of the Jewish nation, everything started from the Bush of Horeb that was burning without being consumed.[243] It was an incomprehensible fact that made Moses, the son of the

---

239 Jon 2:1-3; 1Sam 2:6; Hos 13:14.

240 As it used to happen according to different mythologies.

241 Here, the Fatherhood is considered relative to the Incarnation of the divine Son. It is not referring to the anthropomorphic meaning of Ex 4:22 where God is supposed to be contrasting His firstborn with Pharaoh's firstborn.

242 Despite His will and His teaching, He will continue to be considered God of gods.

243 Ex 3:1-6. *N.B.* "In that period, the XV century B.C., the concept of 'sun – king' was quite widespread in humanity, from China to the Aztec's land. With the fact that the king was also god, therefore, he was to be considered sun-god. Moses was well acquainted with this idea. He had it as an integral part of his Pharaonic education and expected to become one himself." (*Cf.* Taok, p. 175.)
     According to the biblical commentaries that compare what Moses sacrificed of his worldly glory to serve Yahweh's salvific Economy and what Jesus sacrificed, titles like "sun", "fire-flame", "light" should be considered legitimate. (*Cf. Bible Dictionary, s.v.* "Moses") Why did Moses abandon them while Jesus did not? Jesus rather, at a certain point, declared Himself the "Light" of the world. Accordingly, the symbolic vision of

Pharaonic palace and of comprehensive education,[244] decide out of curiosity to make a detour to see what kind of fire it was.[245] The book says, "The Angel of Yahweh appeared to him in a flame of fire from the middle of a bush."[246] The angel was neither the flame nor the fire. It was Ehieh Himself Who, before making known that His name was Ehieh, seeing Moses making the detour, called to him from the midst of the burning bush, which remained unconsumed, and made him understand the following words:

> "Moses! Moses!"... "Do not come any closer," God said. "Take off your sandals, for the place where you are standing is holy ground." Then He said, "I am the God of your father, the God of Abraham, the God of Isaac and the God of Jacob." At this, Moses hid his face, because he was afraid to look at God.[247]

This text is crucial for our endeavor. This scene has now been tamed thanks to the Incarnation, and we can approach it without taking off our shoes. It has become for those who have the *Râzâ* in our hearts, our intimate space.

Before beginning the exegetical study to try to understand the secret of the 'spark' that caused this fire and its flames, we will continue to enumerate the most striking references that describe the God-Father as fire, and all that revolves around this concept. In the Book of Samuel we meet the following plea:

Mount Horeb becomes much more significant. If Yahweh is the fire or flames that burned the bush (*Cf.* Augustine, *De Trinitate*, II, XIII), the light that Moses saw could only be the reflection of His Voice, even the very Word that he heard and which, despite all his Pharaonic education, he could not resist.

This scene seems to prepare prophetically for that of Paul's conversion on the road to Damascus. In both cases, under the impact of the divine Light, the individual project suddenly changes. The eyes that see a certain reality open up to a "differAnt" predictable and secure one which they can no longer resist. They lose their own self-orientation. Even their bearings change to become the ones of the "differAnt" reality that we called 'râzâtic'.

244 Archaeological and ancient religious studies have proved that there were mythological-cultural exchanges between the peoples of the great civilizations of yesteryear. They shared the same gods but under local names suitable to their local dialect. (*Cf.* Taok, Prologue.)

245 Ex 3:3.

246 Saint Augustine confirms that it was God Himself. *Cf.* Augustine, *De Trinitate*, II, XIII.

247 Ex 3:5-6. It is of great importance to point out that in this encounter the Ephremian genius would find its enchantment. The reason is that, to take the very meaning of the name of Moses, "the one who was removed from the water", or as it is henceforth translated according to the latest etymological explanations "the son of water", we understand why Moses feared to fix his gaze on God. If God were only 'Light', why would 'water' fear Him? The importance of Epiphany resides, as Ephrem pointed out, in the fact that 'fire' has married 'water'. *Cf. Al muḥit al jameᶜ fil kitab al muqaddas wash-sharq el qadim.* (المحيط الجامع في الكتاب المقدس والشرق القديم)

> In my distress I called to the Lord;
>> I called out to my God ...
> Smoke rose from his nostrils;
>> consuming fire came from his mouth,
>> burning coals blazed out of it.
>> bolts of lightning blazed forth.
> The Lord thundered from heaven;
>> the voice of the Most High resounded.
> He shot his arrows and scattered the enemy,
>> with great bolts of lightning he routed them.[248]

Under the pen of David, Psalm 18 echoes such language, saying:

> In my distress I called to the Lord;
> ...smoke rose from his nostrils;
>> consuming fire came from his mouth,
>> burning coals blazed out of it.
> The Lord thundered from heaven... etc.[249]

And Isaiah also confirms this phlogistic nature several times, the most striking of which are Is 29:6 and 30:27-33 that call on God or His tongue or His Breath to be "devouring fire" or "similar to devouring fire".

What is to be noted here is that God's jealousy of other gods is not always what drives the sacred writer to describe Him as 'fire', but above all, His compassion for humans.[250] This will even allow reaching the infrastructure of the divine compassion deep down to the depth of God's jealousy, caused this time by the 'double belonging' of humans. The heroes of the Song of Solomon express the apex of this jealousy:

> Place me like a seal over your heart, like a seal on your arm;
> For love is as strong as death,
> Its jealousy inflexible like Sheol.

---

248 2Sam 22:7-15. Any verse in the Bible that describes fire and any information that belongs to the realm of fire, *e.g.*, lightning, is vital to our subject. Lightning, to our amazement, will inspire Ephrem with an analogy to describe the birth of Christ. "It was widely thought in antiquity that the pearl (*margânitâ*) came into existence when lightning struck the mussel in the sea. This birth through the conjunction of two disparate elements is seen by Ephrem as a symbol of Christ's birth in the flesh from the Holy Spirit, the Fire, and from Mary, 'the watery flesh'." (*Cf. TLE*, pp. 107-108.)

249 Ps 18:9-14.

250 Jealousy, here, is of God in relation to other gods. Yahweh, according to the first commandment, was to be the God without equal.

It burns like blazing fire, a flame of Yahweh.[251]

And the author of Deuteronomy declares it frankly: "For Yahweh, your God, is a devouring fire, a jealous God."[252]

What we find very interesting at this point is the link that could be made between these two facets of divine Fire, 'jealousy' that consumes and 'compassion' that inflames. Which Fire did Jesus intend when He said one day: "I came to cast fire upon the earth; and how I wish it were already blazing!"?[253] Is it the fire that consumes or the one that inflames? We have decided to introduce this problematic, in the service of exegetes.

1.4.1. The Fire that consumes or the Fire that inflames?

If Luke 12:49 contains a dilemma, it is precisely because of the difficulty in discerning whether Jesus meant the fire that consumes or the one that inflames. Ephrem mentions it saying, "This divine Fire has a double aspect: it can sanctify, but it can also destroy."[254]

If we admit two persistent and contradictory types of fire, we will fall into a kind of Manichean dualism. To admit that Jesus meant both at the same time would be absurd because He knew that both should have the same source, His Father. Ephrem, like many other Syriac Fathers, describes Jesus in the Jordan as 'Fire', so would it not be possible to say that the very entry of Jesus into cosmic history represents the wish He expressed from the depths of His Heart? As a matter of fact, by meditating deeply on the meaning of this verse, it turns out that the anguish of Jesus was not related to the act of sowing fire. It seems that this act was already done: mission accomplished by the Incarnation. His anxiety was

---

251 Song 8:6.

252 Dt 4:24.

253 Lk 12:49; The fire that the angel of the apocalypse threw on the earth. (*Cf.* Rev 8:5.)

254 *Cf. H Epiph* 3, 10; *CSCO* vol. 186, p. 148. At this point we introduce a remark to be used later. We feel that a certain understanding of this element, of its causes and effects, wants God to bring out either 'from His anger' a kind of fire that destroys or 'from his compassion' another kind of fire that revives or purifies. This feeling embarrasses us as it prepares the ground for dual principles from which the Church has suffered and still suffers. It is often interpreted with a distinction between the God of Old Testament and the God of the New Testament, Yahweh and the Trinity. If we admit that in God there are two contrasting kinds of fire, these can only be seen as belonging to two 'gods', because two incompatible genres of fire could not spring from the same source. Consequently, we tend, like the Pythagoreans, to support the uniqueness of the principle, and that from God comes only one Fire, as a single Word comes out of Him. At the level of creatures, the effect of this Fire differs according to Its receivers: those who are compatible with It are purified and inflamed (vivified), and those incompatible, are burned up (exterminated). The Fire 'speaks' of God and for God. (2Pet 3:7; Heb 4:12)

somewhat related to seeing this 'fire' spreading out and the people starting to be inflamed, to light up.[255] Everyone has experienced a candle that refuses to catch fire from the match brought close to it.

At this level, we notice a quite obvious distinction between the Fire as a being in itself that does not seem to belong to cosmic space-time and which (almost) needs no one to light it, and other beings which need a fire to be lit but do not all inflame to the same extent. There may also be among them individual beings that will never catch fire, hence the anguish of Jesus. The Gospel of John confirms this point of view with the following words: "And the light shines in the darkness and the darkness did not extinguish it ... The Word was the true Light ... He came to His own people, and even they rejected Him."[256] So, one must not think of two 'fires' at the same time, because this would lead either to a contradiction in the One God or to a belief in two gods, one the principle of good and one the principle of evil.

Even though this verse is followed by the quarrel between parents and relatives, there is nothing to suggest that the 'fire' Jesus desired to inflame everything, including languages, hearts, nations, even the Jewish nation, contradicts His mission as Prince of Harmony and Peace. Therefore, we agree with Ephrem that since there is only one God, there can be only one 'fire' in the visible and invisible universe, and that the 'fire' which Jesus speaks of can be none other than the Word of His Father, Himself and/or the Spirit who 'proceeds from His Father'. It is the Fire of the Father, or the 'Father-Fire', Who generates Him, the Light, and from Whom the Holy Spirit, 'Energetic Wave' and Carrier of the Light, proceeds.

But from where then, does the 'fire' that punishes, consumes, annihilates, and more precisely the 'fire' of hell, come? The answer to this question will be evident later. Let us return to the Father.

Based on the clues of the tongues of fire at Pentecost, the baptism of the Son in the Jordan, and the Baptism by the Holy Spirit and Fire, affirming that God the Father is fire remains insufficient if we do not clarify what this fire is. After all, it is from the nature of this fire that the nature of the 'grafts' between Science and Religion, as well as between the *Râz* and creation, will be discernible.

---

255   Einstein's discovery of the quantum of light (photon) is very significant at this level. We will shed more light on it in Chapter Five.

256   John, The Prologue.

2. What is Fire?

Jacques Derrida, the contemporary French philosopher,[257] helps us find a convincing answer to this question. Believing the intuition and the analytic power which perfect the human mind allows us to say that the 'fire' in question should be something like a non-fire, which emits a non-light, and so on. Is this applicable in our case?

2.1. The deconstruction of fire

By introducing this Derridean 'non-', we do not wish to indicate the negation *nihil* of the universally recognized concept. Instead, we want to make clear that if a word, or any term, has a clear and precise root, and is therefore un-deconstructible, we submit it to its reciprocal, or to its opposite, or to this ontological 'non-', to bring out the best of what it can provide to human understanding.[258] If, according to Plato, each sign has its signified of the same nature,[259] the word-sign 'fire', once written, literally means fire according to its own reality and its own nature. Yet there are signs which, although they signify what they mean, refer at the same time to a non-existent and still-existent reality, something totally different from what the simple linguistic construction means. For example, Man, this image and likeness of God, means man and refers to God. He is then called 'god'. The reciprocal is applicable and 'God' is also called 'Man' or 'Son of Man'. It is just then, at this level, that the words of Karl Rahner, "When God wants to be what is not God, man comes to be", reach their apex.[260] How could this reality be, once

---

[257] Bou Mansour resorted to Derrida who shared with Ricœur the same deconstructive philosophy. (*La Pensée,* p.32)

[258] Ephrem, for example, uses this distinction to express so many things that are inexpressible and have caused enough troubles, for example, the knowledge of the Son of God in Mt 24:36. Was His knowledge less than that of the Father? Feghali, by his translation of the 25th paragraph of the 77th *Hymn on Faith,* says: ᶜ*ariful kulli sara la* ᶜ*arifan*, which means: "He who knows everything has become a non-knower." The translation could have been *sara ghayru* ᶜ*arifen*, that is to say, "ignorant", which is not the case at all. It is this 'non-' that makes all the difference. In the first case, there is a kenosis of what is obviously known to just keep it in a certain background. In the second case, there is total loss of precedent information, which contrasts with the verb "to know". If, for example, this distinction was within the reach of Origen when he made his treatise based on Jn 14:28, on the scale of greatness between the three divine Persons, he would have found it a perfect solution.

[259] Augustine, *De Maestro*, VIII, 23. "We signify the things we speak of, and what comes forth from the speaker's mouth is not the thing signified, but the sign by which it is signified. We make an exception for signs that signify themselves." (*Cf.* https://archive.org/details/TheTeacherTheFreeChoiceOAugustineSt.RussellRobe5162/page/n3/mode/2up?view=theater [Accessed June 2019].

[260] Rahner, Karl, *Foundations*, p. 225; *Cf.* Roger Haight SJ, *Jesus Symbol of God*. Orbis Books.

deconstructed? Jacques Derrida speaks of "differAnce".[261]

By his Semitic audacity, Jacques Derrida, who is of Sephardic Jewish origin, and who tries to tame his own religion or to become familiar with it,[262] regenerates Ephremian symbolism in a contemporary mode, *i.e.*, defended the same cause, without having known Ephrem. In addition, he appeases Augustine's anxiety to make it clear that, [263] as far as the Bible is concerned, Plato's theory of signifiers is not sufficient because one must know the referent, in advance, in order to grasp what the signifier refers to beyond the signified. St. Augustine, who like Ephrem was taken by the *râzâtic* reality and who once wrote, "*Credo ut intelligam*", profoundly influenced Derrida who will return him this favor through his book *Circumfession*. By this book Derrida aims at circumcision by Faith without knowing that, at the same time, he will serve the symbolism of Ephrem. What do we mean by this?

A study done at the University of Quebec in Trois-Rivières, deduced the following:

> The Derridean conception of the Sign is then always linked to the structure of Western philosophy. The schema "signifier = signified" (direct relation between signifier and signified) is therefore reviewed. Let us take the example of water:

N.Y. 1999, p. 326.

261   Moreover, this understanding of what seems different and not different at the same time, perceptible and imperceptible at the same time, largely supports the Ephremian genius according to the analysis made by Bou Mansour under the title *A Symbolic Cosmology* where the whole article is about *'kasyâ'* and *'galyâ'*, this being 'veiled' and 'unveiled' at the same time. The approach, seen under the concept *'râzâ'* which stands behind the meeting of perceptive minds in a reality that goes beyond time, space, and even Religion, is very significant since it belongs to the cognitive universality. Bou Mansour advances this example: "The privileged Home of the Son, the 'Womb' of the Father, is also for Creation, the difference being essentially that of a dwelling of Nature and another of Grace. The proof is that, in the two texts of *H Fid* 4, 17 and *H Fid* 11, 4, the two passive tenses of the verb 'to place' *(semâ, semeen)* emphasize the divine initiative in the act of receiving the world in His 'Womb'." (p. 133)

262   He who was circumcised but who refused to circumcise his two children, not by revolt but by preference to free them from this sign *(remzâ)* that refers to the rigidity of the law. He would open their eyes to the circumcision of the Heart, by the Love that opens it to the 'other' instead of locking it up and withdrawing into oneself.

263   Augustine, *De Maestro*, X, 34. "What I am trying most of all to make you see, if I can, is this, that we learn nothing from signs which we call words. For, as I have pointed out, it is rather a question of learning the sense of the word, that is, the meaning hidden in the sound, from a previous knowledge of the reality signified, than it is of perceiving that reality from a sign of this kind." (*Cf.* link in footnote 259)

$$\text{Signifier "water"/ signified} \longrightarrow H_2O \begin{array}{l} \text{Drop of water} \\ \text{Swimming pool} \\ \text{Rain} \\ \text{Glass of water} \end{array}$$

> When reading the word water, one can think of drops of water, a lake, the chemical symbol H_2O, etc. We do not necessarily think of a fixed image of water, a universal mental representation. So, every concept (signifier) that water can refer to, sends to another signifier. This infinite chain, from signifier to another, results in an endless game and opens the text, repositions it, makes it a moving one.[264]

In other words, it is an example of polysemy. Would it not be moving even more, and with synergy, if we build anew the steps that one day the original biblical observer built? Let us take an example:

> And God said, Let there be a firmament in the midst of the waters, and let it divide the waters from the waters. And God made the firmament, and divided the waters which were under the firmament from the waters which were above the firmament: and it was so.[265]

In addition, we also find in the same firmament the sun that originates from a given primordial 'fire' and the 'light' of the first day.

The only way to stabilize what is moving and to give believers – of Nisibis and of Edessa – a little more certainty and predictability as to their fate was to have someone who knows in advance, someone of the type of the Freed Slave mentioned by Plato in his *Myth of the Cave* or someone such as John of the Apocalypse or Paul the Apostle. Could these last two have been so sure of themselves, if they had not seen, tasted, and therefore known these things in advance? What about Ephrem?

We thus reach a conviction that the same signifier may have a signified proper to its visible natural reality, another specific to some apparent biblical reality (*galyâ*), and also a third one beyond the *galyâ*, a veiled one (*kasyâ*). With the quantum reality of today, the same signifier can also have a referent beyond the *kasyâ*, whether it is physical, metaphysical, or symbolic. If sometimes the signified becomes confused, as with the word 'womb' (ܪܚܡܐ), it is up to the experts or to the visionaries like

---

264 Guillemette, Lucie and Josiane Cossette. *Cf.* http://www.signosemio.com/derrida/ deconstruction-et-differance.asp; [Accessed June 2019]. *Cf.* more specifically the article "Some General Characteristics of Deconstructive Readings". [http://web.utk. edu/~misty/Derrida376.html] [Accessed June, 2019]

265 Gen 1:6-7.

Ephrem to have the possibility to discern its various uses and to point with precision at the referred object: 'womb' of the Father, 'womb' of water, 'womb' of wood, 'womb' of the ear, womb of Mary, womb of Eve, etc.

It is when this variety of use joins two different realities that the Derridean keyword "differAnce" applies, because it is neither right nor logical to compare two realities that do not share the same nature. Between the 'womb' of the old Eve and that of the 'new Eve' there is no difference, there is rather "differAnce", and this "differAnce" is due only to the intervention of the divine Word Who has made everything new, "differAnt", in a "differAnt" reality that could from then on resemble only itself and refer only to itself. We say again that this is what Ephrem intended when he adopted the concept '*râzâ*' and what we aim at by our contribution.

What would happen then to the concept, or rather, to the signifier 'fire'?

Yet Nature contains fire, just as the Book contains it. Ephrem experienced both at the same time: the fire of the Persians who attacked his hometown to occupy it and the fire of the heretics who attacked his religion to make him lose it.

> Swords have openly killed (the body),
> And tongues in a hidden way. [266]

But a third 'fire' seems to be on the horizon of Ephrem, promised by the Scriptures, and which, according to the psalmist as well as to Daniel, stands as a shield, because nothing can protect from unfriendly 'fire' like a friendly one, the 'fire' of Faith, when there is no possibility of extinguishing any 'fire' by the water of a flood.[267]

2.2. The Fire of Faith

So a third 'fire', or rather Fire, inflamed the heart of Ephrem without anyone being able to see it. Everyone knows the Fire we are talking about, as well as to which Fire we are referring. Those who like us sincerely admit that indeed "everyone knows" prove first of all that they share with us the same educational background. They admit, like us, a kind of Fire which belongs to a category that they all consider *râzâtic*, which means visible and invisible, mystical and non-mystical, real and non-real, etc., at the same time. For those who do not admit this statement or have not grasped our idea, we say clearly that it is the Fire of Faith, of Pentecost, of Mount Horeb, of the prophet Elijah, etc. In short, the Fire of God as mentioned above.

---

266 *Les Origines*, p.215.
267 Gen 6:17; Ps 18:8-15.

Is it not poetic what we are saying here? Maybe! But working with Ephrem requires that we do it, we would say, especially to deal with enigmas. A time will come when, from our perspective (under a certain angle), we will show that poetry is also one of the ultimate means of expression of quantum reality. We allow ourselves here to make a detour, similar to that which Moses made at Mount Horeb, which aims, even if not closely, for a so-called positive scientific method, to answer as much as possible the aforementioned sub-question: what 'fire' are we talking about?

In fact, if Science wants to identify a kind of fire, a category of fire, or the nature of fire, like the one of the sun, for example, without having means to reach it, then Science studies it according to its effects on the experimental instruments. Science succeeds in teaching us when the sun, this star of fire, with its heat, is good or bad for human life. In short, all the civilizations and the mythologies teach us that the moon has been a closer 'friend' to humans than the sun; the evidence is that the majority of the first calendars were lunar.[268] What would that mean? In our opinion, and this is what will be proved in the scientific part of this book, it would seem that fire is measurable according to the margin of predictability and certainty that it provides to human reason. If we find nowadays that electronic thermometers are everywhere in cities, it is not to indicate experimentally what the present temperature is, but rather to inform us clinically to what extent the prevailing climate remains healthy for life.

These are the kinds of stepping stones that Ephrem sought during the years of attacks that his city was undergoing: security, certainty, predictability. However, it was out of pastoral responsibility, and not of a civil one, that he sought them. The fact is that when a fire exceeds the norms, like the one that annihilated Sodom and Gomorrah[269] or that of September 11, 2001 in New York, the 'rational animal' becomes alarmed and falls into disarray due to uncertainty and unpredictability. Psychology proves that this is the worst situation that a person, or a group of people, can go through. Humans come out traumatized because they are not created for such atrocities. Recall here the wife and daughters of Lot.[270] This subject is so crucial that it is enough to surf the sites of electronic libraries to see that a good number of publications in Science and Religion today deal with what could return certainty to humans and predictability toward their future, that of their loved ones, as well as that of the universe.[271] Is this not the essence of Christ's mission to the world and

---

268   Taok, pp. 154-157.

269   Gen 19:24-26.

270   Gen 19:30-38.

271   Ford, Kenneth W., *The Quantum World: Quantum Physics for Everyone*. Cambridge, Massachusetts - London, England; Harvard University Press, 2004, Introduction. From now on, we will refer to this book by *The Quantum World*.

His invitation to believe in Him in order not to die anymore?[272] It is enough to take again the analogy of the hen and its chicks which He used while considering the perilous situation of Jerusalem.[273] *Ad hoc* Churches were founded for this purpose.[274] The impact of scientific discoveries and the tremendous progress of the twentieth century have only increased this concern among almost all peoples, especially those in industrialized countries. Who better than Ephrem, given the continual attacks on his hometown Nisibis[275] followed by the famine that reigned over Edessa, could psychologically and somatically have experienced the truth of the derisory constitution of all that is created?[276]

> In his collection of hymns entitled *Carmina Nisibena*, the world is called the "inhabited world" (*ʿumrâ* ܥܘܡܪܐ) and its ephemeral character is expressed by the two verbs "to disappear" (*tleq* ܛܠܩ) and "to cease" (*bṭal* ܒܛܠ).[277]

Moreover, it is by contrasting the wickedness of man and the goodness of God that a large part of the hymns (*memrê* ܡܐܡܪܐ) of his *Carmina Nisibena* have formed a theological treatise, in good and due form, on the descent of Christ into the world of the dead, the martyrs, to give believers certainty and predictability as to the fate of their loved ones who fell every day. The signs and symbols of his people's sufferings inspired this theme. Thus, the second prayer that the Maronite Church raises during the night of Sundays, dedicated to the martyrs, calls them to advance toward God with their hands full of their bones like 'pearls' (*margâniâtâ* ܡܪܓܢܝܬܐ) and to intercede for the living ones. The prayer says:

> To you the pleasant incense;
> Your bones are pearls.
> May the blood that has flowed from your necks
> Be compassion for the whole universe.
> Here, out of your bones emerges relief for humans.[278]

---

272 Jn 11:25. And Jesus added: "Do you believe this?"

273 Lk 13:34; Mt 23:37.

274 For example, Christian Science Church.

275 *TLE*, p. 16; *Cf.* also, *Actes du Colloque XI-Alep 2006;* CERO, 2007, Brock, "In search of Saint Ephrem", p. 18.

276 Eastern Christians, especially Lebanese, Syrians and Iraqis, would benefit from re-reading Ephrem from the perspective of what they have endured in recent decades.

277 *La Pensée*, p. 127. *Cf. C Nis* 77, 1-2.

278 Translated by the Author from the original. It is not necessarily Ephremian. The original text.

A pure theatrical twist unfolds before us: the Fire of Faith causes the divine Environment to go beyond all time and space. It even surpasses all reality and makes the Church both a 'reality' of the 'here' and the 'beyond', each 'wing' in continual intercession for the other, all alive, thanks to the One who has conquered death and has taken control without rival of the two worlds of adversity divided between the mythological gods.

Ephrem, thanks to the poetic force as well as to the *ḥaylâ* of the Holy Spirit Who has infiltrated his hymns, eliminates, as an expert, any aspect of awaiting and separation. It is in the here and now that the truths of Faith of which he speaks become real and take place. The Light is here, no doubt, but for Ephrem it is instead the Eucharist that is here, this *"Râzâ* full of *râze"*, this Forgiving Ember, *(Gmurtâ Mḥasyânitâ* ܪܡܚܣܝܢܝܬܐ ܓܡܘܪܬܐ), this Pearl (*Margânitâ* ܡܪܓܢܝܬܐ). The Eucharist is the Mirror (*Maḥzita* ܡܚܙܝܬܐ) that makes one of the two faces of the Icon of the Church see Her other Face as if watching Herself in a mirror. The sacrament confirms that Jesus Christ, in Whom She fully trusts, controls the 'world of death', especially the world of martyrs whose blood is the seed of the same Church, His *Râzâtic* Body.[279]

Who better than Ephrem to have recognized the importance of the Faith that restores the derisory and unpredictable to make out of them a bridge to what should be euphoric certainty and stability? Would this criterion of the Eucharist be applicable to the friendly and unfriendly fire of which we have spoken?

2.3. The friendly Fire and the hostile one

Since the 'fire' (*nurâ* ܢܘܪܐ) present in nature, in any form, can be beneficial if controlled and harmful if out of control, it facilitates our understanding of its symbolism contained between the two covers of the Book. As a matter of fact, the

---

ܠܟܘܢ ܗܘ ܓܪܡܝܟܘܢ
garmaïkun margâniâtâ
ܕܠܘܬܟܘܢ ܥܛܪܐ ܚܠܝܐ
dalwotkun ʿeṭrâ ḥalyâ
ܕܡܐ ܕܪܕܐ ܡܢ ܨܘܪܝܟܘܢ
dmâ dardâ men ṣawraykun
ܗܘܐ ܚܢܢܐ ܠܟܠܗ ܬܒܠ
hwâ ḥnânâ lkuloh tibêl
ܘܗܐ ܢܒܥܝܢ ܡܢ ܐܪܡܝܟܘܢ
w hâ nâbʿin mên armaïkun
ܥܘܕܪܢܐ ܠܒܢܝܢܫܐ
ʿudrânâ labnaïnâšâ

279   *Cf. La Pensée*, pp. 61-62.

relationship between the 'fire' of the Scriptures (said to be of God) and that of Nature (said to be of Man) reflects a perfect anthropomorphism as regards their impact and their 'behavior' in a theocratic system.

This anthropomorphism has even pushed allegories and analogies to lend to God the sophisticated fire invented by the artisans of wars and as means of extermination have: fire and sulfur together, as fell on Sodom and Gomorrah. The combination of fire and hail, about which we can say little, except that it is possible that this allegory was taken from the fall of a meteorite on a very cold night. This attribution makes it difficult to distinguish between the 'fire' of God in a rage and that of a king in a rage. With regard to Israel, both 'fires' would seem to be the same, an unfriendly, terrifying 'fire'. However, one could even say that it is instead God's own 'fire', which is often at the disposal of the kings of Israel. The 'fire' of the prophets (*e.g.,* Elijah) or that of the different theophanies (*e.g.,* Horeb) – we mean the friendly 'fire' which is supposed to be the source of both divine justice and compassion – is "differAnt". Wherever we find it, it is a simple, reassuring, expressive, efficient bearer of its Master's message and perfect executor of His will. It is a salutary one because it accomplishes the Salvific Economy that is in the making. Furthermore, this kind of 'fire' does not belong to our natural categories. Neither allegories nor analogies adequately capture it. The Song of Songs, speaking of Love, gives us an idea of this impossibility. To what is Love compared? It is compared to death, to passion, and at best, to Sheol. As for the traits of Love, it compares them to a flame of Yahweh that even the great waters cannot extinguish nor the rivers submerge.[280]

This 'fire' of zeal toward the House of God which inflamed the hearts of the patriarchs, prophets, disciples, martyrs, saints, and especially of Jesus, enjoys in addition a specificity which projects a better light on its nature: it is a 'fire' that inflames, even set ablaze, but never consumes,[281] especially the free will of humans.[282] On the contrary, the fire of nature only burns and consumes. It is one of our ecological problems. The 'fire' of the Book, as Ephrem described it, sometimes consumes but other times does not; it depends on its 'source', anthropomorphic or purely divine, and its target, good or bad. If the 'fire' of faith, of zeal, or of the "Comforter", *i.e.,* the Holy Spirit, had been the same as that of humans, neither Pentecost nor Cornelius' baptism by St Peter would have taken place.[283] Jesus promised this 'fire' to His disciples without withholding it from non-Jews who believed in Him. While respecting the human person's will, this 'fire' inflames the heart and the spirit without consuming them (*e.g.,* the disciples travelling to Emmaus). This 'fire', to which Ephrem refers

---

280   Song 8:6-7.

281   Jn 2:17.

282   *H Epiph* 3, 10; *TLE,* p. 38.

283   Acts 10:44-48.

as primordial, resides in *šekinah* in the heart of Adam since YHWH blew it into his nostrils, just as the 'fire' that Ephrem supposed was created, by implication, with the creation of Heaven and Earth. Its Christological activation always awaits the personal encounter, in the style of the *Emmaus walk*, and the acquisition of the grace of Faith, as Peter and Thomas encountered it after the resurrection.

It was the Ephremian contribution that put an end to the controversy of the origin of the 'light' which God ordered to be, whereas nothing in the Scriptures had yet been said about the creation of fire.[284] What would be interesting to emphasize here is that Ephrem, despite his insistence on using the "sun-light-heat" type to prove the divine "Trinity", had not yet given a clear answer, nor a definitive explanation, of the mode of being of the 'fire' which was always, eternally, at the source of the *râzâtic* light.

This position of Ephrem inflamed in us sacred jealousy to complete what he has once begun and to unveil the friendly 'fire' that is the source of light which Jesus spoke of when He said, "I am the Light of the world." It is here, we would say, that the *râzâtic* spark (*šalhebitâ* ܫܠܗܒܝܬܐ), like Ephrem's radiant Ember, comes into effect. How can this *Gmurtâ* so dear to Isaiah and Ephrem, conservable and transportable, 'fire' and 'non-fire' at the same time, help us?

2.4. Can 'ember' (*gmurtâ* ܓܡܘܪܬܐ) serve as a unit of measurement for the *râzâtic* reality?

We recall here that the One Who created Adam from the mud blew something "differAnt" into his nostrils, one thing that made him a rational being, an image (*ṣalmâ*) of the One who created him, and God cannot contradict Himself. The *râzâtic* here, while trying to interpret Ephrem as best as possible, specifically wants to make it clear that the graft is there, hidden, and that this hidden matter there, refers to a more advanced hidden being, beyond the "hidden" of sacramental or exegetical everyday life. The graft is the *râzâtic* ember, more precisely, the "purifying and sanctifying ember full of *râzê*".[285] At creation, the divine 'graft' was accomplished in a general way in the whole creation and in a specific way in the human being. Saint John Paul II calls this breath 'the spark'. He says about it, meditating on Michelangelo's fresco of creation:

---

284 *Les Origines,* p. 58: Thus, the works of the first day were created from nothingness (*nihil*); the sun was created from fire, and the other works were from the first elements of creation: "Heaven and earth, fire, wind, and water were created from *nihil*, according to the testimony of Scripture. But the light that was made from the first day, and the other works that were made afterwards, were from something." (*C Gen.* 1, 14).

285 *Gmurtâ mḥasyânitâ w malyat râzê* ܓܡܘܪܬܐ ܡܚܣܝܢܝܬܐ ܘܡܠܝܬ ܪܐܙܐ

The first man, Adam, became a living being. Michelangelo's immortal genius depicted on the vault of the Sistine Chapel the moment when God the Father communicates the vital energy to the first man, making him a living being. Between the finger of God and that of man, stretched in a way they almost touch each other, an invisible Spark seems to spring: God puts in man a shudder of His own Life, He creates him to His image and His likeness. In this divine breath, lies the origin of the singular dignity of the human being.[286]

At the Incarnation, the opposite took place: the human graft, by divine will, was welcomed by the Trinity in general, and by the Son in particular. The *fiat* of Mary was the approval of the humanity that was groaning, awaiting this event, that of achieving the exchange with Heaven between divine Spark and human flesh (*sarx*).

The Syriac liturgy, based on a hymn of an unknown author (but attributed to Ephrem due to its Marian theme), chants words addressed by Heaven, at Christmas, to the Earth saying:

> Congratulations! Happy you are because you possess the
> most precious thing in existence! You have a mother. Give me
> a mother, and I will give you God.[287]

This proposal lays down the element of the great 'chiasmus': it is when the spark of creation bombards the flesh of Mary that the "divine Ember" (*sic*) becomes a unit of measurement for a soteriological reality that is measured only by the succession of these bombardments, *i.e.,* of this exchange. The more humanity, no matter how sinful it is, follows the model of Mary and opens its 'vaginal ear' to the divine Word to let the *Gmurtâ* multiply by *râzification*, the more it is accomplished and perfected, and with it, Creation. And since humanity has no concrete 'womb', it is merely up to the 'ear' of each person, the *râzâtic* ear, to assume this role. The Old Testament proves that God always addressed His people Israel by inviting him to open wide his 'Ear' to listen to Him: "*šama͑ Israël* שְׁמַע יִשְׂרָאֵל".[288]

---

286   *Cf.* https://www.vatican.va/content/john-paul-ii/en/messages/urbi/documents/hf_jp-ii_mes_20001225_urbi.html, #2 [Accessed June 2019].

287   *Maronite Breviary* (Arabic), Christmas Time, Tuesday Morning prayer, USEK, Jounieh, Liban 1987.

قالتِ السَّماءُ للأرض: هَنيئاً لكِ ما أسعَدَكِ، لأنَّكِ تَملكين أثمَنَ ما في الوُجود! عِندَكِ أمٌّ. أعطيني أمّاً فأعطيَكِ الله.

[*Kalate Ssama'u lel 'Ard : Hanee'an lake ma as͑adaki, li 'annaki tamlukeen ahhma ma fel wujud! ͑indaki ummun, a͑tiee umman fa-a͑teeke Allah*] (Transliterated by Google translator). *Cf.* http://church.marantoniosalkabir.com/AF_4-24371 [Accessed June 2019]

288   Gen 49:2; Ps 81:9; Bar 3:9; The writer of Deuteronomy and the prophets use it as a

The *Gmurtâ* that results from the chiasmus between the Word of God and the human 'ear' can ably serve as a unit of measurement of the *râzâtic* reality, as the 'pearl' that results from the bombardment of seawater by lightning does in the wealth of jewelry. Nature offers us another phenomenon of the kind that Saint Ephrem would have appreciated. It is the case of the *camā*, or *kamā*, that sort of desert truffle from the Middle East that is said to grow only under the effect of thunderbolt hitting the desert soil.[289]

Saint John Paul II was right in his interpretation of Michelangelo's fresco, "The Creation". If Creation is an infinite process, so is the creation of humans, and therefore the 'Spark'. It is repeated in the manner of the spark that is produced by a dynamo and a distributor.[290] What remains to be prepared is enough 'yes' (*fiat*) from humanity, otherwise the dynamo would no longer emit and the synergy, as we said above, would fail.

At the Incarnation, the current passed from the finger of God to that of the New Eve, (*Ḥawâ* ܚܘܐ), Mother of all living beings, and the 'Spark' glittered in our cosmos. The lightning happened and the 'Pearl' was fertilized in the womb of the obedient Mary. It was from that moment that everything started to be "made new".[291] And Mary was the first to sing:

> You have united, O Lord, Your divinity to our humanity
> and our humanity to Your divinity,
> Your life to our death, and our death to Your life.
> You have taken what is ours, and you have given us
> what is Yours...[292]

---

keyword. In Deut 27: 9 it joins silence to listening. Jesus, also, recalls it. (Mk 12:29) Ezekiel even calls the places where Israel lives to listen: "Thus says the Lord Yahweh to mountains, hills, ravines, valleys." (Ezek 36:4).

289 Terfeziaceae or desert truffles is a family of truffles (Berber: *Tirfas*, Arabic: كمأ *Kamā*). *Cf.* https://en.wikipedia.org/wiki/Terfeziaceae. The Arabic article underlines the popular thought that links the growth of this legume to thunder :the more there is thunder over the desert in winter, the larger will be the harvest of the season. Accordingly, this kind of truffle is called in local dialect "the plant of the thunder" or "the daughter of the thunder". *Cf.* https://ar.wikipedia.org/wiki/كمأ. [Accessed May 2020]

290 This brings us back to the Aristotelian theory of the unmoved mover or prime mover. It is advanced by Aristotle as a primary cause (or first uncaused cause) or "mover" of all the motion in the universe.

291 Rev 21:5.

292 *Cf.* Footnote 162.

ܣܘܝܬ ܡܪ ܐܠܗܘܬܟ ܒܢܫܘܬܢ،
Ḥayedt mâr alohutoḵ bnâšutaan,
ܘܢܫܘܬܢ ܒܠܗܘܬܟ.
wo nošutaan balohutoḵ.

Would the Big Bang be an event similar to the moment when, one day, the current passed between the finger of God and that of Adam? How could we understand this unification (*ḥayedt* سحيده) to have happened without lightning or thunder? Would the Maronite priests who sing these words after the consecration be aware of what is going on between their hands? Is it not that the spark, or the thunderbolt, is fire, yet without being so until it meets a flammable substance to ignite it? What ignited the water in Saint Charbel's lamp? For people like Charbel, the answer comes from the Bible: "For here is the strangest: in the water, which extinguishes everything, the fire had only more ardor; the universe indeed fights for the righteous."[293]

Whereupon, we can well affirm that the primordial Fire which caused so many headaches to the theologians of the first centuries, and which is the object of our inspection, was always there. It was present as the letters of the verb "to be" were present in all beings, as the letters of the verb "to create" were found in each of the creatures, and as the letters of the verb "to generate" were there, in every gender and generator, and in fertility. If this fire were not there, from the beginning, as it was in the aforementioned *naphtha* of the prophet Nehemiah, nothing would have allowed the spark to ignite the lights, to illuminate the eyes of the creatures and to warm their hearts. It is more or less the same phenomenon which, according to the new anthropological theories that refuted Darwinism, also accompanies the definition of *homo sapiens*: if humans were not *sapiens* from the seed of their kind, they could never have become so. The proof is that other kinds of apes have never become so. This theory is true and supports predictability.

Can we now, as the idea of 'God-Fire' has been admitted, synthesize what kind of 'fire' we are talking about?

We dare to say yes. Saint Ephrem will not disappoint us, because we need a foot on the ground, somewhere, to be able to advance in reconciling Religion and Science. It is a kind of *râzâtic* 'fire', a hidden one (*kasyâ*) that leads us to foresee a "differAnt" hidden one, more advanced than the one the Prophets predicted, a 'fire' that the women of the Old Testament foresaw in the DNA of the bloodline from which the Savior Jesus will be born, as mentioned by Sebastian Brock.[294]

It is also a hidden 'fire' like the one the Immaculate Mary felt at the Annunciation. It is the *râzâtic* 'fire' that her Son, Jesus Christ, the Son of Man, saw and desired so much that everyone would begin to experience. It is the Fire

---

ܗܘܝܬܟ ܒܡܝܬܘܬܢ ܘܡܝܬܘܬܢ ܒܗܘܝܬܟ

Ḥayutâk bmitutaan, wo mitutaan bḥayutâk,

ܫܩܠܬ ܕܝܠܢ ܘܝܗܒܬ ܠܢ ܕܝܠܟ...

šqalt dilaan wo yhabt laan dilâk...

293   Wis 16:17-19.
294   *Actes du Colloque*, "Saint Ephrem on Women in the O.T."

that has acted in eternity so that all other beneficial, fertile and reassuring 'fire' exists, and not the opposite kind of 'fire'.

Nothing prevents us from calling this 'fire' Love, Passion, or divine Zeal, as long as its primary property is to be one and triune at the same time, as God is, after the Christian perspective. Whoever feels it, or can see what it hides, rushes to accept any kind of kenosis, even to give his own life, to possess it, or to be possessed by it. In the case of this 'fire', to contain it is to accept being contained by it, as Khalil Gibran mentioned in *The Prophet* saying: "When you love, do not say, 'God is in my heart', but say rather, 'I am in the heart of God.'"[295]

Conclusion

This 'fire' phenomenon, of which Gaston Bachelard made good use, has allowed psychiatry to reach the bottom of the abyss of human psychology and fertility, even of life, the only fruit of Love. In honor of Saint Ephrem, Doctor of the Church, we baptized it, as seen before, *râzâtic*.[296]

In accordance with this 'Fire', the three Persons of the Holy Trinity, Father, Son and Holy Spirit, can be said to be 'Fire' as They are called Love. God is 'Fire', and what comes naturally from 'Fire' is Light and Heat. The Light is, as we have described in the first chapter, the corollary that gives the eye its reason for being. This 'Fire' inflamed *râzâtically* the Scriptures, and the divine 'Light' illuminated it. The 'synergy' between the two warmed the whole thing and made it a *râzâtic* 'Sun' that Ephrem would tame as a type for his Trinitarian theology.

We have thus succeeded in unveiling the identity of the 'Fire' we are talking about and in taking a stand in Ephrem's polemic *vis-à-vis* those who worshipped the created fire as a deity, whereas the *râzâtic* Fire exceeds this step by far.

We have reached with Ephrem a consensus about the existence of the divine Fire that generates a divine Light, and out of which, as well as out of its Light, proceeds the divine Heat.

But to the displeasure of the authorities of other religions, all Three are of a nature that refers to a "differAnt" reality, inaccessible and incomprehensible to those who do not possess the 'Luminous Eye', the One enlightened by faith in Jesus Christ.

It is time to explore this "differAnt" *râzâtic* reality, already mentioned several times and to which the uncreated Fire belongs. This is the challenge that will confront us in the next chapter.

---

295 Kahlil Gibran, *The Prophet;* On Love. *Cf.* https://poets.org/poem/love-8. [Accessed June 2019]

296 And in honor of Jacques Derrida, the "differAnt" fire.

PART II

Interpretation and the *Râzâtic* Reality

Chapter III

The *râzâtic* reality

To assist reason in its effort to understand the mystery there are the signs which Revelation itself presents. These serve to lead the search for truth to new depths, enabling the mind in its autonomous exploration to penetrate within the mystery by use of reason's own methods, of which it is rightly jealous. Yet these signs also urge reason to look beyond their status as signs in order to grasp the deeper meaning that they bear. They contain a hidden truth to which the mind is drawn and which it cannot ignore without destroying the very signs that it is given.[297]

Saint John Paul II

Reality, like Truth, represents an ontological challenge to the epistemological consciousness, properly scientific, as well as to the religious consciousness. In Book VII of *The Republic*, Plato had advocated an uncompromising idealism which negates the existence of empirical reality. This introduces us to a first reality, the world of Ideas that prepared the ground, or better stated the underground, for the use of signs, types and symbols, as well as of the power of dialectics. He borrowed from the Orphics their famous Allegory of the Cave that represents for him the image of the possible human illusion of reality. He developed it, thanks to his maieutics, into a real art of building realities. We reproduce a synopsis of that allegory to consider.

> ... Prisoners are chained to a place where, cut off from the world, they see only the shadows cast by a fire, far behind them; what reality can they be aware of? Plato makes Socrates ask Glaucon... And the dialogue continues: "If they could talk to each other, do not you deem they would think they were naming real objects, naming the shadows they would see?"[298]

In this chapter, we will focus on the contribution that Ephrem has left us regarding the determination of an inevitable reality which has been neglected for

---

297 John-Paul II, Encyclical letter, *Fides et Ratio*, §3. From now on we will refer to this Encyclical by *Fides et Ratio*.

298 Plato, *The Republic*, Book VII, p. 373. *Cf.* http://www.idph.net/conteudos/ebooks/republic.pdf [Accessed June 2019]. From now on, we will refer to this book as *The Republic*.

so long and which we have qualified as *râzâtic*. We will develop our presentation around three axes:

I.   The "angles of view" (*zewyâtâ dḥezyân* ܙܘܝ̈ܬܐ ܕܚܙܝܢ). It is an axis that would support a theory similar to that of Hans-Georg Gadamer's Horizons and which would deal with "perspectives" under which humans often look, analyze and judge ...

II. The *râzâtic* reality, angle and description

III. The verb "to *râzify*"[299] from the Syriac verb *râz* (ܪ̇ܐܙ), indispensable for linking Religion and Science mainly at the quantic level.

We will limit our input to the following plan:

1. The "angles of view" (*zewyâtâ dḥezyân* ܙܘܝ̈ܬܐ ܕܚܙܝܢ)

   1.1. How to distinguish between the angles of views

   1.2. How to differentiate between intermediaries (*meṣʿâyê* ܡܨ̈ܥܝܐ) and interpreters? *(fušâkê* ܦܘ̈ܫܩܐ)

      1.2.1. The problematic of interpretation in the universal sense

      1.2.2. The role and responsibilities of the interpreter as intermediary

      1.2.3. The criterion of the good intermediary

      1.2.4. Ephrem's originality

         1.2.4.a. Ephrem's interpretive method

         1.2.4.b. Originality in Ephrem's interpretation

2. The *râzâtic* reality, angle and description

   2.1. Would the *râzâtic* reality be a mixture of subjective and objective?

   2.2. Would the *râzâtic* reality be a compromise or a utopia?

   2.3. What can this reality be and to which category would it belong?

      2.3.1. The becoming

      2.3.2. The relationality

   2.4. Describing the relationality through behavioral analysis

3. The verb *râz* (ܪ̇ܐܙ) and the verb "to *râzify*"

   3.1. The requirements that demanded the creation of the verb "to *râzify*"

   3.2. Clarifications that would reduce the challenges raised by the various applications and difficulties of this new verb

      3.2.1. Christological challenges and difficulties

      3.2.2. Linguistic challenges and difficulties

      3.2.3. Challenges and difficulties in practical applications

      3.2.4. Soteriology of the 'hymen' and the 'Navel' (umbilicus)

      3.2.5. The virginal Birth of Jesus by Mary

Conclusion

---

299 In consensus with Professor Paul Feghali, we have coined this verb to free the contemporary cultural spirit from the weight of the pejorative deviations which now weigh on the verb "to mystify" in both its profane and sacred meanings.

1. The "angles of view" *(zewyâtâ dḥezyân* حزين وا تاوهي *)*

In French culture, describing someone as having a compass in the eye means that he/she is able to draw free hand a perfect circle. Analogically speaking, we can describe persons as having a protractor in the eye who are able to draw angles with precise measures, 45° or 60° or 90° etc., without the help of a protractor. If these two expressions lead to some deduction, they confirm that there are people more able than others to narrow the fields of vision, to fix angles of view for humans and to abuse the natural goodness of groups to push them to see and understand only through the angle drawn for them. They are mediators or intermediaries who are there, even if Plato does not mention them in his Cave. They could be demons, demiurges or simply philosophers like Socrates. The problem that comes up here, as Ephrem once perceived it, is to answer the questions that any believer would ask: How does one distinguish between angles of view? How does one distinguish between interpreters?

1.1. How to distinguish between angles of view

By analyzing the introduction of this point in light of the words of John Paul II that we have adopted in the epigraph of this chapter, we are referred back to the issue of the eye-light relationship raised in the first chapter, to launch a comparison between the contribution of the Greeks and that of Ephrem.

Almost eight centuries before Ephrem, we find Socrates using types and symbols, by means of Plato's maieutics, in a very eloquent dialogue, even if not poetry. He used it on purpose to push his interlocutor to see with a new eye, from a new angle of view, reality as the interest of the Republic requires it.[300]

Socrates, according to Plato, aimed at the Republic; Jesus, according to Ephrem, aimed at the Kingdom. These are two types *(ṭûpsâ* توبسا *)* which, at first glance and despite the very different angles of view which outline them, suffer sufficiently from idealism and virtuality to paralyze their implementation. While Socrates voluntarily committed suicide before seeing his dream realized, Christ likewise died voluntarily on the cross, both accused of having disturbed the circumscriptions of the compass of the system, particularly the cults and the circumcision of the Mosaic Law, respectively.[301]

Their accusers, therefore, are the people of the system who determine the fields of vision and limit their angles so that any reflective person conforms their

---

300 Glaucon, relatively speaking, was one of the prisoners of the Cave.

301 *Cf.* Justin(Saint), *First Apology*, 64; *Cf.* Gemayel Boutros, *Salat al Mo'men* (Arabic), Volume 3, Imprimerie Catholique, Beyrouth 1967, p. 622.

way of thinking to the coordinates of the reality of the *fait accompli*. This is the most dreadful of realities since, to believe Qoheleth, it is secreted frequently and continuously in the geopolitical 'here and now' by the materialistic dialectics related to the interests of the strongest.[302]

This infernal circle has allowed Qoheleth to say, "Nothing is new under the sun." Such things had already happened long before Marx's dialectic. These people of the system, whom Socrates opposed, as Christ did later, used to play the occult role of manipulating the universal reality. They were clever enough to place before the eyes of the inhabitants of the Cave the filters for their interpretations, the shadows that the inhabitants saw and the ideals to which they aspired and that they felt in the beating of their heart.

Socrates describes these people of the system, with the following words:

> Whereas if they go to the administration of public affairs poor and hungering after their own private advantage, thinking that hence they are to snatch the chief good, order there can never be; for they will be fighting about office, and the civil and domestic broils which thus arise will be the ruin of the rulers themselves and of the whole State.[303]

Christ does the same with a list of "woes" from which we choose two that were addressed to the experts in the Law:

Jesus replied, "And you experts in the law, woe to you, because you load people down with burdens they can hardly carry, and you yourselves will not lift one finger to help them... Woe to you experts in the law, because you have taken away the key to knowledge. You yourselves have not entered, and you have hindered those who were entering.[304]

From these two descriptions temporally distant and belonging to two completely different situations, a universal moral truth arises before every eye anxious to find the right angle of view for better predictability and certitude. But, although the protractors and compasses are universal and invariable instruments, once they are incorporated into the eye of the human being, they become colored by the colors of his/her proper 'eye' just the way it happened in Eve's 'eye'. The words of Christ, "The lamp of the body is your eye. When your eye is healthy, your whole body is also luminous; but as soon as it is sick, your body too is dark,"[305] apply to any instrument that becomes like an integral part of the 'eye' or of the

---

302   Eccles 1:1-11.
303   *The Republic*, p. 379.
304   Lk 11:46-48.
305   Lk 11:34.

body. This explains the unlimited diversity of angles of view among humans, even among those who belong to the same group. Hence the need for a guide whose eye is luminous (*šafyâ* ܨܦܝܐ) and who is as much as possible disinterested; otherwise other words of the Nazarene would come into effect: "Let them alone. They are the blind leaders of the blind."[306]

Diversity, especially when directed by systems, can by abuse of maieutics (brainwashing) reach angles of view and fields of vision opposed to the common good. The Socrates of the Republic invited the inhabitants of the city to look at reality in one way, whereas the people of the system invited them to do it in another, totally opposed to the common good. In the Book of Ecclesiastes as well, a group invited reading reality in a way opposed to that of Qoheleth. And Christ? Did He not upset the system as He wanted to submit everything to the Will of His Father? Therefore, one could well say that the maieutics would be a double-edged sword and that it is in the interests of the intermediary, and not in his eye, that we must seek the compass as well as the protractor of his interpretations. It is in the enigma of serving two gods at the same time that Christ has located evil *par excellence*. It is according to this criterion, whether in Science or in Religion, that the angles of view of the various intermediaries and interpreters should be analyzed and judged.

Ephrem had become hypersensitive to this socio-politico-religious phenomenon where Mammon's share is inevitable. That is why he was able to unveil the angles of view of his opponents, as well as their myopia; therefore, he knew how to thwart and counteract them. To answer the basic question of how to distinguish between angles of view, asked by the aspirants to the truth, Ephrem saw that it was not enough to show them the right or correct angle, as if he was passing them a fish for their 'dark night'. Instead, by all that he wrote, he tried to teach them how to fish for the answers in the water of Epiphany. He taught them to identify themselves in the Mirror of the Church, to determine their origin in the light of the "Origin", in order to know which fishing they are worthy of, – would it be that of the 'Pearls' (*Margâniâtâ* ܡܪܓܢܝܬܐ), for example?– and in addition, to determine which angle of view to follow in order to succeed, since this angle is often not necessarily one of the widest of the world. Fortunately, thanks to the Incarnation, this angle of view always has one of its two sides in the divine Plan. Ephrem taught his parishioners to recognize the true Light, the true Sun, and to recognize that their freedom to deliberate, choose, and decide is sacred.

In Nisibis, which was the scene of all these circumstances, he was able, thanks to the model of Daniel's friends in the furnace, to make them focus on the 'type' that would best suit their situation. Daniel will teach them that the salvation of

---

306   Mt 15:14.

their souls is besieged by the fire of the Persians from outside the walls of their city and by that of heresies from within these walls, and moreover, that the only solution is the one that he, Daniel, and his friends adopted, the one that made them join the equation: $1 + 2 = 4$.

It is this unexpected fourth Participant who is the sole mediator and guarantor of the discovery of the best angle of view that they search for in the beating of their heart. But this fourth Participant, invincible and reassuring, is for Daniel only the *Râz* Himself. He reveals Himself through angels such as Gabri-El and Rapha-El, witnesses such as Mary His Mother, John the Baptist, the Magdalene, disciples and collaborators such as Peter and Paul who, like their Master, are living signs, intermediaries of the Economy of Salvation. Once the angles are well sharpened, let's see with Saint Paul[307] how to distinguish between intermediaries and their interpretations.[308]

1.2. How to differentiate between intermediaries (*meṣ῾âyê* ܡܨܥܝܐ) and interpreters (*fušâkê* ܦܘܫܟܐ)?

According to Plato, everything starts, ideologically, with the light emitted by the rays of the primordial Fire from the background of the Cave and ends up in the ear of the one who listens to the interpreter. Plato was not the only one to write that everything starts with fire and light. The Bible did too. How did Ephrem interpret the beginning of this whole matter, he who fought, among other things, against a certain Greek school of logic, that of Aristotle? Starting from this question, we shed light on the following:

- 1.2.1. The problematic of interpretation in the universal sense
- 1.2.2. The role and the responsibilities of the interpreter as intermediary
- 1.2.3. The criterion of the good intermediary
- 1.2.4. Ephrem's originality

1.2.1. The problematic of interpretation in the universal sense

Ephrem, relying on the Bible, speaks in turn of the 'primordial fire', considered hidden, placed in the background of creation. He supposed it to be included in the concept of 'Earth' which was created together with 'Heaven' from 'nothingness' on the first day of Genesis.[309] This consideration is already an interpretive effort, whether on his part or on Plato's. Moreover, subjectively speaking, Plato mentions signs and symbols similar to the trees of Paradise, just as he also mentions a

---

307   1 Cor 3:4.

308   Gal 1:8.

309   *Cf.* Footnotes 180 and 285.

certain 'free will' and a 'comprehension force' (*noûs voυς*) which invites humans to leave their servitude and to open themselves to the absolute that they feel in their 'heart'. Ephrem mentions the same faculty under the name of 'calculating force' (*maḥšabtâ* ﺧﺴﻌﺒﺘ). According to the author of Genesis, God has locked humans behind the closed door of death with regard to the knowledge of good and evil: to know is to die.[310] However, Socrates invites them to go beyond this restriction and to know just what is needed to satisfy the faculty of self-awareness: know thyself.[311] The consequences of any neglect of these two dictates, as well as of any revolt against them, are well emphasized in the scene of the temptation of Eve, and even better in that of the Annunciation.[312]

There are enough various and confusing interpretations here to plunge us into an inexhaustible dialectic between Faith and Knowledge. As Reason once compelled the Church to ask for the criteria to consider a book of the Bible canonical, the same Reason demands today to ask what the requirements are to discern a correct interpretation. With that, it turns out that it is worth knowing that the verb "to die" forms part of human knowledge and that it worth knowing too that God took a human Mother, even though it would cost Him the price of death on the Cross to deify the human race. Ephrem underlines this in his *De Fide* hymns:

> If your cognitive faculty
> Does not succeed to know itself

---

310 Gen 2:17 and 3:3.

311 *Fides et Ratio*, Preface title.

312 "Eve, know yourself." As she was ontologically invited to recognize her limits in the light of God's counsel, her fault was to have understood the good and the bad relative to the senses and tendencies, before having acquired the capacity of analyzing what is the truly good, the truly beautiful, and the truly just. These are, according to Plato, the criteria that determine the scale of scientific priorities to be taught in the Republic. The comparison between the characters of *The Republic* and those of the Biblical Paradise reveals a very significant symmetry. On the one hand, Socrates invites the prisoners to a conversion of 180°, to turn away from the outside world towards their respective selves, or to turn away from primitive observation to qualitative reflection on the process of intrinsic knowledge (*ad intra*). On the other hand, Eve accomplishes the opposite, undergoing an invitation of the Serpent to the extrinsic knowledge that contrasts with the enlightening Wisdom of God's Counsel. However, God's Counsel does not conflict with Socrates'. This knowledge reveals the symmetric reflection between the centrifugal revelation that the Creator put at the disposal of Greek philosophy and the centripetal one that He put at the disposal of the nomads of the Bible. This description sends us, clearly, to the two traditional Christological currents recognized by the kerygma of the Church: the Christology from below (that of the Synoptic Gospels, especially Luke) and the Christology from above (with John). For the Greeks, it was the revelation from below, from Nature, out of which Aquinas found inspiration for his five proofs of the existence of God. For the Hebrews, illiterate and adopted by divine Wisdom, it was from on High.

How dare it murmur what is of the nativity
Of the One who knows it all?
You, [created one], who are unable to know yourself,
How can you "stare" at the One Who made you? [313]

Then there are the 'eyes' of the prisoners, those instruments of knowledge of which we have spoken in the first chapter and which we find back in the biblical allegory of Adam and Eve. What is ethical in Plato's Myth of the Cave is that he allowed these "images and likenesses" of humans to have eyes that see. What is not ethical is to have blocked their necks – hence their eyes– so that these can see from only one well-defined angle. Here lies the problematic of interpretation at the level of the ethics, justice, and goodness to which the intermediary hidden between the shadows and the real objects is subjected. The angle of view here is not measured by mathematical degrees, but rather by grades of education and honesty that assure a disinterested and liberating interpretation which allows an opening to the Absolute. Ephrem applied this theory to his parishioners by telling them about paradise and the importance of continence and chastity to acquire it:

To the one who abstains from wine by separation,
To him/her the 'vineyards' of paradise will be made
available,
And he/she is delighted to overcome, one after the other, the
traps put to him/her.
And if the one who frequents it [abstinence] is a solitary
It [abstinence] touches the strong core that this chaste
solitary needs
To avoid falling into the 'chasm' of the 'bed of copulation'.[314]

313 Feghali, Paul. *Anasheed fil Eemān*, 1 – 40; (Arabic), *Yanabeec Al Eeman* # 15, Paragraph 16, p. 12, Antonine University Publications, 2007. *Cf. CSCO* 154, *H Fide* I, 16; p. 5. (Translated by the author).
Cf. https://drive.google.com/file/d/1oBLu4MBhRqNdf1G8x5xmvMiX4EoSldXS/view [Accessed May 2019]. The original text:

ܐܝ ܗܘ ܡܚܫܠ ܕܢܒܚܫ ܗܝ ܠܗ ܕܝܕܥ ܠܐ ܢܒܚܫ
ܐܝܟܢ ܬܬܒܩܐ ܒܗ ܒܡܕܥܗ ܕܗܘ ܝܕܥ ܟܠ
ܒܪܝܐ ܕܗܘ ܠܗ ܠܐ ܝܕܥ ܐܝܟܢ ܫܪܝܐ ܠܥܒܘܕܗ

314 *Cf. H Par* 7,18. *Cf. CSCO* vol. 174, p. 29; (Translated by the author). The original text.

ܐܝܢܐ ܕܡܢ ܚܡܪܐ ܨܐܡ ܗܘܐ ܒܦܘܪܫܢܐ
Aynâ dmen ḥamrâ ṣâm wâ bfuršânâ
ܠܗ ܣܘܚܢ ܝܬܝܪ ܓܘܦܢܘܗܝ ܕܦܪܕܝܣܐ
Leh sowḥân yatir gufnao dfardaysso
ܘܚܕܐ ܚܕܐ ܣܓܘܠܗ ܡܘܫܛܐ ܕܬܬܠ ܠܗ
w-ḥdâ hdâ sgulâh mušṭâ dte-tel leh

In short, each "reality" has an angle according to which the compass draws its circumscription. Paradise, although invisible (*lâ mêthaziân* ܠܐ ܡܬܚܙܝܢ), also has its own. However, who or what determines the best opening of the angle and its best orientation, as well as the sufficient extent of its sides, to reach the reality that human mind qualifies as heavenly? Ephrem replies: the Scriptures, Nature, and interpreters like the Evangelists, the Fathers of the Church, and the holy Bishops he praised. Actually, he was able, thanks to the teaching of the latter, to expose the errors of the Arians and Manicheans who had distorted the message of the Scriptures.[315] As for him, he considers himself the least of these Fathers, deserving only to graze around the fence of Paradise and to eat only the crumbs from their sacred table:

> And if it is not possible for me to enter Your paradise,
> Allow me to pasture outside, by its enclosure:
> Within is the table of the diligent,
> But let the fruits of its enclosure drop like crumbs outside,
> So that through Your grace sinners may live. [316]

ܘܐܢ ܕܝܢ ܒܬܘܠܐ ܗܘ ܛܘܒ ܐܥܠܝܗ
wên dên btulâ hu tub aᶜlih,
ܝܚܝܕܝܐ ܕܡܛܠ ܕܟܢܐ ܥܘܒܗܝܢ ܠܓܘ
lhidâyâ dmetul da-knâ ᶜubhên Lgao
ܠܐ ܢܦܠ ܒܓܘ ܥܘܒܐ ܘܟܪܣܐ ܕܙܘܓܐ
Lâ nfal bgao ᶜubâ w karsâ dzuogâ

315 Feghali, Paul. *Anasheed fil Emān*, 1-40; (Arabic), Introduction.

316 *Cf. H Par* 5,15; *CSCO* vol. 174, p. 19. *Cf.* Sebastian Brock, *The Harp of the Spirit*, San Bernardino, CA. USA, 2018, p. 34. (From now on, we will refer to this book by *The Harp of the Spirit.*)

ܘܐܢ ܗܘ ܕܠܝܬ ܦܪܘܣ ܕܐܥܘܠ ܠܦܪܕܝܣܐ
wên hu dlayt frus dêᶜul lfardyssâ
ܐܫܘܢ ܐܦ ܡܢ ܒܪ ܠܪܥܝܐ ܕܣܝܓܗ
Ašwân âf mên bar lrᶜyâ dbsyâgueh
ܒܓܘܗ ܢܗܘܐ ܦܬܘܪ ܟܫܝܪ̈ܐ
bgaweh nehweh fṭour kašrê
ܘܦܐܪܐ ܕܣܝܓܗ ܡܢ ܒܪ ܐܝܟ ܦܪܟܘܟܐ
w-fîrâ dbasyâğêh men bar aḳ farkuḳâ
ܢܛܪܘܢ ܠܚܛܝ̈ܐ ܕܢܚܘܢ ܒܛܝܒܘܬܟ
neṭrun lḥatâyê danḥun bṭaybutâḳ

1.2.2. The role and responsibilities of the interpreter as intermediary

It is, therefore, the intermediary who, as mentioned above, plays the role of the third eye (*'aynâ dre'yânâ* ܥܝܢܐ ܕܪܥܝܢܐ).[317] Enlightened by the good light, he/she would be simultaneously the midwife of maieutics and the interpreter who has the compass of Truth in his/her eye. Ephrem emphasizes this point by insisting that this messenger must possess particular qualities at the level of the 'reality' of which he/she is the intermediary and the interpreter. The essential qualities of the one who mediates between the human eyes and the shadows are transparency (*šafyutâ* ܫܦܝܘܬܐ), justice (*zadikutâ* ܙܕܝܩܘܬܐ) and chastity (*btulutâ* ܒܬܘܠܘܬܐ). Anyone else would shade the truth according to his/her own colors.[318] But are there no other criteria for the interpreter's reliability?

For Ephrem, with regard to the sacred reality and its signs and symbols, the indisputable criterion of the interpreter's reliability is the conformity of his/her interpretations with the joint information of the Scriptures and of Nature. Ephrem delights in describing these last two together as the interpreter *par excellence* of the things of the Creator. They form a well-polished Mirror and are potentially able to reflect the divine Reality for the one who looks at him/herself in it. Still, no one can actually see the ontological reality in this Mirror, because the eye of the one who scrutinizes it (*bsâ* ܒܣܐ) must also be clear, chaste and bright to see the reflection of the Truth:

> The Scriptures are laid out like a mirror:
> And he whose eye is lucid sees therein the image of Truth.[319]

For the intermediary, to be good it is not enough to see oneself and to know oneself in this Mirror, nor to know how to know oneself. His/her role, by vocation, is to transmit the know-how to others and to convince them of the effectiveness of the Scriptures and of Nature as means of knowledge of the divine Reality, then to leave them freedom of action. Despite the authority of the kerygma, the role of the intermediary remains crucial, because he/she is the one who guides the maieutics, and it is to him/her that one lends the eye and the ear, or the ear-eye as mentioned in the first chapter.

He/she is also the one who establishes the rules of the dialectics and its

---

317 Our Syriac prayers are full of supplication to God to illuminate this specific 'eye' of our mind (conscience). Ephrem writes: "With the eye of the mind, I saw Paradise." *Cf. Les Origines*, p. 117.

318 Lk 11:34-36.

319 *H Fid* 67,8. *Cf. TLE*, p. 47; *Cf. CSCO* vol. 154, p. 207.

directionality. Regarding the past, in Religion, for example, we can name some persons like "Socraton"[320] of *The Republic,* Moses of the Israelite nation, Jesus of Nazareth, Paul of the nations, our guide Ephrem of Nisibis etc., while in Science, we can point to Pythagoreans, the Greek atomists, Aristotle with his empirical physics, astrologists, etc. Regarding our modern times, we find among the intermediaries the theological, scientific and charismatic authors quoted in this book. They have been of great interest to our entire study of Creation and the harmonization between Science and Religion.[321]

Here, the rigors of the ethical and ontological relativity of the 'angles of view' that the interpreters represent and some *ad hoc* consultation between them impose themselves on their role as well as on their responsibility. The relative vertices of these angles must at least start from the center of the same circle that remains to be determined. Could the center of this circle be the same as the vertex of the Angle determined by the Stone "that the builders rejected and that has become the Cornerstone"?[322]

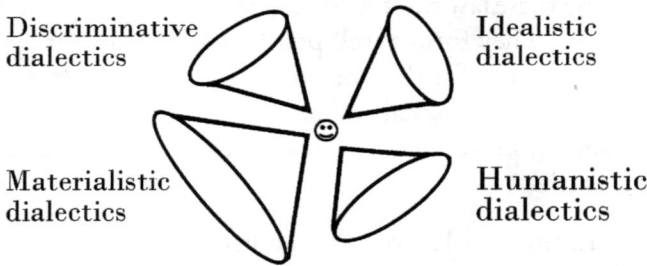

Discriminative dialectics

Idealistic dialectics

Materialistic dialectics

Humanistic dialectics

Figure 7: Diversity of orientation of the angles of view.

1.2.3. The criterion of the good intermediary

If the intermediary is a thief or a bandit[323] who knows himself and, at the same time, knows well the nature and inclinations of the prisoners of the 'universe', he will play on the field of vision of these chained ones in order to possess their souls. He will enslave them to all kinds of realities, except the one that frees them.Why would he manipulate these humans without freeing them? Well, 'Socraton', pointing out

320 A compound word resulting from apocope and apheresis of the names of the two famous Greek philosophers Socrates and Plato.

321 This could be extended, in future studies, to cover other religions, civilizations, and even the philosophical currents which have influenced modern times, *e.g.,* Marxism.

322 Mt 21:42; Lk 20:17; Mk 12:10.

323 Jn 10:1-10.

the risk that even a good reformer would run, explains it by the following:

> S. And if any one tried to loose another and lead him up
> to the light, let them only catch the offender, would not
> they put him to death?
>
> G. No question, he said.[324]

Having said that, how can we recognize the good interpreter so that he does not risk the same fate as the good reformers who preceded him and whose righteousness has been proved a little too late?

Jesus taught saying, "The good shepherd gives his life for his sheep."[325] Even if a 'thief' guides them to their own interest, would he be ready to die for them?[326] Only the true 'shepherd', faithful to their true good Master, unchains them and guides them to the true source of the Light that liberates them, at the cost of entrusting to them his own neck, knowing in advance that they will make him a scapegoat and sacrifice him. Does he not know them as he knows himself, and as he knows the One Who charged him with this mission? If he knows the prisoners well, he will also know that, if they are prisoners, it is because for the most part they want to be. They are not entirely blind, because their mother Eve had already eaten from the tree of knowledge of good and evil. They fed on her milk and are proud of her, and there is no more use in describing their mother – the Woman (*Îšah* ܐܢܬܬܐ) or Eve (*ḥawâ* ܚܘܐ) – as a vain woman who, by the foolishness of an ill-calculated curiosity, went against the natural law and caused them this miserable situation. To believe Plato, they do not consider their situation miserable. They have no other choice and, moreover, they are unconscious, and unconscious of being unconscious, of being prisoners. That is why it is dangerous to wake them up. However, as human beings, their unique problem comes from the same milk of that mother. It is a vain self-esteem, which blocks them from the duty of humbly recognizing the favors of another person to their dignity and freedom. In other words, they cannot bear the idea of being 'indebted' to anyone, not even to God, and, if necessary, they would kill their liberator. They would even kill God, their creditor, so that their debt of gratitude would no longer be real, that is, present day and night before their eyes.[327] This debt must become 'moral' and pass to the archive of a fanciful reality.

Moreover, the 'shepherd' faithful to his Master is the one who knows,

---

324   Plato, *op. cit.*

325   Jn 10:11-12.

326   In this lies the core of the Exodus.

327   *Cf.* Ps 52:11.

wisely and equally, the Book and the Nature of created things and holds, by a well-experienced hand, the art of the relational dialectic that feeds the two parts. This 'servant' should possess his Master's 'luminous Eye' in order to foresee the consequences that flow from this dialectic, especially the problematic of evil.

In fact, his major duty, as intermediary, is to awaken the inhabitants of the Cave to the truth of their state of being prisoners. They have to be conscious that the fault of their imprisonment is not totally their own, that the One Who created them in His image and likeness has endowed them with an imagination that can only be satisfied with the most beautiful. He has granted them a heart that rests only in the most Just and most Unifying Being who, to believe Saint Augustine, reassures and protects from all self-division of the heart, thus from all self-destruction.[328] As long as they are created, and it is not exclusively their own fault, the range between the most and the least remains vast relative to each of them. It allows an unlimited number of choices of 'realities' in the middle of which they find themselves locked up. Consequently, they may well recognize that their own state of being prisoners is due either to the total lack of choices due to too much light relative to the darkness of the Cave, or to the large number of choices caused by the lack of light. In both conditions, the major reason is the absence of the 'guide' ready to be killed by them, for them.[329]

Therefore, the fatal criterion of the interpreter's reliability is, at the same time, self-knowledge, fidelity to the Scriptures and Nature, and the total and voluntary readiness to sacrifice oneself until signing with one's own blood the will of one's own interpretations. For humans, this kind of 'signature' makes its signatory immortal, even if it becomes an 'archive'.[330] It is this melodramatic situation of the absolute reality of creation and of human–human relationality that Plato, the novice of Socrates, expands like a dining table in front of humanity.

Ephrem reveals another 'dining table' that the Creator has extended through Nature and the Scriptures and that Christ, the Incarnate Word, served during the Last Supper and whose completeness He signed by His blood on the Cross. Bread and wine became Body and Blood of a Reality written with a capital R, called the Kingdom, in which believing becomes necessary for understanding, not the other way around.

Intermediaries as interpreters specialized in this Reality, who stand between its Light and its Shadows, do not have much choice, as they must respect the rule of

---

328   *Cf.* Mt 12:25.

329   *Cf.* Phil 1:21-24.

330   Key word of the Derridian philosophy.

'equality' between Master and disciple, between 'Table' and 'table'.[331] They must be capable of a good discernment because the welfare of the Kingdom and of its inhabitants depends on it. Truth and Justice also depend on good discernment. This seems seriously applicable to the reality of the smallest Platonic city, to ancient Rome, to the City of God, *i.e.,* the Church, as well as nowadays to the Super Powers, like the USA after its claims to be able to establish the "New World Order", etc. How does Ephrem of Nisibis tackle this issue of the 'Signed Reality'?

1.2.4. Ephrem's originality

Under this subheading, we will deal specifically with the following:

1.2.4.a. Ephrem's interpretive method

1.2.4.b. Originality in Ephrem's interpretation

1.2.4.a. Ephrem's interpretive method

Ephrem, our guide, was not a stranger to what was happening in the two cities of Nisibis and Edessa which were affected by the following three lights: Gnosticism, Torah and Gospel. He was not a stranger to what was happening in their churches to which he offered his knowledge and his life. Similarly, he was familiar with the 'light' of the Cave of Plato, dialectically approachable, as well as with the 'light' of Genesis imposed from on High at the risk of dazzling to blindness. Most specifically, he was not a stranger to the 'light' of the Tomb of Jesus Christ, which is the result of a chiasmus between life and death. The characteristic of this latter one is that it stands just at the edge of death and life, and invites those who wish to acquire it to approach it according to a definite path, the way of Golgotha.

This *sine qua non* condition is determined by the precepts of the Master Who made it clear that a disciple can never be better than his Master, that he has to accept willingly to drink from the cup his Master drank from and to be baptized with the same baptism his Master passed through.[332] Accordingly, the waves of any interpretation aiming at this Light must be 'in phase' with its Source and follow its Way. It is that Way which is sensitive to only one of all the dialectics ever known, the one of Love.

So, Ephrem – without considering the grace of holiness he enjoyed – was in a historical and cultural situation that allowed him to start a quite particular method of interpretation. He saw that if the Greeks, and specifically Plato, laid on a dining table, as it were, the condiments necessary for all kinds of realities, Jesus, through

---

331   Jn 13:13- 17.
332   Mt 10:24 and Mk 10:39.

the reconciliation He accomplished between 'heaven' and 'earth', dispelled any annoying shadow between the two and opened to its full extent, 357°, the angle of vision for all humans.[333] He challenged them to respond to a unique reality, that of the Kingdom, to which He invited them by His anguished question, "Do you love Me?"[334] The anguish behind this question finds its roots in the ever unsatisfactory and disturbing materialistic reality that the Greco-Latin world developed through the centuries before Christ, and which even affected the people of the Book. Jesus, quoting Isaiah, described it in the following words:

> For this people's heart is waxed gross,
> And their ears are dull of hearing,
> And their eyes they have closed;
> Lest haply they should perceive with their eyes,
> And hear with their ears,
> And understand with their heart,
> And should turn again,
> And I should heal them.[335]

This reality of the Kingdom of Love is what Ephrem tried to elucidate through his interpretations, at the cost of his life, at the very price of becoming a priest. He insisted that people train their human eyes to see as large, or to acquire the 'Luminous Eye' capable of covering this angle of 357°, that of the maximum of Love and Trust in the 'Other', avoiding confusing it with the angle of 3°.

Brock quotes Robert Murray SJ to speak of a comparison between Ephrem and Dante Alighieri.[336] It is precisely from this theatrical worldview where signs, types and symbols represent the means of rescue with which adventurers should arm themselves to bring light into the eyes of those attached to the dark abyss. Plato hid himself in the world of ideas, under Socrates' signature, and Aristotle resorted sometimes to the empiricism of physics, sometimes to the unreachable metaphysics. Finally, even nowadays, where freedom of expression exists, Jacques Derrida found his refuge in his brilliant theory of "differAnce" thanks to which he was able to join the BEING, the SIGNATURE, and the ARCHIVE, in an inseparable way, into an indivisible reality.[337] Between these three builders of

---

333   The remaining 3° represent the domain of sin, which has remained apart from all reconciliation, as evidenced by the prayer of the Maronite liturgy: "No one is without sin except our Lord and God Jesus Christ." Although the number 3 is minimal, it remains perfect just as the number 357 which is a multiple of both 3 and 7.

334   Jn 21:15-17.

335   Mt 13:15.

336   *TLE*, p. 173.

337   Bennington, Geoffrey. *Derridabase,* pp. 74-75.

relative realities,[338] other creators of 'reality' have followed through the centuries, especially the founders of the quantum reality, to reveal the truth about what "is", what "will be", and what "could be", without ever perfectly succeeding. If the latter can boast of having achieved anything, it is rather, to believe a very serious criticism advanced by the *Encyclopedia Universalis*, a compromise of reality.[339] It is a good mixture of "objective" and "subjective" information, of which humanity has always been victim, since no one has been ready to defend the Truth at the cost of the extreme sacrifice.[340]

In declaring "I am the Way, the Truth and the Life", the Nazarene, the Son of Man, takes any complexity of interpretation upon Himself, as He takes upon Himself, in Person, the responsibility to shake all human logic.

At this point, our reader would be excused if he/she reacted by asking whether Jesus Christ would not be part of this series of philosophers and whether this Christian reality of the Kingdom of God would not be the same mixture of "objectivity" and "subjectivity" in reading the state of being. Let us explore how Ephrem could answer this question and inform his audience that this is not the case, indicating what is objective, predictable, and reassuring of that Reality that we have baptized *"râzatic"*.

1.2.4.b. Originality in Ephrem's interpretation

The interpretations represent the trap into which Ephrem led the heretics. It is in this domain that he excelled in order not to lose the Light that was continuously re-energizing him. With a protractor and a compass in the eye, he knew how to move towards the most limpid and most necessary Source of Light for his mission.

His originality is that he aimed straight at the concept of 'womb', the center of all conception. It is to the 'womb' of creatures that he went to draw the hidden Light, as he did with the One hidden in the womb of Mary. But this is only part of his contribution. Let us look at everything he has added to this concept.

To be able to distinguish the difference and nuances that exist between one light and another, one reality and another, with the aim of achieving an interpretation toward which the inhabitants of the 'City' would be invited to turn and convert,[341]

---

338 We have already suggested that each of them is the pioneer of a different reality: idealist for Plato, empirical for Aristotle and "differAnt" for Derrida.

339 *E.U.*, Concept de Réalité, Introduction. Article written by Jean Hamburger.

340 *Ibid.*

341 S. Whereas, our argument shows that the power and capacity of learning exists in the soul already; and that just as the eye was unable to turn from darkness to light without

is not a simple activity. It is rather an activity that depends on the most serious responsibility towards the absolute good of these people, rather than towards the interpreter's or any demiurge's. Book VII of Plato's *Republic* also testifies to this.

The Myth of the Cave, in light of our reading of Ephrem, inspires us with a nuance in the understanding of the biblical reality. We consider this nuance so interesting that we find it appropriate to mention it at the outset. It is simply the issue of cognition by inductive mode and/or by deductive mode.

Who among educated persons did not ponder over the clever Platonic maneuver that requires not only that the necks of humans be stiffened, but that the globes of their eyes be fixed in their sockets? The purpose of this maneuver is to subject them perfectly to the demands of maieutics. Those who have meditated on this trick must have found *ipso facto* that the Myth of the Cave is of a different paradigm than the perception of Creation according to the text of Genesis. Let us explain.

The fact that the necks and eyes of the nomads –of whom Abraham is an illiterate representative and Moses a highly literate representative– have full access to all that surrounds them and, implicitly, to the entire cosmos, contrasts with the necks and the eyes of the heroes of the Cave.

The nomads' necks and eyes were free to search and gaze (*bṣâ* ܒܨܐ),[342] day and night, in nature, in books and in the Book.[343] They were driven by the need for self-sufficiency, for protection of their kind and its survival beyond death, to deduce a knowledge useful for predictability and certainty. Thanks to their analytical mind, these nomads analyzed everything that happened to them, like everything that happened within their field of vision, that is to say "under the

---

the whole body, so too the instrument of knowledge can only by the movement of the whole soul be turned from the world of becoming into that of being, and learn by degrees to endure the sight of being, and of the brightest and best of being, or in other words, of the good.
G. Very true.
S. And must there not be some art which will effect conversion in the easiest and quickest manner; not implanting the faculty of sight, for that exists already, but has been turned in the wrong direction, and is looking away from the truth?
Cf. The Republic, p. 377.

342 This term is borrowed from Feghali, used in his last translations of the Ephremian texts. It is nowadays much more of a popular use in Egyptian dialect (*bassa, boss*) than in the North Mediterranean Arabic countries. In the latter dialects verbs like (*ḥaddaka, tafaḥḥassa, maḥḥassa, tafarrassa*), are more literarily and dialectically familiar despite the differences in their pronunciation. We notice that Ephrem uses it frequently to express the idea of "to scrutinize", while he uses the verb "to calculate" (*etḥšêb*) and its derivatives (*maḥšbâtâ, ḥušâbâ*) etc., to express the act of analyzing.

343 The books of other civilizations and religions in comparison with their Book.

sun",[344] without ever considering the need to know themselves. They consulted together and interpreted the discoveries of their primitive observations in the light of the intuition that came to them from the phenomenon of their mother Eve who wanted to know as much as God. They synthesized the information and passed it down, one way or another, in the hope that their descendants might one day reach the Truth that resides beyond utilitarianism. They hoped, as well, to acquire the necessary expertise to reach it on their own, or at least to recognize the good guides who will help them to reach it.[345]

These nomads left to their descendants words such as: "In your unfailing love you will lead the people you have redeemed. In your strength you will guide them to your holy dwelling".[346] Concerning the guides, they wrote to them: "My people, your guides lead you astray; they turn you from the path."[347]

For Plato's heroes, things were happening in the opposite way. Being stiff-necked, a demiurge brought them knowledge by passing before their eyes the signs and symbols of what is supposed to be reality.[348] These were virtual beings that imposed themselves on their visual memory, their phobias and philias, by penetrating their eyes, as described by Gaston Bachelard.[349]

---

[344] *Cf.* Eccles. 1:1-11.

[345] But to their great disappointment, these ancestors could not have predicted that one day their descendants would no longer be primitive and that they would lose the dream, for the experience. With this, Bachelard continues to say: "... The same act working the same matter to give the same objective result, does not have the same subjective meaning in mentalities as different as those of the primitive man and the educated man. For primitive man, thought is a centralized reverie; for the educated man, reverie is a relaxed thought. The dynamic sense is inverse from one case to another. (Bachelard, p. 48.)

[346] Ex 15:13*ff.* Taking into consideration Bachelard's analysis, we comment on this point by adding that "it is right and good". Man, in the making, would never succeed to know himself. Only the One Who created him and Who "probes his heart and his loins" (Ps 139: 13, Rev 2: 23-24) is able to guide him. The ancestors found in this faith the solution to the limitation of the cognitive heritage transmitted, sacramentally, from ancestors to their descendants.

[347] Is 3:12.

[348] This phenomenon is not so strange for today's civilization. It suffices to look at the people sitting in front of screens, especially children between 1 and 3 years old. The addiction to television is alarming, except if we take into account the dream and imagination phenomena that every child enjoys from birth. It seems that Plato's prisoners did not enjoy it. This is what makes the myth of the cave inhuman.

[349] Bachelard, p. 18.

Figure 8: Concertation

Because of their limited view, Plato's heroes were obliged to open up, to discover themselves, to discover what was going on in them and in front of them, to get to know each other better, but only individually and without consultation or exchange of experience and analysis with neighbors. This inductive method was supposed to push these prisoners to realize what was happening and to become conscious of being conscious of their being and of what was going on before them, that is to say of a probable reality with its causes and its effects, without fantasy or reverie. Finally, this method was supposed to help them to acquire a truth relative to what they saw. It was meant to help them to gain in certainty and predictability – no doubt, subjective – the common sense that allows everyone to clear the shadows and to penetrate this truth, without ever knowing what truth it was, or if it was the same for all of them who share the same situation and destiny. It is evident that in their case it was of no use to seek, without a guide or without a director, an objective and universal Truth. Socrates affirms this by saying to his novice Glaucon:

> Why, yes, I said, and for two reasons: in the first place, no government patronizes them [subjects of study]; this leads to a want of energy in the pursuit of them, and they are difficult; in the second place, students cannot learn unless they have director. But then a director can hardly befound.[350]

---

350 *The Republic,* p. 87. *Cf. Les Origines,* Note #197 p. 80. Feghali comments on Ephrem's defense of the deity of the Holy Spirit and His role as Director of the Church. Ephrem, indeed, criticizes in his hymns the heretics saying:

> Nobody has ever seen a boat on the sea -
> To walk alone and without a seaman,
> To govern oneself and to direct oneself!

So, it seems that the eyes of Platonic necks are instruments on which depends, primordially, all human knowledge, nourished by a need which is not personal, but rather the need of the Republic in this case. And, as mentioned above, this knowledge is accidentally nourished by the curiosity of someone, whether male or female, a born philosopher who knows how to dream, as in the case of Socrates.

Figure 9: Brainstorming

This sheds light on this determinative nuance of the two visual methods of becoming aware of the true reality in the absolute, either by deduction or by induction. However, a third typical instrument of learning, a corollary, emerges from the biblical side: the ear. It is the gift of hearing. This is the originality proper to Ephrem.

In the case of the prisoners of the Platonic Cave, the ear is, indeed, almost non-existent. It is an instrument that stays in potency until one of them breaks the chains, goes to the real sources of "Light" and comes back to tell the others about what he saw and learned. But would they believe him if they did not see with their own eyes?

In the case of the Bible, it is the opposite. We find the 'ear', in act, from the first verses of Genesis, associated with the freedom of the eyes and neck. Long before the making of Adam's and Eve's ears, a hidden 'ear' of the universe heard the voice of God, and at His command, the 'raw material' of the universe, existing in potency, reacted.[351]

As we explained in the first chapter, the synergy between the eye and the ear is a

---

Like boats, every being needs direction:
The soul of freedom, the creature of a Creator,
The Church of a Savior and the Altar of the Holy Spirit.
(*Cf. Hc Haer.* 5, 2; *CSCO* vol. 169, pp. 22-23.)

351 This is what the Son will prove during the storm. *Cf.* Mk 4:39-41.

sine qua non condition for the conceptualization of all that elucidates the absolute Truth about Light and Life. It is the ear of the apprentice, the 'child Israel'. It is through the ear that one could have the eyes and the mind guided and directed. It is through the ear that we start with the apprentice, or the initiate, a dialogue that develops into dialectics and cognitive material. Moreover, Plato will base his method on the ear, even if it has been overshadowed in Book VII of *The Republic*. Indeed, Plato affirms that knowledge is present in the mind of human beings, and to educate them means only to remind them, through the ear, of what they already have in their core.[352]

> S – Now, these are to be apprehended by reason and intelligence, but not by sight.
>
> G – True, he replied.[353]

It turns out that Plato's maieutics, indispensable for the freeing of prisoners, is accomplished through the ear.[354] Would it then still be strange to admit the Ephremian theory that Mary conceived the divine child, the Word of God, through her ear?[355] Would it be so strange to talk of the womb of the ear?[356]

And once conceived by the ear, from where would one give birth? Is it not from the mouth?[357] How beautiful would it be then to meditate on the words of the Lord which refer to the ear-mouth relationship as Jesus described it by His answer to the devil saying: "It is written: One does not live by bread alone, but by every

---

352   'Core' is taken here in the Platonic–Augustinian meaning, the base of reminiscence and not of memory.

353   *The Republic*, p. 389.

354   We hold to this position until the contrary is proved despite the fear of Ephrem noted by Bou Mansour on page 112 of his book, *La Pensée*, that the hearing does not favor scrutinizing information or its source: "... In relation to listening, it asserts itself as a priority. Starting from the symbolism of numbers hidden in the bosom of the human mind, Ephrem makes it clear that neither its language nor its silence are addressed to the ear; it reveals its hidden symbols only to the eye *(H Eccl* 35, 15-16). In fact, this privilege granted to the eye finds its ultimate reason in the fear that hearing does not favor scrutinizing. Ephrem attests to it when it comes to the Marcionites ... "

355   *La Pensée*, p. 248: "In the same way, he compares the Word in Mary's womb with the word spoken in the lobe of the ear." *(H Fid* 83, 11-33)

356   *Cf. TLE, H Eccl* 49, 7, p. 33. "So through a new ear, that was Mary's, life entered and was poured out."

357   There follows a very interesting prayer, which Ephrem places on his own lips assuming the role of a sterile woman, which highlights this allegory: "Sterile is my spirit, Lord, to give birth again, make it fertile and give it a child just as for Anne; so that the voice of the child when it comes out of my mouth is offered to you as the child of the sterile." *(H Eccl* 30,1; *Cf. TLE*, pp. 113 and 172.)

word that comes from the mouth of God."³⁵⁸ It would also be nice to meditate on the word which is "digested" by the ear: "Do you not see that whatever goes into the mouth enters the stomach, and goes out into the sewer? But what comes out of the mouth proceeds from the heart..."³⁵⁹ And, the door of the heart is the ear: "For Ephrem, Adam and all men are subject to a law inscribed in their 'hearts', which means, well rooted in their ears."³⁶⁰ One would also enjoy the meditation on the vision of Ephrem as Brock describes it with the following words:

> Particularly beautiful, and readily capable of being appreciated by a modern reader, is the story of his vision of a vine sprouting from his tongue, reaching up to the sky and producing myriads of clusters and bunches of grapes - his *mîmre* and his *madrâshe*.³⁶¹

Then, for an additional reference in this sense, it would be useful to emphasize that the education of the People of Israel began one day with the order, "Listen, Israel –*šama Israël*",³⁶² and not at all, "Look, Israel".

Eye and ear were and will always be the target of Christ's messianic role as foreseen by Isaiah, "Then will the eyes of the blind be opened and the ears of the deaf unstopped,"³⁶³ and all this so that they may convert. And in order to convert, it is necessary to have beforehand a compass, like that of the blind man of Jericho, which indicates the Orient from which the Light points.³⁶⁴ It is also necessary to possess two eyes that see and two ears that listen to the director who guides the research on this Orient of Light, as well as on the path of the Way that leads to Him. And finally, it is necessary to have a 'heart' capable of knowing itself, of understanding, of getting along with the other, and of converting.

---

358 Mt 4:4. The word "mouth" appears sixty-five times in the New Testament, and "ears" fifty-two times.

359 Mt 15:17.

360 *Les Origines,* p. 193.

361 Brock, Sebastian. "In Search of Saint Ephrem", p. 17.

362 Ps 49:7; 77:1. Although the Bible, in general, is dominated by this reciprocal invitation to listen, – God invites humans to listen to Him and humans beg God to lend them His Ear, – Jesus sealed by His words a 'truth' which emphasizes that any communication with the Father, through the eye or through the ear, is impossible apart from Him Who is the Word and the Light.

363 Is 35:5.

364 Why not say about the blind man of Jericho, like all the other blind persons whom Jesus healed, what Brock said about the women of the Old Testament? A tacit dialogue (*kâsyâ*) existed prior to any miracle between Jesus and those who had received Grace from his Father, long before the event became apparent (*galyâ*) and subject to criticism from the Pharisees, *e.g.,* the blind man "saw" before seeing, and that is why he rushed the steps or jumped. We will recognize this phenomenon in the disciples of Emmaus.

Jesus joined, as Master, the two methods mentioned above to prove that He came to perfect all the disciplines of research and cognition of the divine that preceded Him, and not only those of the Old Testament. He did not exclude any teaching method that could serve His mission. That He was Himself the Sign, the Signified, and the Referent, however, posed a problem. What made His mission more crucial was His refusal to be a sign of Himself, and to refer only to Himself, the 'Visible'. He insisted in a way that bothered the people of the "system" to refer everything to His Father, the Invisible.[365]

It was of primary importance for Him to make sure that the reality of the Kingdom, of which He declared the institution at Caesarea Philippi, is not taken for a mixed reality, not only mixed between divine and human, but specifically between political messianic and soteriological messianic. If He succeeded inHis mission, it is because He did not expect to see its realization during His lifetime. His words to the disciples of Emmaus revealed this fact.[366]

To stand at an equal distance between the Greeks (*yawnâyê* ܝܘܢܝܐ) and the Hebrews (*ᶜebrâyê* ܥܒܪܝܐ), and to marry the ear, the eye, and the mouth together by their relative *râz*, was the strength of Ephrem's interpretation.[367] To see exhaustively from Nisibis, and especially from the "mystical pulpit" on which he had risen, was the synergistic element that allowed the application of the formula $1 + 2 = 4$, where (1) is the *Râz* and (2) represents the two methods of deduction and induction mentioned above. Here, the fourth unexpected element is the new 'heart' that once guided by divine Grace is able to see exhaustively, even before the eyes see or the ears hear.

Ephrem's originality is to have recognized that any interpretation is trapped by the *kasyâ – galyâ* dichotomy (that of veiled – unveiled) and that despite all the assurances given to us by Faith and Science, we always feel the awkward position imposed by the dialectics of choice and deliberation.

Our freedom that also comes from our mother Eve is what challenges us. Being created makes our hearts always uncertain, worried. At night, the inductions of dreams worry us, and during the day, the deductions of our awakening alarm us – not to forget the current uninterrupted role of the media.

Ephrem's originality was, and still is, to reveal that the interpreters are themselves the intermediaries and must be conscientious directors specialized in probing the 'wombs' of reality and dreams, of what is visible and of what is invisible, to reassure humans about their true value and true fate, as Christ did with

---

365   *Cf.* Jn 5:17 and 37-38; 8:18-19 and 41-44; 10:30.

366   Lk 24:19*ff.*

367   *Cf. H Nat* 15, 5.

the disciples of Emmaus. With this, let us return to Ephrem's texts to elucidate the secret of the *râzâtic* reality, the title of this chapter, which now has enough background to enable us to write the prelude.

2. The *râzâtic* reality, angle and description

If we admit, with Ephrem, an entity called *râzâ* and a *râzâtic* fire that emits a light of its own nature, a realm should exist where these elements naturally belong. Would it be audacious or pretentious to admit the existence of a *râzâtic* reality? To which category would it belong? This is the angle from which Ephrem targets the domain of this reality to provide the answer. We will elucidate the enigma of this reality by proceeding, first, with the elimination of the doubt aroused by the two following questions:

> 1- Is this *râzâtic* reality a mixture of subjectivity and objectivity?

> 2- Is this *râzâtic* reality a compromise or a utopia?

Then, by answering the question that flows from the result:

> 3- What could this reality really be and to which category would it belong?

2.1. Would the *râzâtic* reality be a mixture of subjective and objective?

Ephrem, who takes the angle of the Incarnation as a key in all that concerns his *râzâ,* inspects it through this 'theodolitic' Pearl and analyzes it, if we read him well, under a specific perception. We may summarize the latter by the following: if, in the Person of Christ, the Incarnation of the Divinity in the humanity were a mixture, then the *râzâtic* reality would be too.

Ephrem graciously proclaimed to the heretics of his time the absurdity of implicitly introducing Christological theories based on a mix or a mixture in the Person of Christ. The risk in doing so would be to allow, at a certain moment of the history of salvation, a re-separation of the Divinity and the humanity (*prash meneh*). He even urges the 'sun' to testify to the indivisible unity in the hypostatic union of the three Persons of the Trinity and to the 'non-separability' of Jesus Son of Man and Jesus Son of God.

It is true that the rational soul learns by experience that "Man is a temporary admixture of two substances: spirit and matter, good and evil, light and darkness",[368] but it also learns, "... that the deliverance from this admixture

---

368 *Les Origines*, p. 32.

N.B. In the texts of Ephrem, the word "mix", as translated by our privileged authors, presents a difficulty that any interpretation would only aggravate. Indeed, a mixture

means a complete rupture between the two."[369] According to Ephrem, the error lies in the very fact of comparing 'substances' belonging to two substantially different realities. If the spirit is a substance, matter cannot be one in the same sense. Consequently, to believe Einstein, it is quite absurd to compare good and evil or light and darkness. Ephrem has drawn to the attention of heretics that, even in terms of WORDS and NAMES, they have no right to make a comparision without applying the motto that warns, "Give back to Caesar what is Caesar's and to God what is God's." He wrote:

> Let us confuse heretics in the manner of thieves.
> Because the wealth they stole shouts against them, it speaks (eloquently).
> They stole the names and dressed something that does not exist
> With the name of God; they have always been dressing the idols they honored and
> Thus, by their denomination, they became essences (*ityê* ܐܝܬܝܐ), they, that do not exist.
> Their names were adored ... they became famous.[370]

---

may derive from the verb *ḥlat* (ܚܠܛ) "to mix", specific to materials whose substances are incompatible, *e.g.,* mixing the various ingredients to make food, or mixing oil and water etc. It can also be derived from the verb *mzag* (ܡܙܓ), also "to mix", but specific to materials whose substances are compatible – *e.g.,* mixing two oil-based colors, – or *mzag*, this time in the sense of mixing a small quantity with a much larger one, as in the Liturgy mixing water and wine in the Cup. To these two verbs is added a third which implies much more demanding dimensions: "to unite" *ḥayed* (تبد). And here, as we say in Lebanese, "The Rosary is long." It goes from union in the fertilization of cells to the hypostatic union within the Holy Trinity. The Incarnation is its apex and we recall the aforementioned formula of consecration in the Maronite Church which says, "You have united, Lord, your divinity to our humanity..." On this, taking into consideration the level of understanding of the Ephremian *râz* in the most delicate of its applications in the Book of Daniel, we say that Ephrem did not only attack the Platonic duality and the necessity of breaking with its two components to return to the absolute Oneness, but he also strongly supported the impossibility of breaking a hypostatic union, as the Pythagorean theories of indivisibility and of "unity in multiplicity" show. Ephrem, by his *râz*, reveals the incapacity of the Manicheans and Bardaisanists to see the third intermediary, who first of all makes it possible to join the perfect number of three, and hence the perfection of unity in the Trinity. Therefore, Ephrem highlights the unifying, fermenting, fertilizing, life-giving element, which never tolerates stagnant parallelism (grains of wheat on the rock), asphyxiating confusion (grains of wheat among thorns), nor undesirable juxtaposition (grains of wheat on the road), three cases of different duality. As we develop this thesis, we will have to go back to this triune conception which seems, even today, not to have been surpassed by Religion, Science or, in particular, by quantum physics.

369 *Ibid. Cf.* Gal 4:14; Rm 7:24
370 *Ibid.* p.49.

Today, the Gestalt theory would even question the use of upper and lower case letters. Clarity and distinction demand it, as Descartes would say.

Having seen with Ephrem that this *râzâtic* reality could in no way be a mixture of subjectivity and objectivity, either in form or in matter, or even linguistically, let us see if it can be a compromise or a utopia.

2.2. Would the *râzâtic* reality be a compromise or a utopia?

Ephrem also rejects this possibility by resorting to the Fire that discriminated mercilessly between the offerings of Cain and Abel and between the God of Elijah and the God of Jezebel. (1Kg 21) It was also the Fire that distinguished between the three young people in the fiery furnace and the rest of the Babylonians who carried them there. "The king's command was so urgent and the furnace so hot that the flames of the fire killed the soldiers who took up Shadrach, Meshach and Abednego."[371]

This Fire knows no compromise; either it inflames and preserves as a sign of acceptance, or it burns and consumes as a sign of refusal and annihilation of everything that has to do with the compromise. The Apocalypse bears witness to this *categoricity* by emphasizing the words of the 'Amen':

> These are the words of the Amen, the faithful and true
> witness, the ruler of God's creation. I know your deeds,
> that you are neither cold nor hot. I wish you were either
> one or the other! So, because you are lukewarm—neither
> hot nor cold—I am about to spit you out of my mouth.[372]

If the reality Ephrem proposes, even being largely poetic, is considered a compromising or utopian reality, the same answer would be valid.

However, epistemologically speaking, Science considers 'real' what comes in concepts and categories that have rational coordinates that positively delimit its space and time and help the intellect to measure it. It is under this aspect that Einstein, opposing Kant, was able to introduce space and time into reality and to later add the space-time relation (*Raumzeit*) to this same reality. By this "relationality", Einstein exposed a huge difference between the understanding of what Kant considers '*a priori*' and what Faith reports as invisible and therefore immeasurable.[373] "Once the

---

371 Dan 3:22 and 48.

372 Rev 3:14-16. This allegory finds its source in the words of Christ in Mt 5, 37: "Let your word 'Yes' be 'Yes' or 'No', 'No'; anything more than this comes from the evil one".

373 Commenting on Kant, Einstein says: "The only justification for our concepts and system of concepts is that they serve to represent the complex of our experiences; beyond this they have no legitimacy. I am convinced that the philosophers have had a harmful effect upon the progress of scientific thinking in removing certain fundamental concepts

coordinates are grasped, reality is cooked," declares Derrida.[374] As a result, the unreal becomes real and any compromise loses all support. Without a 'protractor' to define angles and fields of vision, even dreams do not come true. We deduce, therefore, that the *râzâtic* reality cannot be a reality without coordinates; otherwise, it would be taken for utopian or for a compromise between a subjective and an objective phenomenon. Even worse, it could simply fall into the category of 'hustle and bustle' (תֹהוּ וָבֹהוּ) of the 'period' before the ordering of Creation.

Therefore, since this *râzâtic* reality cannot be a utopia or a compromise of anything, since it is alien to 'everything', let us try, always in the Incarnation field of vision, to explain its nature.

2.3. What can this reality be and to which category would it belong?

The *râzâtic* reality, being, on the one hand, neither mixture nor compromise, and on the other, neither fantasy nor utopia, is just a "differAnt" reality, like that of Love.

But even in Love, empirical coordinates are needed and the most important of these is the point 'zero', the intersection of the four dimensions of the Cross that represent the horizontal, the vertical, the depth and the path in space-time of all other coordinates. However, taking into consideration the description of the concept '*râzâ*' and everything that turned around in Ephrem's eyes and ears during his meditation on the Pearl or on the Mirror, without forgetting Nature and the Scriptures, the description of this strange reality can be formulated as follows:

> A "differAnt" reality that joins "differAntly" humanity and
> divinity, Creator and creature, Alpha and Omega. It is in a
> continual becoming, mutually relational with any other kind
> of reality.

John also described it well in his Revelation. We just quote him with this verse: "Do not be afraid, I am the First and the Last, the Living; I was dead, and here I am alive for ever and ever, holding the key of Death and Hades (*Sheol* שְׁאוֹל)".[375]

---

from the domain of empiricism, where they are under our control, to the intangible heights of the *a priori*... This is particularly true of our concepts of time and space, which physicists have been obliged by the facts to bring down from the Olympus of the *a priori* in order to adjust them and put them in a serviceable condition." *Cf.* https://en.wikisource.org/wiki/The_Meaning_of_Relativity/Lecture_1 [Accessed July 2019]

374 Bennington, Geoffrey & Derrida, Jacques. *Circumcision*, series Contemporaries, Seuil, Paris, 1991, Epigraph: "As soon as it is grasped by the writing, the concept is cooked". (Same epigraph in French for the English version)

375 Rev 1:12-20. Msgr. Michel Hayek, a Maronite scholar, introduced poetically these words in the first verse of his accompaniment liturgy for deceased persons: "With the dead-living Christ". (*Cf.* Maronite accompaniment hymn: *Ma' Yassou' el māyt wu hāy*

This concise description of the *râzâtic* reality no longer allows doubting its own existence, because by doing so, we put our own existence in doubt. We will try to highlight in a succinct way two strong points of its "behavior" which are the 'becoming' and the 'relationality'.

2.3.1. The becoming

The becoming of this *râzâtic* reality differs from the fate of any other reality, be it philosophical or scientific, by the very fact that the realities that are familiar to us have the beginning and the end of their becoming on the same plane or on the same orbit. In other words, while the Alpha and the Omega of common realities belong normally to the same reality whose future will one day reach an impasse, the *râzâtic* reality will never reach that impasse because its Omega never belongs to the same plane as its Alpha. Its orbit is synergistic, Trinitarian, open to a beyond that, as shown above, is 'here' and 'not here' at the same time, thanks to the event of the Incarnation. Moreover, the Alpha and Omega of the other realities, those subjected to the same exhaustiveness, are arithmetical and self-sufficient, and the *Fullerian* equation of $1 + 2 = 4$ only results in a fourth dimension of their own nature. But the Alpha and the Omega of the *râzâtic* reality reveal, naturally, a fourth element "differAnt" from the three primordial ones. This fourth element belongs to the beyond, *i.e.,* the symmetric, the other side of the 'Mirror'.[376] It is this fourth element that forms the basis of the becoming of the *râzâtic* reality, of its perpetuity and its nature.

2.3.2. The relationality

Quite simply, it is a relationality that roots itself in the ontological and qualitative coordinates of the Alpha point represented by the letter *A* of "differAnt", with a small nuance, "differAnt" this time, hidden behind this *A*. We take pleasure, thanks to the Ephremian lights, in revealing this nuance, for the first time, by the following:

Some kinds of relationality, perceptible by the human intellect and whose signs and symbols fill the Book and Nature, exist between spiritual and natural realities. As they depend only on the human intellect as the agent of interpretation, they confuse it and lead humans to fall into *idololatry* (*sic*).[377] In the case of the *râzâtic*

(كتاب الجناز الماروني، نشيد المرافقة، مع يسوع الميت وحي).

376 It is of great importance to note that the symmetry between an object and its image through the mirror changes poles, contrary to mathematical symmetry.

377 It is not a spelling error. It is instead the correct word to talk about worshiping Idols. The latter Greek's origin is *'eidolon'* and not *'eidos'*, added to *'latreia'* for adoration should be said *Idololatria*.

relationality, the intellect can do nothing unless it is enabled and seasoned with the salt of the *Râz* of Incarnation, as previously mentioned. The *râzâtic* reality is observable only with a luminous 'Eye', such as that of Mary or, at least, that of Paul. Only a non-flammable 'protractor' can trace its angle and field of vision. To do this, the 'protractor' must be able to measure the opening of the arms of Jesus on the Cross, which covers at the same time, from the moment of the Crucifixion, the four 90°angles of the Cross multiplied by the four poles of the universe and its seven conceivable dimensions. Within the same action, the 3° mentioned above as lacking to the Perfection of Salvation must be borrowed by the 'Eye' behind the 'protractor' from the One Who cried out, *"Eli, Eli, lmâ šbaktan(i)?"*

One unique 'compass' is valid to draw the exhaustive spherical surface of this reality. It should be able to have one point in the 'opening' made by the spear of Man in the side of the God of forgiveness, the God of Love, and the other point on the orbit of the 'Heart's Ear' of Him Who hears His Son saying, "Forgive them, Father ..." This nuance is what has to be unveiled. It is to this Love (*Amor*) that the *A* of the concept "differAnce" refers and from which the relationality of the *râzâtic* reality takes root. However, the Love we evoke here, as we mentioned above in the case of the Song of Songs, is never absurd or romantic. It is of the divine reality that creates, vivifies, reassures, affects every being, and finally *râzifies*.

The coordinates of this relationality, distributed over open planes between divinity and humanity, and once established on either side of the intersection of the two boards of the Cross, become inseparable (*lâ praš* ܠܐ ܦܪܫ), even at the level of the space and time of the universe. In this case, time (*chronos*) the basic instrument of realities, is transmuted into the eternal (*ta eschata*) just like the Love that indelibly marks it. The eternity of the *râzâtic* reality comes transmuted into 'time' by its very nature. Furthermore, the 'chiasmus' meant by the words of the Maronite Liturgy, "Lord, You have united Your divinity to our humanity and our humanity to Your divinity...", has its birth from this relationality and allows us to declare it without risk of preaching confusion, elimination or substitution. This leads us to wonder if the rules of grammar and conjugation can satisfy this relationality! It is just as one day the question was whether the rules of Newtonian physics could contain quantum physics. Would it be logical to convince students that the verb "to unite" of the above-mentioned prayer, whether it means "unify" (*ḥayed* ܚܝܕ) or "mix" (*mzag* ܡܙܓ), can be conjugated under the same moods and same tenses as in the empirical or common cases? We assert that if this happens, it would be the worst misstep an interpreter can commit, because the sun in this case would continue to rise in the same way these students get up from bed.

That said, it turns out that it is impossible to apply categories that can be grasped by our pure reason or by our intellect, no matter how analytical they may be, to the instances of this relationality of Salvation. Getting out of this embarrassment requires us to approach it from a behavioral perspective. And, since there are no categories that can determine the Love we have spoken of, and since the *râzâtic* relationality suffers from the same problem, we will try to describe it on the basis of its behavior, causes, conditions and effects.

2.4. Describing the relationality through behavioral analysis

We humans 'instinctively' observe the behavior of everyday reality, through its smallest details, in order to describe it objectively *a posteriori* after having described it *a priori, i.e.,* by prejudice. For this reason, we are compelled to question its behavior in order to be able to discern its nature, its dimensions and the basis of its ethics with greater certainty. Would this relationality involve believing Qoheleth who asserts that nothing is new under the sun? Would it be an archival relationality or a living one? Could it not be a dialectical relationality in the making?

We purposely omit the discussion of the first two of these last three assumptions to aim at the analysis of the third, because thus we will reply exhaustively to all of them. The path of realization includes them all.

Being 'in the making' could be understood as if the making is absolute, without end, while being in process of realization means being in the path of taking place, of seeking perfection in a creation where nothing can be said to be perfect. Therefore, each 'being' including the *Râz* and, by the same token, the *râzâtic* relationality must be subjected to the fatality of the two philosophical movements, 'Passion' and 'Action'.

Practically, Ephrem discusses this point with the same energy that he has invested in the discussion of the imperfection of Creation.[378] For him, it is obvious that if Plato's concept of God (*Unum, Bonum, Verum*) and Aristotle's Causing and Uncaused Cause mean and refer at the same time to a perfect Being Who performs only Action, his opponents would have prevailed over him as well as over the Gospel.

The theologian and Bible scholar Feghali, who spares two titles and several pages for this point in his book *Les Origines*, says that the Bible chorus "and God saw that it was good" is cited by Ephrem only once in his commentary on Genesis. This is due to the simple reason that between the verb 'to form' (*taqên* ܬܩܢ) and the verb 'to adorn' (*sabêt* ܨܒܬ) there was 'to wait'.[379] Who had to wait? Obviously God, said Ephrem, since between these two stages of divine action – before delivering the universe to Man to make it his habitat – God waited a good

---

378 *Cf. Les Origines,* p. 62*ff.*

379 *Les Origines,* p. 87.

while before expressing His satisfaction. Quite simply, the work created and/
or made was not yet adorned, thus imperfect and incomplete.[380] It is from this
imperfection that relationality has its source.

But, would it not seem absurd, for those uninitiated to Ephremian thought,
to see time and eternity joined in a *râzâtic* time at the service of this relationality?
Yes, of course. What else can be done, would Ephrem say, if the divine activity
according to the two previous stages of creation is directed towards an end that
is not of its own reality? Although the formation (*tuqânâ* ܬܘܩܢܐ) begins in
eternity, perfection (*ṣubâtâ* ܨܘܒܬܐ) is accomplished in temporality, at the
service of mankind, the first motive towards which everything has been motivated.
Consequently, the absurd becomes logical, but according to the logic of the Love
that wants, "differAntly" from Kantian Pure Reason, everything to get always
better, so that every person "may have life, and have it to the full".[381] Thus a 'bridge'
of relationality rises between eternity and temporality, of an indescribable nature
for some, but for Ephrem, as for 'mystics' like him, it is the exhaustive Cross on
which rest at the same time protractor and compass, time and eternity, God and
Man, body and bread, blood and wine, death and life. At the same time, this Cross
assumes the role of the astrolabe and the compass of the *râzâtic* reality which
direct and guide from the least sacred to the most sacred, from the least perfect
to the most perfect, until reaching the perfection of the heavenly Father.[382] This
is what the Maronite Liturgy teaches Her children to wish, by singing in Arabic,
"Lord, let Your Cross be a bridge for the souls...",[383] and in Syriac, "Your Body
and Your Blood that we have received, may they be for us a bridge, a passage from
darknesstolight."[384]

A 'bridge' is what the two prayers emphasize, and it is what proves that
relationality is indeed, at least for the Churches that enjoy an Ephremian
spirituality, a mission in the process of realization.

The Maronite Church, in view of the infinite persecution it has endured since
its foundation, as we have been told by the late scholar Chorbishop Michel Hayek, is
"the Church of Holy Saturday, the waiting Church *par excellence*".[385] Accordingly,
She should be the 'Bridge Church' between Good Friday and the Glorious Sunday.

---

380   *Les Origines*, p. 86.

381   Jn 10:10.

382   Mt 5:48.

383   A traditional Arabic Maronite hymn, *Salibuka Jissran yakunu lahum* ...

384   *Manual of Divine Liturgy*, Bkerke, 1995.

ܦܓܪܟ ܘܕܡܟ ܕܩܒܠܢ ܗܘܘ ܠܢ ܓܫܪܐ ܘܡܥܒܪܬܐ
ܘܡܥܒܪܬܐ ܕܢܥܒܪ ܡܢ ܚܫܘܟܐ ܠܢܘܗܪܐ

385   He used to repeat it in almost all his conferences as his own leading idea.

Indeed, this Church enjoys the hope inspired by the continual realization of the Crucifixion, Resurrection, and Eucharist, as foreseen by Her Lord. She *râzâtically* reads these events and admits, *râzâtically* as well, that the role entrusted to Her as *râz* of Fire and Spirit is to assume (hierarchy and community at the same time) the continuation of the realization in the *hic et nunc* of the Economy of Salvation.

Thus, between an *a priori* perfection and a promised one, the relationality of the Incarnation can only be in progress. Its two movements would be the WAITING for the 'Passion' and the ANAMNESIS for the 'Action', which is translated by self-feeding with *râzê* (through the Sacraments), while waiting to join in act the perfection with which the heavenly Father is qualified.

At this point, a compelling problematic faces us. It comes from the fact that Christ, the *Râz par excellence*, Who invites us to perfection, does so in His quality of Way, Life and Truth. He says that perfection is in His Father and not in Him, at least as long as He is in this world. So, we have the right to ask: what does He invite us to through this relationality? To an end or to a means?[386] Everyone recognizes that between the end and the means exists a persistent conflict that influences Science and Religion at the level of both morals and ethics. This conflict, which has reached its apex with everything related to genetic manipulation and cloning, seriously hampers the dialogue between Science and Religion. It prevents their nuptial entrance – so long awaited – under the Tent of Meeting between God and humans. It compromises the ontological value of the *râzâtic* reality that does not seem concerned with these scourges. Chapter VII of this book will explore this in enough depth to prove the opposite.

For the moment, let's move on to the development of the verb *râz*, as presented by specialized dictionaries, and try to find an English-language equivalent that, as with every verb, should be the key to the reality with which it is radically affected.

---

386 He invites us to take the 'way' to the 'Way'. It is obvious that Jesus, even if He presents Himself as Truth and Life, meant that He is the Way to His Father. Why, then, must we always seek the Father? The answer would be: He insists on leading us to the One Who loves Him and with Whom He realizes a different synergistic formula where: $1 + 1 = 1$, and once the concept Love is involved, the formula becomes: $1 + 1 + 1 = 1$, always *râzâtic*.

3. The verb *râz* (ܪܐܙ) and the verb "to *râzify*"[387]

We find the verb *râz* in two written forms. One refers to the past, *râz* (ܪܐܙ, ܪܙ, ܪܙܝ), and the other to the continuous present, *étérêz'* (ܐܬܪܐܙ / ܐܬܪܐܙ), as explained in the Syriac-English dictionary.[388] This verb derives from the name *râz* whose substantive is *râzâ* and its linguistic applications are similar to the ones of the verb 'to love' that derive from the substantive 'love'.

Râz and *étérêz*, are the two forms of the Syriac verb often encountered and widely explained by dictionaries through the nuances of their applications. But as we notice through the various sources, the first use is transitive. In English, it can be translated by the verbs 'to reveal', 'to declare', 'to teach', 'to pass' a secret, 'to initiate' someone. Its second use, as we have deduced, has the sense of 'conspiring' when it is intransitive. These two definitions are what we will retain after consulting the Syriac-Latin encyclopedic dictionary.[389]

As we have done above when dealing with the requirements that prompted Ephrem to borrow the word *râz* from Persians, we will proceed under this heading in highlighting the following:

1-   The requirements that demanded the creation of the verb "to *râzify*"

2-   Clarifications that would reduce the challenges raised by the various applications and difficulties of this new verb.

3.1. The requirements that demanded the creation of the verb "to *râzify*"

What obliged us to forge the verb 'to *râzify*' for the English language stems, first of all, from the fact that the verb 'to mystify', which is considered to serve the field of 'mystery' (and also of 'mystics') that we are targeting, refers to a completely different field. In English it means "to abuse someone's credulity to have fun at his/her expense; to mislead by giving reality a seductive, but false idea".[390] The need for a different word is also due to a gap that the Western reader should have already

---

387   By the advice of Prof. Feghali, an expert in Syriac and ancient languages, the consensus was to create the word '*râzify*' so that it comes to be used as a desired transitive verb that will serve the requirements of the *râzâtic* reality.

388   *Thesaurus Syriacus,* (Syriac Dictionary); R. Payne Smith; The Clarendon Press; Oxford, 1957, p.524. *Cf.*http://www.tyndalearchive.com/TABS/PayneSmith/index.htm [Accessed July 2019]

389   *Encyclopedic Syriac-Latin Dictionary*, Convent *Mar Ašaia* Library, Antonine Order, Lebanon, column 3875.

390   *Cf. Larousse.*

noticed. It can be summed up as the impossibility of expressing in English all the applications of the Ephremian *râz* and, consequently, the failure in any attempt to erect a 'bridge' between the Ephremian theology of Nature and the quantum analysis of the universe. The limitations of language will block humanity from crossing from the idea of *Creatio Ex Nihilo* to that of *Creatio Ex Amore, i.e.,* from being created out of nothingness to being created out of Love.

We consider that if the quantum reality has the verb 'to quantify' that is a key to the problematic of Creation and its consequences, the *râzâtic* reality should have a proper verb too.

A third component of this necessity is the fact that a reality of the scope of the 'Mystery of Salvation' that we have described as *râzâtic* requires a verb that complements the verb 'to synergize' proposed above. This verb will have to manage the transmutative action (*mšaḥlaftâ* ܡܫܚܠܦܬܐ) from the simply spiritual to the 'mystical', and also beyond the 'mystical' (1 + 2 = 4), in order to satisfy the direct objects that exceed common sense as in the case of the Burning Bush in the account of Moses. For example, what verb do we use to deal with this bush which suddenly changes and attracts the intellect to a hierophanic condition? It is up to humans to find the necessary expressions to describe the event. A verb whose direct object is completely and radically outside the category in action, whatsoever its subject, will fall into the absurd. We allow ourselves to write *râzify* (the bush is *râzified*) and to claim, according to Rev 21:5, that the One who sits on the throne would have said, "Behold, I *râzify*' the universe," and that would have better expressed His mission. After all, is it not up to humans to coin words relating to the action undertaken by God?

3.2. Clarifications that would reduce the challenges raised by the various applications and difficulties of this new verb

The challenges and difficulties that would accompany the applications of the verb *râzify* are spread over three levels: Christological, linguistic and practical. Let us proceed, point by point, to clarify these different levels.

3.2.1. Christological challenges and difficulties

Whether according to the Synoptic Gospels (Christology from below) or to Saint John's Gospel (Christology from above), there has been, in a certain fullness of time,[391] a certain chiasmus long-awaited and sensed by civilizations, and so much anticipated by the prophets of Israel. By this chiasmus, a hypostatic union between divinity and humanity took place, God became Man, the Word became

---

391 Gal 4:4.

Flesh, since, as said above, everything should have been *râzified* by this *minima naturalia divina*,[392] which is the Word-God, the Photon of the divine Light, the Spark of the divine Fire and the living *Râz* of His Father by divine DNA.

Since then, speaking of the deification of Man or of Nature has always represented a challenge, and in the best of cases, a difficulty that has made several philosophers and theologians fall into the trap of pantheism.

Speaking of the insertion of divinity into humanity or of divinity that now wears (*lbêš* ܠܒܫ) 'human dress' (or vice versa) would run the risk of providing a foundation for a dualism or a parallelism which would sin against the Ephremian intuition. The latter admits only one relationship between the divine and human realities, very different from any expression of relation and communication common to the created world and to common linguistic practices. The natures of the two parts are essentially incompatible with each other. However, they are made compatible by the act of *râzification* which takes place at the same time in the created 'realm' as in the uncreated one. They are made compatible by a third 'element', a "differAnt" intermediary which is neither of them, but is rather, synergistically speaking, the one, the other, and Himself at the same time. This third 'element' is not a product of passive emanation but rather of a free and loving Will that forms the synergistic element of *râzification*.[393]

If we can join Ephrem in his amazement at this 'Element' Who is none other than God in fetus in the womb of Mary, we will have to ask ourselves, with him, about the nature of the 'tongue' the angel used when addressing Mary. Ephrem wrote:

> Mary bore a mute Babe
> Though in Him were hidden all our tongues.[394]

3.2.2. Linguistic challenges and difficulties

What language was objectively used during the dialogue between Mary and the Archangel Gabriel which triggered this *râzâtic* union? It is certainly not one of the *Babelian* languages. Ephrem would say that, since Mary is conceived

---

392   We will not repeat the proof already provided by the paradoxes of Zeno of Elea which prove that in the absolute - as in the divine case - the *minima* and the *maxima* - are not distinguished; they remain the indivisible and non-multipliable One. As for Ephrem, he prefers the symbol of salt as in his hymns to the Nativity. (*H Nat.* 15, 10).

393   *Les Origines*, p. 47. *Cf. H Fid* 26,1-2, *CSCO* vol. 194, pp. 88-89.

394   *H Nat* 4, 146, *Cf. CSCO* vol. 186, p. 38; *Cf.* Kathleen E. McVey, *Ephrem the Syrian Hymns*, The Classics of Western Spirituality, Paulist Press, New York, Mahwah, 1989, p. 100. Henceforth, we will refer to this book by its title.
         Tᶜinâ wât Mariam ᶜulâ šêlyâ   ܠܒܝܟܐ ܗܘܬ ܡܪܝܡ ܥܘܠܐ ܫܠܝܐ
         kad beh hu ksên waw kul lêšânïn   ܟܕ ܒܗ ܗܘ ܟܣܝܢ ܘܘ ܟܠ ܠܫܢܝܢ

immaculate, it should be the very 'language' God used with Adam and Eve in Eden. Obviously it does not resemble any of the Hebrew, Aramaic, Syriac, or Arabic tongues. Therefore, we deduce that it is a *râzâtic* tongue similar to that of the One Who "intercedes for us in wordless groans".[395] Ephrem says precisely:

> Blessed be He Who has fitted the senses of our minds
> In order to sing on our lyre
> Something that the mouth of the bird
> Is unable to sing in its melodies.[396]

A *râzâtic* reality that Ephrem meditated on through the Mirror,[397] the Pearl,[398] or even through the Ember (*Gmurtâ*) that touched Isaiah's lips requires a language that suits the same nature of what results from the divine-human chiasmus.[399] All Ephremists would call it the language of types, signs, and symbols that form its basic resources. In fact, Ephrem tried his best to give each of these three elements of his meditation – of daily use in both Nisibean and Edessean societies – a cornerstone value in the light of the primordial Fire and under its burning Heat.[400] To succeed in this mission, in the service of faith, he 'bombarded' them with the

---

395   Rm 8:26; Hence the importance of silence and listening in the relationship with God. Nowadays, a digital user would suggest the so-called emoticon language as an adequate answer.

396   *Cf. H Nat* 3,16; *CSCO* vol. 186, p.23.
N.B. We noticed that McVey, in her book *Ephrem the Syrian Hymns*, p. 86, get confused by saying: "He is He Who constructed the senses of our mind" while the real meaning should be as mentioned above: "He who makes it His own..."

397   The use of the mirror symbol and its corollaries light and eye allows Ephrem a way to explore the 'dioptric' of spiritual perception. *Cf. Les Origines,* p. 168. "Man is then like a mirror that reflects the divine glory in the manner of the son who reflects his father." *(Hc Haer* 5,12. *Cf. CSCO* vol. 169, p. 21)

398   Here, what is to be emphasized is Ephrem's comparison between the birth of pearls in the sea-water and the birth of Christians through baptismal water: "Lightning penetrates the water to give birth to pearls, as the divine Fire enters the water to give life to the baptized..." The allegory of the pearl covers, according to the *Hymns of Faith,* a large part of Ephrem's theology.

399   *TLE,* p. 104: "The first begetting of Christ by his Father and the second by Mary are put, at this level, in balance with the human birth of Mary and her second birth (in other words her baptism) through the pregnancy chiasm. (Ephrem considers in fact that the baptism of Mary had already taken place as soon as Christ was in her womb.)"

400   Analogously, one can refer to the reality that shaped the construction of the Tower of Babel, as well as the one that stood behind the Pharaohs' Pyramids. Each of these two constructions, which symbolize even today a *râzâ malyâ râzê* (mystery full of mysteries), marks the evolution of human rationality and its ability to conceptualize a reality. The "Tower of Babel" allowed Man to go higher, the "Pyramids of Egypt" to go broader and more perfectly. What was left to the Greeks was to build deeper, introspectively, in the very psyche: "Know yourself."

concept *râzâ* just as physicists do with an atom to unveil the subatomic reality. In other words, he seasoned them with a more familiar symbolism, as a mother seasons food with salt to give it taste:

> The hundred and fifty psalms he sang were flavored by You
> Since all the words of prophecy are in need of Your seasoning.
> All the words of the prophets take from Your condiment
> because, without Your salt, all wisdom is tasteless.[401]

Thus, the same necessity mentioned above requires the consideration of the verb "to *râzify*" as a grammatically complete verb like "to love" and "to create". Like them, it is transitive without limit, conjugated in all tenses, and under all moods, but in addition, it enjoys a particular mood and several related tenses that it acquires specifically from the *Râz.* It is the Mood and its Tenses which join Creator and creation, which means that through our intellect, we humans can sometimes discern the act through the 'subject' and the 'object', but not by its effects, or we can grasp it through its effects, but not by its 'subject' or 'object', regardless whether the object is direct or indirect.[402] We can conceive it only at the *râzâtic* level. Some examples would help us adopt Ephrem's thought without puzzlement:

> Jesus said: "Take and eat ... this is My Body ... this is My Blood."
> A believer would say: "I eat ... the Eucharist (Body and
> Blood of Christ–God)."

As the believer actually eats something, the act is discernable by the subject and object but not by its effect. The latter remains in the realm of faith.[403] The direct object of the act cannot be subjected to any of the categories of human logic except the one of faith, and "Faith has its logic that logic doesn't understand".[404] The converse is applicable to the other Sacraments, also called 'Mysteries', such as Baptism, Priesthood, Reconciliation... It is especially the latter, in which sins are forgiven, that highlights the discernment of the act by its effects and not by the subject and the object. In this act, the subject is "God" who forgives, and the direct

---

401   *H Nat* 15,10; *Cf. CSCO* vol. 186, p. 83. *Cf. Ephrem the Syrian Hymns*, p. 147.

ܘܡܐ ܘܚܡܫܝܢ ܙܡܝܪ̈ܐ ܕܐܙܡܪ

      W mâ w ḥamšïn zmirê dazmar

402   Derrida was able to unveil this overlapping, at the epistemological level, by writing one day, "I am dead."

403   It is by the Holy Spirit that we take communion, by the Grace of Faith.

404   A pastiche of the Blaise Pascal's famous aphorism: "The heart has its reasons which reason knows nothing of... We know the truth not only by the reason, but by the heart." *Cf.* https://www.goodreads.com/quotes/559339 [Accessed Oct 8 2021].

object is the sins. The effect of the forgiveness that is hidden (*kasyâ*) is made visible (*galyâ*) by a miracle such as the one granted to the paralytic at Capernaum who stood up, carried his mat and went home.[405] This is what is meant by the grammatical art that Ephrem preached through the verb 'to *râzify*', because the healing of the paralytic, with the invitation to the noetic transmutation that Jesus made to 'people of the system', is the wonder that supports our grammatical assertion.[406]

In any case, without faith, the situation could be said to be either simply bizarre or, to borrow the qualification used by Jean Guitton, absurd. Despite this typical dismissal, we continue to use these sacramental expressions because they are the ones that reassure us the best and that allow us to foresee the unexpected. They also allow us to see the Invisible, to eat ambrosia, to drink the Water of Life and Christ's Blood if, and only if, we have faith. And, they work (*sic*).

They work, yes. However, at what price? Is it not at the price of making oneself aware, and aware of being aware of eclipsing one of the two realities, often the theological one, sacrificed to pagan, materialistic, pessimistic habits and customs?[407]

3.2.3 Challenges and difficulties in practical applications

We have actualized through what has preceded some of the challenges presented by the application of the verb 'to *râzify*', while counting, on one side, on divine Grace and, on the other, on the preceptiveness of the human mind to help us overcome any difficulty. Ephrem did not admit any compromise, in the particular historical situation of his time, as a director of seminarians, as a pedagogue and deacon, all fired up by his love for the Mother of God and for the House of his Heavenly Father.

By his *râz*, he made possible the intermingling (*mzâgâ* ܡܙܓܐ) of the two tenses, eternity and time, and the two moods, divine and human. Thereby, it became possible for humanity to mingle (*étmazag* ܐܬܡܙܓ) with the divinity and vice versa, without ontological union, confusion, exclusion, or overlapping.[408] In another sense, it is a lyrical shore that Ephrem was preparing for the relationship between the Book and Nature which translates for us between

---

405 Mt 9:1-8; Mk 2:1-12; Lk 5:17-26.

406 Mk 2:3-12.

407 Is this not what the Church suffers whenever there is a ceremony of Baptism, First Communion, Confirmation or Marriage? Ordination and Extreme Unction hardly escape this paganization. Confession, the only sacrament that does not admit camouflage, is condemned to continual confinement. As for burials, let's not even talk about it!

408 The example of the intersection of two rays of light, or of the interlacement of musical notes in harmony.

Religion and Science.[409] Thus, baptism, communion, and all other sacraments become '*râzâtic* meeting places' where 'human grammar' and 'divine grammar' agree to conjugate the verb "to *râzify*" in the same tense and under the same mood, both *râzâtic*.

We afford ourselves a last practical, although unusual, application of the verb "to *râzify*" through an example which will later serve as a cornerstone to build the *maškanzabnâ* (ܡܫܟܢܙܒܢܐ) so much desired by Science and Religion. We call this application, which joins the hymen of Mary and the navel of Adam, the "Soteriology of the Hymen and the Navel". Let us see how this soteriology depends on the eternal virginity of Mary and the perfection of the humanity of her Son.

3.2.4. Soteriology of the 'hymen' and the 'navel' (*umbilicus*)

The Second Council of Constantinople (553 AD) dogmatized the perfect divinity of Jesus Christ, born of Mary, as well as His perfect humanity. However, it did not exclude the fact that, like every human being, He possesses a navel. Consequently, we deduce that every perfect human being must have a navel.

Let us apply to this case one of the Aristotelian syllogisms to highlight the contrast of realities and of the positivist's reductionism:

> Every perfect man must have a navel
> Now, Adam had no navel (since he was not born of a woman)
> So, Adam was not made a perfect man.

It is shocking, especially if we consider this syllogism inversely where the man without navel, according to the Bible, is taken for perfect. The case is also applicable

---

409 We consider it interesting to mention the philosopher Blondel in one of his references which considers this meeting place as the hymen: "Only the actual practice of life contrasts, for each in secret, the question of the relations of the soul and of God. Thus, from the very meaning which it was necessary to attach to this hypothetical necessity, it appears that the legitimate scope of philosophical conclusions stops at the threshold of the real operation in which only the human act and the divine one, nature and grace, can unite. From then on, philosophy remains, and can only remain, forever, beyond this mysterious hymen: it is this reserve that remains to be understood, with all that it implies. (*Cf.* Blondel, Maurice; *Les Exigences de la Pensée Contemporaine*; 1899; p. 133). Then Blondel explains his comment by adding: "Where did this hybrid hymen come from? It is without a doubt that, thanks to the double manifestation by Reason and by Revelation of the same divine λογος, the sharing and the agreement of Science and Faith seemed equally natural. It is above all because, beyond all that ancient thought could conceive and dominate, a new domain was discovered. It goes beyond the Man purely human and seems to extend its horizon infinitely, without limiting or diminishing its former pretensions. (*Ibid*. p. 141)

to Eve and Mary at the level of the hymen, whether intact or not. Thereupon, we wonder if this discovery, reached thanks to the 'Luminous Eye' that pierces the norms of common sense by *râzâtic* synergy, would not shake the foundations of human logic? Will a shift take place or, as happened following the discovery of Galileo, will our contribution be condemned and the sun continue to 'rise' and 'set' as in the biblical story, even if this is illogical?

The sun indeed continues to accomplish what it has always accomplished, that is to remain immutable, incandescent and calorific, insensitive to human divergences.[410] The problem of sunrise and sunset is not its problem. Similar will be the case of the human race with its navel and Eve's kind with their hymen. The problem is rather that of the 'luminous eye' with the angle from which the very narrow vantage point from which humans see, interprets and describes reality. Therefore, the problem is *par excellence* cognitive, linguistic, etymological and above all a problem of reductionism.

We are quite sure that even if one of the Greek philosophers once raised this issue of Adam's navel, the famous painter Michelangelo was not aware of it. We notice in the following picture of the ceiling of the Sistine Chapel of the Vatican that he even provided God with a navel so that Adam as 'Image' does not differ from his 'Origin'.

Figure 10: The Creation by Michelangelo.[411]

Sincerely speaking, a boy who was preparing for his First Communion raised this question and thus inspired us to consider this idea further.[412] Why did he

---

410 Mt 5:45.

411 Courtesy of the Vatican website.

412 Andrew Ghassan Khreich (8 years old), in preparation for his First Communion; Native of Aïn-Ebl, Southern Lebanon, resident in Mississauga, Canada.

care so much about Adam's belly button? We do not know, but it seemed like a providential message for our research, and we welcomed it positively. Since then, we could not read things the same way. The Man after Eden, Cain for example, possessed a navel which is only a natural consequence of natural conception and birth, and yet he must have been as perfect a man as his father, Adam of Eden, who was without a navel. But what can we say in the case of Christ Jesus several thousand years later? A new soteriology took shape in front of us based more on Eve's hymen than on the absence of a navel in her and Adam.

Even though created 'Perfect', Eve could not shy away from the temptation to commit the offence of disobedience, the result of which was fully intended by her. Ephrem said, "The devil poured poison into her ear and she gave birth to death."[413] A first and different hymen seems to have been deflowered by the devil, and it is that of Eve's ear. This is why Ephrem insisted that it be by the same ear of the new Eve that salvation begins, as a sign to remedy the hymen of the human ear.

By this allegory, he pointed to the rehabilitation of the hymen of motherhood, which instead of continuing to procreate for death, will begin to procreate for life. This is also why Jesus, the First-born of this new motherhood, compared bread and the Word of God in His first confrontation with the same devil of Eden just before He started His mission. He, Himself, will be that Word Who will fertilize any ear prepared by its own hymen to believe in Him.

The post-Edenic Eve should therefore be, despite her navel, a perfect woman, the navel being only a mark, a signature with which God marked her as He marked her descendants so that no one would kill them for the offense that their 'mother' committed.

The imperfection sits then in the fact of having lost the chastity of the ear, and by the same token, of the heart, and of having made Adam lose it, resulting in both of them losing all protection. This explains the act of the preparation by God of the garments of skin to provisionally remedy the deflowering of their ears and their hearts. They will have to wait for the total renewal of chastity by the Paschal Lamb Who is offered by His Mother, the new Eve, by the power of the Holy Spirit, to be immolated for the total redemption and protection of the 'hearing' and the 'heart' of His Mother's kind.

This protection was more palpable in the case of Cain, the eldest son (*bukrâ* ܒܘܟܪܐ), by whose birth the natural defloration of Eve is accomplished. As matter of fact, Cain was the first of Creation to bear the navel as a signature, but also the first to commit fratricide. For his protection, God provided him with another sign-signature, worse than the first, since it will be a visible scar and not simply a

---

413   *Cf. TLE,* p. 33; "Just as from the small womb of Eve's ear death entered in and was poured out..." (*H Eccl 49, 7*)

scar that various clothes can veil.

Can we consider this post-Edenic perfection the same as the Edenic one? And if Adam and Eve of Eden had desired to have children during the Edenic period, would Eve have suffered from physical defloration and have had children with relative navels? Would the same terminology be applicable to the activities of mating and childbirth, before and after the fall? Who could prove that Eve who did not have a navel needed a hymen? An equation such as "navel = hymen" would normally be more bearable in the case where the hymen would be considered something "differAnt", that is to say, of a different functioning that allows continual virginity. Hence, all the terminology of physical virginity would be affected and would make any mythology as well as any symbolism of defloration unimaginable.

Let us move, soteriologically, to the practical application of this theory to the case of the virginal birth of Jesus by Mary.

3.2.5. The virginal Birth of Jesus by Mary

Mary, who had been elected from all times by the divine Trinity and conceived free from original sin, had the privilege of a 'perfect' body, with the same perfection as that of 'Eve' before the fall. This caused her to be called "full of grace", despite her belly button. The same applies to her Son Jesus, the New Adam, especially that He was, as the Maronite Liturgy states, the only One in the world without sin.[414] Ephrem was able to highlight the chastity of Jesus (the symmetry of the virginity of Mary, hence, *râzâtic*) by presenting Him to believers as the pre-eminent chaste Person *par excellence (Btulâ)*.[415] Moreover, in the Arabic language, the adjective 'chaste', in the sense of never-married, is *'batul'* (بتول), and it is non-gendered. We can qualify a monk as well as a nun as *'batul'* (بتول).

For Ephrem, chastity is the basis of the Edenic perfection and, therefore, of salvation and redemption; the navel is not at all a sign of imperfection. So it is fair to say that Jesus is a perfect man but of a perfection "differAnt" from that of other humans. The difference is in the *forma* and not in the *materia,* both *râzâtic*. The Maronite Church, obedient to the dogmas of the first six Councils, affirms this truth daily in her Eucharistic prayer as mentioned above, by repeating: "No one on earth is without sin except our Lord and God Jesus Christ."

Likewise, Mary's hymen, hidden under her belly button which is, as mentioned above, the indelible signature of the *râzâtic* umbilical cord linking God and

---

414 Maronite Missal 2005, p. 831. Supplications for the dead. (Translated by the author)

415 *Cf. TLE,* "The Ascetic Ideal", pp. 136-137. *Cf.* also *La Pensée,* p. 389 *(sgulâ btulâ* = the virgin cluster*) (H Nat* 8, 8); *Cf.* also *Ephrem The Syrian*, p. 120.

humans, should be of the same perfection as Eve's hymen in Eden.

So, if we have to scrutinize (*bṣâ* حصا) the type of this hymen, we say it is of Edenic nature, *râzified* by the Holy Spirit from all time by the predilection and the maternal-filial union. Nothing prevents it being called transparent and luminous (*šafyâ* عصفا), 'a luminous hymen' as Ephrem might say.[416] It transcends in a "differAnt" way the post-Edenic hymen. We will say that in the case of God who is made Son of Man, Who is the *Râz par excellence*, a *râzâtic* hymen never gets deflowered.[417] It is true that the Incarnation was a post-Edenic event, but

---

416   We witnessed in Canada a psycho-spiritual case in which a very religious Christian immigrant woman of Lebanese origin, the mother of three children, was tested by the revolt of that "transparent hymen" that we consider present, latent, in the subconscious of every woman.

That woman had for a long time been subjected to indignation and marginalization exerted on her by her husband and her parents-in-law, and subjected as well to blackmail on the part of her husband, for having hidden from him a serious flirtatious affair before their marriage. Not having been able to convince him of her honesty as well as of her detachment from all that is past (while the husband did not have to detach himself from anything), she was seized by a crisis of revolt provoked by this oppressed *râzâtic* element of her subconscious. This crisis took place during a very cold winter night in Quebec City, about -10o C. She left her house following a dispute with her husband, to undress completely outside and run naked through the street, traumatically silent. She said that she wanted to declare in the open, once and for all, through her screaming silence, that she was of pure heart, innocent, and transparent. She ended up in the hands of the public security agents, then in a psychology clinic.

What amazes in this story is that the psychiatrists to whom that woman was submitted and whom she had to consult later found her very lucid, conscious, 'normal': she said that they wrote a "perfect woman".

In our judgment, this is true, and the story is believable. But what makes a difference is that, for a moment, the intellect of that woman, atrociously suffering, transmuted to function at the level of a reality that psychiatrists cannot grasp. We say it is "differAnt". Only people able to transcend intellectually to this level and who can have access to that Edenic realm, *e.g.,* Christian 'mystics', can understand it. This woman entitled her experience "the aromatic expression".

We put forward this testimony for psychologists and psychiatrists as proof that the human mind, in spite of the opacity of the body, can join over the temporal margin the Edenic state composed of *râzâtic* space-time.

417   Psychologically speaking, it is well accepted that a sudden transmutation in the human person from a certain situation to another can shock and traumatize. The trauma is indelible and causes a kind of schizophrenia that allows living the new situation without ever losing the 'reminder' of the old one that resurfaces from time to time. In our case, and based on the previous story that the woman in question gave us permission to share, we would like to propose a sub-concept, which will be of the domain of the subconscious, and that is, sub-hymen. By sub-hymen, we mean the Edenic hymen situated in the *râzâtic* subconscious, far from any human intervention. It must belong to both sexes, and be compared to the door of the Freudian "secret garden" that every human instinctively protects.

This sub-hymen is part of the conscience that Rousseau described by saying,

it comes from Faith itself, and not only from Science, to affirm that everything has been prepared for the Mother to conceive and give birth in the Edenic reality.[418] It is the condemnation of Adam and Eve and the loss of their luminosity (*šafyutâ* ܫܦܝܘܬܐ) by the infraction committed, which made the hymen become vulnerable, part of the suffering foreseen for Eve by God, breakable by the natural sexual intercourse between two bodies that have become "opaque" by what has been perpetuated as original sin.

When God said to the woman, "I will make your pains in childbearing very severe; with painful labor you will give birth to children. Your desire will be for your husband, and he will rule over you," (Gen 3:16) the breaking of the hymen by the violent intervention of the man should be included. That is what all violated women, even legally married ones, remember, traumatized, and that is why, we can say, the act of defloration took on a sacred meaning and entered into cults during which the woman voluntarily offers the sacrifice of her virginity and the suffering of her defloration to the gods. It is, moreover, what has always been held in the infrastructure of the vow of chastity in the churches of apostolic traditions, where the nuns are supposed to offer their virginity, with all the sacrifices that this vow will require, to their divine fiancé, the *Râz*, Jesus Christ.[419]

It is not needlessly that Ephrem so praised chastity, Noah's, Mary's and the Daughters of the Covenant's (*Bnât Qiâmâ*).[420] To say that Jesus was a perfect man, who took charge of our humanity which is perfect in nothing, and that Mary was a perfect woman, who has fully assumed her maternity like any other woman,

---

"Conscience, conscience, instinct divin." It is on this maternal sub-hymen that depend the depths of honor and human dignity, feminine *par excellence*, especially that of the son-mother relation. Otherwise, how can we understand the struggle of a woman, deflowered legally and canonically by her husband, for the safeguarding of her dignity and honor, as if she were always virgin, intact? What is it then of a woman that a man deflowers, even in the marriage bed (Heb 13:4), and what is it of a woman that sin deflowers, *e.g.*, adultery, even the sacred adultery of which the prophets talk?

418 Mary was *râzified* by the divine *Râz*, her Son Who is the divine Light born from the divine Fire and realizes Himself in the secular reality thanks to the fertility and warmth of the divine Love, the maternal Holy Spirit (*Ruaḥ*), procreator and founder of generations, be it human generations, *râzê*, as well as intuitions.

419 Canaanite worship that took place in the temple of Afqa, Lebanon, and the gods were El and Baal, the gods of fertility. A practice that will pass later, even under Christianity, to the feudal traditions where women were considered property of the feudal lord, and their defloration one of his first rights.

420 The Daughters of the Covenant (*Bnât Qiâmâ*), invited by Saint Ephrem to glorify, along with their brothers in baptism, the Sons of the Covenant (*Bnay Qiâmâ*), the One Who, thanks to His Mother, regained for them the Edenic dignity, were all female parishioners, even if Ephrem used to separate the married ones from the celibate ones for socio-cultural reasons.

remains real and true; however, soteriologically the expression "with the exception of sin" should never be overshadowed. Having said this, both enjoyed (action) and suffered (passion) from all that could be called human except the sin that had made and always makes humanity lose its brightness (*šafyutâ* ܚܦܝܘܬܐ), its purity (*dakyutâ* ܕܟܝܘܬܐ) and its holiness (*qadišutâ* ܩܕܝܫܘܬܐ). Hence, to reduce Jesus and Mary to mere test-objects is not worthy of Them nor of the humans who must recognize that, if humanity enjoys a little constant dignity, it is thanks to this *râzâtic* conception. The merit also extends to the verb *râzify* that Ephrem forged to serve the clarity and precision of all that concerns Evangelization and Redemption. This *râzâtic* conception is a guarantee to the Incarnation, to the Birth, to the Suffering, to the Death, to the Resurrection and to the Ascension, as well as to the Assumption.

Hereupon, thanks to the Syriac tradition that has preserved with zeal the "Hymn to the Light" and thanks to the strings of the Ephremian lyre which support the letter 'A' of the perpetual "differAnt" virginity of the Mother of God, we were able to dot the 'i's of a *râzâtic* reality. This reality is the natural product of the verb 'to *râzify*' which connects what is divine to what is human and which is at the base of a *râzâtic* language or, at least, a *râzâtic* dialect. This dialect has allowed and still allows calling the sign light, Light, without ever fearing to offend its referent the Fire that generates it, nor the Heat twinned to it, and vice versa. Here, the horizons are of a "differAnt" nature. Here, no signifier stops at what it signified in the concrete, no symbol symbolizes something created and no type represents a palpable category, although everything continues to be called real, concrete, and palpable, up to the point of confusion and of heresy.[421] Under this 'angle of view', God speaks from heaven and the effect of His Word is realized on earth. Man speaks from the earth and his words resound in God's 'Ear', in 'Heaven': "Blessed is He who has become immeasurably small to make us grow without measure!"[422] Do we write this today *post mortem* on behalf of Ephrem, or do we do it *per vitam æternam* on our own behalf?

---

[421] *La Pensée*, Note 32, p. 45.

[422] *H Nat* 21, Chorus. *Cf. Ephrem the Syrian*, p. 174.

Conclusion

Calling light "divine Light", without fearing to commit an infraction against Its divine Fire and Its Heat, is the result of a synergy that crosses realities to enrich us with an additional, exceptional reality, which we have named *râzâtic*. It is a reality garnished with angles and fields of view adequate for the Luminous Eye, an undefinable reality recognizable only through its own behavior, activated by the verb *râz*, thanks to which it breathes the traditional *Mysterium Salutis* and the *Sacramenta*, foundations of the now *râzâtic* and no longer mystical Body of Christ. The verb 'to *râzify*', with the nature of the reality that emanates from it and that has its source in the eternal-temporal chiasmus of the Incarnation, forms, as long as men believe and deliberate, the keystone of the bridge of relationality between Science and Religion, between Faith and Reason. As for audacity or pretension in proposing a *râzâtic* reality, our feeling, as well as our conviction, reassures us, at this level of its unveiling, of the good Way we have discovered. We leave it to the "other" to judge if we are, objectively, on the right track. It is a reality that we felt deep in our heart, that was there, hidden in the Book and in Nature, and that the spirituality of Ephrem of Nisibis helped us to unveil. Its unveiling, as is the case with the formula of Fullerian synergy ($1 + 2 = 4$), led us involuntarily to an unexpected soteriology called "of the hymen and navel" which affirms the dogma of the eternal Chastity and Virginity of Mary. It will serve in the last chapter to defend the dogma of Mary's Dormition and to support her role as *Co-Redemptrix*.

What follows will reinforce our certainties since the understanding between Religion and Science on the unique and absolute Truth serves the good of humanity, from its Alpha to its Omega. However, what remains to be done is to determine, as far as possible, the unit of measure of this exceptional reality, *i.e.,* what makes it possible to tame it, to better subject it to our concepts, as required by scientific rigor, in order to make it amenable even to the extremists of scientific positivity.

Ephrem and Syriac literature in general present the concept *'yât'* to us (ܝ݂ܬ). This *'yât'* covers the being, every being, from its substance to its essence, and we can see its traces in the air, in the water, in the matter as well as in the fire may it be 'ember', 'visible sun' or 'invisible sun'. This *'yât'* seems to propose itself, by its infinite smallness as well as by its infinite immensity. It propounds itself as the unique desired unit of encounter, the musical 'key of sol' that will ensure the development of the common 'symphony' of the two supposedly incompatible disciplines. What is this *'yât'*?

Chapter IV

The Substance (*yât* ܝܬ), from Ember to Sun

בְּ/רֵאשׁ/ית/ בָּרָא/ אֱלֹהִים/ אֵת/ הַשָּׁמַיִם/ ו/ אֵת/ הָאָרֶץ

B/ riš/ ït/ bara/ Êlohîm/ êt/ h'šamaïm/ w/ êt/ h'êrêtz

In the beginning, God created the heavens and the earth

ܒ/ ܪܝܫ/ ܝܬ/ ܒܪܐ/ ܐܠܗܐ/ ܝܬ/ ܫܡܝܐ / ܘ / ܝܬ / ܐܪܥܐ

B/ riš/ ït/ brâ/ Alâhâ/ yât/ šmayâ w [yât] arcâ

Gen 1:1

Following the consecration and the epiclesis, the Maronite Liturgy directs that the 'Body' be lifted over the 'Cup of Blood' to be broken, even slaughtered. The 'immolator' begins this ritual with the following words:

ܗܝܡܢܢ ܘܩܪܒܢ، ܚܬܡܝܢ ܘ ܩܨܢ

Haïmennan wa-qrebnan, ḥâtminan w qâsenan

ܐܘܟܪܣܬܝܐ ܗܕܐ ܠܚܡܐ ܫܡܝܢܐ

ukarestia hodê laḥmâ šmayânâ

ܦܓܪܗ ܕ ܡܠܬܐ ܐܠܗܐ

faghreh d_Mêltâ Alâhâ

ܘ ܟܣܐ ܕܦܘܪܩܢܐ ܘܕ ܬܘܕܝܬܐ ܪܫܡܝܢܢ

w kâsâ dfurqânâ wad taoditâ rošminan

ܒܓܡܘܪܬܐ ܡܚܣܝܢܝܬܐ ܘ ܡܠܝܬ ܪܐܙܐ ܡܢ ܪܘܡܐ...

bagmurtâ mḥsyânitâ w malyat râzê mên raomâ...

Which means :

> We have believed (*hayminnan*), and we have offered (*qarebnan*), and now we seal (*ḥotminan* +) and break (*qosenan*) this Eucharist, the heavenly Bread, the Body of the Word-God... And we sign (*rošminan*) the Cup of Redemption and Thanksgiving with this purifying Ember full of mysteries *malyat râzê* (ܪܐܙܐ) from above...[423]

---

423 *Cf. The Book of Offering*, Saint Maron Publications, Brooklyn, New York 1994, p. 226. As for the plural '*râzê*' (*malyat râzê*), we reconfirm our conviction that this plural term is taken here for what is immeasurable, and not for what is multiple. A good example

With the following reference, we join the Prophet Isaiah in his meditation in the Temple:

> Then one of the seraphim flew to me, holding an ember which he had taken with tongs from the altar. He touched my mouth with it. "See," he said, "now that this has touched your lips, your wickedness is removed, your sin purged."[424]

We consider this introduction quite sufficient for this chapter in which we will deal with the *'yât'* under the aspect of the above-mentioned 'ember' which symbolizes the substance that is one and triune at the same time in the realm of sacredness and sacralization. This approach allows us to make the necessary flashback to underline the origins and the various applications of this element that can simultaneously play the role of type, sign, and symbol. It mainly symbolizes what is known in the Latin scientific expression as *"minima naturalia"* of a reality, but a *râzâtic* one this time. It is comparable to Plank's Constant, the *"minima naturalia"* of the quantum reality.

This chapter deals with the existent beings according to their substance named *ït* in Hebrew, and not after their contingencies (אֶת) in Syriac, and *ït - êt* (ܐܝܬ) – *yât* recognizable by observation, no matter the kind of instruments used to observe them, whether the naked eye, the most sophisticated microscopes, or telescopes.

Its target is to prove, with Ephrem, that every created existent and being comes out of the same substance, the 'Light', and that in the same 'Light', it shall find its end. We will be guided, in this highly perilous journey by the Light of the tomb of the Man-God, by His Resurrection, and by Ephrem's 'Ember', an 'ember' that played a prominent role in the refutation of the philosophy and the theosophy of the worshipers of the created fire and sun. We will divide the analysis according to the following points:

---

would be the case of 'waters', especially since our theory considers Jesus Christ the *Râz par excellence*.

424 Is 6:6-7; New American Bible Revised Edition (NABRE) *Cf.* https://www.biblegateway.com/passage/?search=Isaiah+6&version=NABRE, [Accessed March 2019]. For us Easterners, this scene reminds us of the famous proverb that says: "The ultimate remedy is purification by fire." While purifying Isaiah's lips, the angel did not tell him that his lips alone were purified, but also that his fault was erased and that, therefore, his sin was forgiven.

1. The symbolism of the ember: etymologic, mythologic, biblical and *râzâtic* dimensions

    1.1. The etymology of the term 'ember' and its mythological and biblical dimensions

    1.2. The *râzâtic* dimensions of the ember

    1.3. The symbolism of the ember

2. The Light in the tomb

    2.1. Incompatibility between 'light' and 'tomb'

    2.2. The enigma of the 'empty tomb'

    2.3. From 'ember' to 'sun': the *yât* (ܝܬ)

        2.3.1. The *yât* (ܝܬ) and the void of the 'tomb'

        2.3.2. The unveiling of the *râzâtic yât*

3. Resurrecting 'light' from matter

    3.1. Distinction between *qiâmtâ* (ܩܝܡܬܐ) and *nuḥâmâ* (ܢܘܚܡܐ)

    3.2. Resurrecting the 'light' from the 'tomb'

    3.3. Descartes and Saint Ephrem (Rationalism vs Exhaustivism)

Conclusion

1. The symbolism of the ember: etymologic, mythologic, biblical and *râzâtic* dimensions

An ember, according to the dictionaries, is a burning residue, the combustion of wood. Mircea Eliade, the author of the article "Gods and Goddesses", states that it continues to be an instrument of purification without failing to be, like ambrosia, an instrument of regeneration. In the Talmud, ember is translated RSPH רְצָפָה and is pronounced either 'ritSPaH', or 'rotSPeH'. According to this last reading it means 'to cleave his mouth'. What does this mean for us?

Following the etymological development of the term, we will subdivide our analysis into two parts: the mythological and biblical dimensions on the one hand, and the *râzâtic* dimension on the other, followed by a focus on the symbolism of ember.

1.1. The etymology of the term 'ember' and its mythological and biblical dimensions

Mircea Eliade, without any intention to give an apology for the Blessed Sacrament, presents to us on a golden platter a proof of the universality of this term and of its pre-evangelical existence in non-biblical civilizations. It is with the goddess Demeter of Greek mythology, who had agreed to take care of Demophon, the newborn son of Queen Metanire, that he reveals this evidence. Indeed, Eliade says:

> ... Having affection for the child, the goddess Demeter wanted to make him immortal: "During the day she anointed him with ambrosia,[425] and during the night she purified him in the fire." [But, adds Eliade:] "On discovering her son on the 'embers', Metanire uttered a cry of terror." It is then, concludes the author, that the goddess revealed herself by declaring: "I am Demeter, the venerated, the one who brings regeneration..."[426]

This is the Greek pre-Evangelium to which the following words of Ephrem correspond:

The Fire of Mercy has now descended and dwelt in the Bread:

---

425 From the Greek ambrosia, food of the gods. According to the ancient Greeks, the food that provides immortality. The gods of Olympus were nectar drinkers and ambrosia eaters.

426 *Cf. E.U.,* "Dieux et Déesses". Article written by Mircea Eliade. *Cf. TLE, H Fid* 10, 8-10; pp. 104-105.

Instead of the Fire that consumed humankind,
We have consumed Fire in the Bread and we have come to
life.[427]

These verses echo in one of Ephrem's hymn dedicated to the Church, where the theologian compares the Deity also to a wet nurse. He says:

The Divinity is attentive to us, just like a wet nurse is to a baby,
Keeping back for the right time things that will benefit it,
For she knows the right time for weaning,
And when the child should be nourished with milk,
And when it should be fed with solid food (lit. "bread"),
Weighing out and providing what is beneficial to it
In accordance with the measure of its growing up.[428]

Biblically speaking, at the semiotic level, the sign 'ember' triggers a series of symbols that it shares with 'fire'.[429] In fact, under Eliade's last reference, the paragraph entitled "The Image of the Son", he asserts that:

... In fact, the sexualized fire involves the symbols of fertility, and more particularly filial symbolism. Logically, the theme that contains *ιγνε - igne* slides towards the theme of the son, the "fruit" of the womb of the mother. The fire is 'son'; it is a natural or industrial product. It produces in its turn, *homeopathically*,[430] birth, rebirth, orregeneration.[431]

On that basis, we find that the concept 'God-Word', the 'God-Verb' that represents the basic unit of the theological reality (Jn 1:1) as well as of the linguistic one, has its origin in the 'God-Fire'. It is from the mouth of analysts of mythologies

---

427   *H Fid 10, 12; Cf. CSCO* vol. 154, pp. 50-51. *Cf. TLE*, p. 112

      nurâ daḥnânâ   ܢܘܪܐ ܕܚܢܢܐ

      blaḥmâ neḥtat wašrât   ܒܠܚܡܐ ܢܚܬܬ ܘܫܪܬ

      ḥlâf hî nurâ deklat nâšâ   ܚܠܦ ܗܝ ܢܘܪܐ ܕܐܟܠܬ ܐܢܫܐ

      nurâ blaḥmâ êkaltun waḥyaytun   ܢܘܪܐ ܒܠܚܡܐ ܐܟܠܬܘܢ ܘܚܝܝܬܘܢ

428   *H Eccl 25, 18; CSCO* vol. 198, p. 57; *Cf. TLE*, pp. 171-172.

                    ܐܝܟ ܕܚܡܝܠܐ ܥܠ ܝܠܘܕܐ   ܐܠܗܘܬܐ ܚܝܠܬܢܝܬܐ

                    ܘܢܛܪܐ ܠܗ ܠܙܒܢܗܝܢ   ܘܝܕܥܐ ܙܒܢܐ ܕܚܘܠܦܐ

                    ܘܐܡܬܝ ܢܬܪܣܐ ܒܚܠܒܐ   ܘܐܡܬܝ ܢܬܬܪܣܐ ܒܠܚܡܐ

                    ܘܡܦܠܓܐ ܘܡܡܢܥܐ ܕܝܠܝܢ   ܡܬܝܬܪܝܢ ܠܗ ܐܝܟ ܕܪܒܝܢ

429   *Cf. E.U.*, "Symbolisme du Feu". Article written by Gilbert Durand.

430   A medical word borrowed from Bachelard. It means: "Therapeutic method which consists of treating a patient with infinitesimal doses of substances which would provoke, in a healthy man, troubles similar to those which the patient presents." (*e.g.*, vaccine)

431   Gaston Bachelard supports this theory. *Cf.* Bachelard, p. 54.

and fire that a different 'fifth gospel' seems to open up to us: the gospel of the Nations (*goyim*).[432] As the 'Fire' gives birth to a 'Son' who is 'Verb', the angel who carries the Word to Mary's 'ear' to be incarnated in her, bears at the same time the 'ember' that will 'cleave the mouth' so that the birth of this 'Word', *i.e.*, His concretization, is secured in due form. However, does the role of this 'ember' end on the lips of the prophet and on those of the enlightened scholar?

1.2. The *râzâtic* dimensions of the Ember

Râzâtically speaking, Ephrem proposes that Mary is also an 'ember' since she has the 'Fire' in her maternal womb.[433] The most sublime union that humans can imagine is that of the fetus with its mother. It lives in her, builds its body in her, developing its own blood from her, takes its form in her, but still is never her.[434] By the fact of this union, the type 'ember' conveys a "differAnt" Fire, which is within humans' reach for the purpose of mutual 'consumption' and not of mutual 'consummation'.[435] Ephrem reinforces this positioning of ember or *gmurtâ* (ܓܡܘܪܬܐ) by his baptismal theology. He wants the baptismal water to be 'ember' too, as every person baptized in it and by it is baptized by the Holy Spirit and Fire. The ember-fire element is necessary, as shown in the prophet Isaiah's case, to cleave the mouth of the baptized person so that the word '*Abba*' comes out of it and inflames the Heart of the God-Father.[436]

To facilitate the assimilation of the *râzâtic* conception of the 'forgiving *gmurtâ*' (*gmurtâ mhasyânitâ*) on which Jesus was broiling the fish at the end of "The Miraculous Catch of Fish",[437] we propose the dynamic parentheses

---

432 By applying to this case Fuller's synergistic theory, we can write $1 + 3 = 5$.

433 *La Pensée*, p. 230. Tanios Bou Mansour comments: "However, even if the term "body" (*pagrâ*) prevails over others, the term "flesh" (*besrâ*) is not entirely absent. Ephrem uses it especially in a hymn (*mimrâ*) on the Lord, where he affirms that the Holy Ember (*gmurtâ*) (Is 6:6) sits (*ettasl*) in the veil of the flesh (*tahfitâ d-besrâ*) (*SdDN* 46, 4-5). In this hymn, the Virgin's flesh plays the role of subject and Ephrem attributes to her the birth of the Son (*SdDn* 2, 18-20).

434 However, he still gives her the qualifiers of 'Mother' and 'fertile', without which she never reaches her fullness.

435 Two verses from a Syriac hymn taken from a book published in India describe the Virgin, saying: "In your arms you embraced the flames and gave milk to the devouring fire: blessed is He, the Infinite, Who was born of you." *Cf.* Babu, Paul; *Veni, Vidi, Vici*; Rabban Benjamin Joseph Publisher, St. Joseph Press, Trivandrum, India, p. 4.

436 The ancient Maronite baptismal liturgy used to require that, for the preparation of water, three embers be thrown there, which, by penetrating the water, would provoke steam and smoke before extinction. Similarly, this happens at the ritual of blessing water on Epiphany Day. The ember is the unique instrument that allows, physically speaking, the introduction of fire into the water.

437 Jn 21:9. The allegory goes so far as to suggest that if the apostles were fishers "of men",

schematized by the following:

(·) This point, between the parentheses, is the Word-Fire. The parentheses that encompass it are the signs of a first kenosis: a first layer of 'ash' whose symbol is well spread throughout the Old Testament; a humble, well-obedient Son, residing in the Paternal Womb, a Word who is 'hidden' under the Wisdom ... So it is a first layer of 'ash' around the Son-Fire. Who would believe in a hidden, humble God?[438]

((·)) The Word-Fire took flesh from Mary. He became Son of Man. He has always affirmed it. There is now a double stratum of 'ashes', a double difficulty to believe in a God hidden under human flesh.

(((·))) This Son-Fire, Who came out of the Father as "God of God, Light of Light", took a mother, was born of Mary, needed her milk, was covered with all the signs of human weakness (hunger, cold, swaddling, risk of death, fleeing, repatriation, etc.): a third layer of 'ashes'.[439] Who would believe in a God who possesses an umbilicus, who suckles, and who needs to be cleansed, protected, warmed?

((((·)))) Moreover, His parents subjected Him to the prescriptions of law and tradition: a fourth stratum of 'ashes'. Is it not God Who establishes the Law and the Commandments? Who would believe in a God Who submits Himself to the human condition and human traditions even if they were sacred? Did the people not once prefer the "golden calf" of Exodus and at another time Barabbas to this kind of God?

(((((·))))) Then, beginning His mission, He took on Himself our sins and our limitations, as Isaiah had anticipated. And since all that humans can see are appearances, they took Him too, as Isaiah had foreseen: "... shunned and rejected by others; a man of suffering and acquainted with infirmity; and as one from whom others hide their faces, he was despised, and we held him of no account."[440] This formed a fifth stratum of 'ashes'. Who would believe in a God Who is a mirror of the ugliness of the human being?

((((((·)))))) At thirty-three years old, during the Paschal Meal that He had desired so much to share with His friends, He did not declare Himself the new David, the long-awaited Messiah. He, instead, projected His Person, Flesh and Blood, into the bread and wine and shared them with the celebrants. Furthermore, as a sign of surrender to those who wanted to kill Him, as He foretold to His disciples, He

---

"men" would be the "fish" that Jesus will prepare by the 'Ember' for the service of the Kingdom and for eternal Life.

438   We will highlight later a first kenosis of God, called primordial, which comes from outside the cover of the Bible.

439   The question is: "Who has ever seen the fire wrapping itself in swaddling clothes?" *De Nat* 5, 4 ; *Cf. TLE*, p. 90 ; *The Harp of the Holy Spirit*, #30, p. 92.

440   Is 53:2-3.

broke the bread, signaling that He will accept death so that His friends may have their sins forgiven and live, and particularly "have life abundantly". (Jn 10:10)

Philosophically speaking, this leads to saying that He immolated Himself in potency before allowing the immolation to happen in act. Since then, He began to speak to His friends as if He were no longer of this world: "Take and eat, it is My body; take and drink, it is My blood, etc." Thus, He instituted, under a *râzâtic* tension, the most delicate stratum of 'ashes', the sixth. This puts the listener, should he/she be Aristotle, in front of one of two deductions: either this person is drunk and delirious, or this person is really God.[441] Indeed, while the disciples, His friends, disappeared furtively following His arrest, two strangers, the good thief and the centurion, spoke their parts well during the realization in act of His immolation by the Roman soldiers. It was the sixth stratum of 'ashes' placed as a capstone in the presence of unbelievers.

((((((((·)))))))) Finally, accepting His fate by accepting death and the tomb, "like a lamb led to slaughter",[442] He sealed the seventh stratum of 'ashes' which brings the 'ember' of His *râzâ* to its perfection: who would believe in a dead God lying in a tomb?

Thus, Jesus made impossible any faith in Him which does not come from 'above' and which is unable to remove these seven layers from the 'door' of the tomb. Similarly, whatever returns to the Father is also made impossible out of the Intermediary Whose resurrection blew away all sorts of layers.[443] The superposition of the seven parentheses seen exhaustively, gives us the following schema:

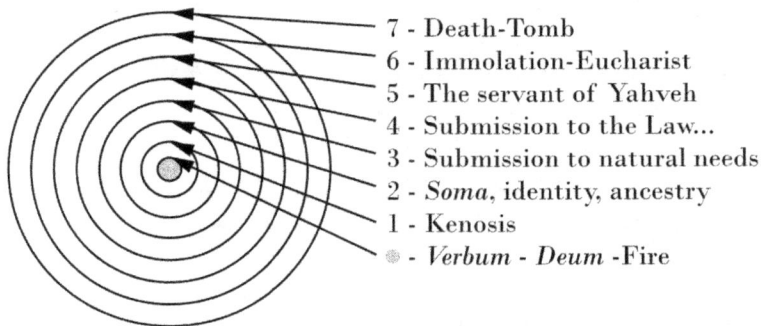

7 - Death-Tomb
6 - Immolation-Eucharist
5 - The servant of Yahveh
4 - Submission to the Law...
3 - Submission to natural needs
2 - *Soma*, identity, ancestry
1 - Kenosis
· - *Verbum - Deum* -Fire

Figure 11: Symbolism of the 'ember' (*gmurtâ*): cross section.

Therefore, starting from the center, we can enumerate the various layers of ash beneath which the Fire-Word veiled Himself so that He is neither heard nor seen

---

441 Mt 27:54; Mk 15:39; Lk 23:46.

442 *Maronite Missal*, 2005. Words pronounced over the bread of the oblations during their preparation, p. 53.

443 Jn 6:44; Mt 11:27.

except by those to whom it will be given ears to hear and eyes to see Him. They will do so even if they were born deaf-mute or blind, or both at the same time. It is a matter of "differAnt" sensations.

What is ***râzâtic*** about all this is that before the Resurrection, and according to the criteria of the "Christology from below", the enumeration of the ash strata was centrifugal, whereas after the Resurrection, as this time requires the "Christology from above", this enumeration has become centripetal. It can even start from the *Ḥaylâ kasyâ*, that invisible 'Power' ***par excellence*** Who has resurrected the Son. Later, we will highlight the third category of Christology originating from the synergistic phenomenon stemming from these strata and from their ***râzâtic*** depth that represents for us, henceforth, this ember. To conclude, let us review the symbolism of the ember.

1.3. The symbolism of the ember

The symbolism of fire prevails over that of embers, as explained in the first chapter. Yet, we still have to confess that the primordial 'Fire' from which the Light of Genesis came and which is the same Fire of Mount Horeb's bush and that of Mount Carmel, remains unapproachable.[444] It is necessary that this Fire and its Light veil themselves to permit access while humans wait for better circumstances to be unveiled.

> That visible glowing fire,
> Pointed to the Fire and the Holy Spirit,
> Which is mingled and hidden in the water.
> In the flame, baptism is figured.
> In that baptism come, enter, be baptized, my brethren, for it looses the bonds;
>
> In it there dwells and is hidden One of the three Persons of God, Who, in the furnace, was the fourth.[445]

---

444 Ex 34:29-30.

445 2 Cor 3:13-15. Ephrem says that the water and oil used in baptism reduce the strength of this Fire with which the person is baptized. *Cf. H Epiph* 8, 6; *Cf. CSCO* vol. 186, p. 170; *Cf. La Pensée*, p. 95; (Translated by the author).

ܗܿܝ، ܢܘܪܐ ܓܠܝܬܐ ܕܠܗ ܢܣܝܟܐ
ܥܠ ܢܘܪܐ ܪܡܙܬ ܘܪܘܚܩܘܕܫܐ
ܕܗܿܐ ܡܚܠܛܐ ܘܟܣܝܐ ܒܡܝܐ ܬܡܢܐ
ܒܫܠܗܒܝܬܐ ܗܘ ܓܝܪ ܪܫܡܗ ܠܡܥܡܘܕܝܬܐ
ܒܗܕܐ ܡܥܡܘܕܝܬܐ ܬܘ ܥܘܠܘ ܥܡܕܘ ܐܚܝ؛
ܕܗܐ ܫܪܝܐ ܐܣܘܪܐ
ܕܒܗܿ ܗܘ ܥܡܝܪ ܘܟܣܐ ܬܠܝܬܝܘܬܗ ܕܐܠܗܐ
ܕܒܐܬܘܢܐ ܐܬܝܗܘܗܝ ܗܘܐ ܪܒܝܥܝܐ

In fact, as soon as the Light of this Fire appeared in the created universe, in response to the imperative order of the word *Fiat*, a bombardment of the state of 'hustle and bustle' (תֹהוּ וָבֹהוּ) happened.[446] It caused the restructuring of the concept 'darkness' and brought out of it a dichotomy called light-darkness. The 'light' and 'darkness' of this dichotomy are in fact only two symbols of the same creative Light which acts substantially in the form of light and reacts accidentally in the form of darkness. This dichotomy, which was at the root of every other dichotomy as well as of Ephremian theology in general, marks the widest possible extension of the power of the divine Word as well as of the divine Fire and its Light.

It is at the height of the information released from this extension that the human intellect will be molded, because reaching beyond this extension is like succeeding at achieving the goal of the Tower of Babel – humans reaching God.[447]

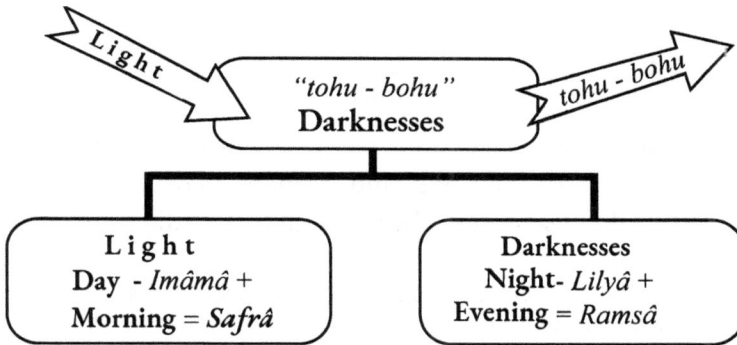

Figure 12: Incompatibility between light and chaos

So, in Genesis, the light (*nuhrâ* ܢܘܗܪܐ) of the first day was encompassed in the term 'day', called *imâmâ* (ܐܝܡܡܐ) in Syriac and *nahar* in Arabic, and the remaining darkness was encompassed in the term 'night' (*lilyâ* ܠܠܝܐ): "God called the light Day, and the darkness He called Night."[448] Thus the divine Word facilitated the taming of these two states by the eyes ('*aynê* ܥܝܢܐ), by the intellect

---

446   Henceforth, for example, we can assert, by simple observation based on new scientific discoveries, that the sun acts by its substantial light on the Earth that orbits around it. Therefore, a variety of shadows and darknesses is accidentally produced on the other side of objects that obstruct the flow of light rays. It becomes obvious, then, that the light is what creates shadow or darkness. In other words, light exists in itself whereas shadows and darknesses do not.

447   This point of view is fully supported by St. Ephrem; *Cf. La Pensée*, p. 114; and *TLE*, p. 41 where Brock comments on *H Fid* 44, 7 in an exhaustive way.

448   Gen 1:5.

(*réyânâ* ܪܷܝܳܢܳܐ), and by human reasoning (*mahšbânâ* ܡܰܚܫܒܳܢܳܐ). Similarly, at the level of the concept 'universe', 'light' and 'darkness' became two well-settled concepts. For the Creator, it is done. Mission accomplished. The primordial 'light' entered to fertilize the universe, and the contrasting contingent, darkness, incompatible with life and which reacts only to the extent of the action of 'light', disappears. It is in the 'ember' and through the 'ember' that this was revealed to the goddess Demeter before the Incarnation, and to Ephrem afterwards:

> Our Lord died and the Sun darkened:
> for the crucifiers, double darkness!
> They are darkened from outside,
> they are darkened from inside.

> For you who have confessed the Crucified,
> double light at the same time:
> Illuminated by the anointing,
> You shine again from the baptismal water.[449]

God, then, called them 'day' and 'night' and distinguished one from the other by the two concepts 'morning' and 'evening'. This transmutation at the level of names, "In the beginning" (*b-riš-ït* ܒܪܷܫܻܝܬ),[450] allows us to analyze this

---

[449] *De Epyph 3, 31. Cf. CSCO* vol. 186, p. 153. (Translated by the author).
   Also, *Cf.* Cassingena, *Spiritualité Orientale*, Série Monachisme Primitif, Abbaye De Bellefontaine, 1997, N° 70, p. 39.

ܟܰܕ ܗܘܳܐ ܡܳܪܰܢ ܘܰܐܚܫܰܟ ܫܶܡܫܳܐ	ܗܰܘ ܚܙܺܝ ܕܥܰܦܝ ܚܶܫܟܳܐ
Mît Moran wa ḥšaḵ šêmšâ	*ᶜâf wo ḥeškâ lzokufêh*
ܘܗܰܦܝ ܡܳܪܰܢ ܚܶܫܟܳܐ ܓܰܠܝܳܐ	ܗܰܘ ܚܙܺܝ ܕܥܰܦܝ ܚܶܫܟܳܐ ܟܰܣܝܳܐ
ḥafi Moran ḥeškâ galyâ	*w ḥafi enun ḥeškâ kasyâ*
ܢܘܼܗܪܳܐ ܚܰܕ ܬܪܶܝܢ ܐܻܝܬ ܠܟܘܢ	ܠܟܘܢ ܕܰܘܕܻܝܬܘܢ ܒܰܙܩܻܝܦܳܐ
Nuhrâ ḥad trên ît lḵun	*lḵun dawditun bazqifâ*
ܬܘܒ ܢܳܗܪܻܝܢ ܐܰܬܘܢ ܒܰܥܡܳܕܳܐ	ܕܶܬܢܰܗܪܬܘܢ ܒܡܶܫܚܳܐ
Tub nâhrîn atun baᶜmâdâ	*detnahartun bmešḥâ*

[450] We have always understood the term *b-rish ït*, which comes from *rosh* in Hebrew and is equivalent to Syriac *rïsh* and Arabic *ra'ss*, as all translations, interpreters, and preachers have suggested, with the meaning of "start", "beginning" : "In the beginning God created the 'Heaven' and the 'Earth'." Today, and thanks to the Ephremian *râzâtics*, we allow ourselves to declare that this is no longer enough. *Rïš'ït* is an Aramaic word composed of two parts: *Riš*, for the head, the beginning, and *ït*. What does this *ït* that we have accentuated with a capital letter mean? To compare it with the Hebrew *Ët* and Syriac *Yât*, which precede the names *Êrêtz* and *Shamayïm* of the same verse, it turns out to be from their root and to refer to the same signified as they do, *i.e.*, the substance. In short, according to our interpretation, the first sentence of the Bible would then be: "In the initial substance of time, God created the initial substance of the invisible (*Ët Shamayïm*)

phenomenon of 'Time' and to consider it as the creation of the substantial unit *'Ët'* of chronological time. In fact, it is the intercalation between day (*imâmâ*) and night (*lilyâ*), thanks to both intermediary concepts 'morning' (*ṣafrâ*) and 'evening' (*ramšâ*), that form together the supposed 'day' (*yaomâ*) of twenty-four hours. Consequently, Time (*Chronos*) and the temporal order (*taksitâ*) will take place, will limit the disorder of the chaos and will allow humanity, henceforth, to reach the "the fullness of time".[451]

As a matter of fact, we will not meet the concept 'darkness' again except in the circles of *Hades* (Ἀδης, *Sheol*, שְׁאוֹל), and all that resembles them, *i.e.*, all that is non-concept, non-being, non-life, non-fecundity, which would resemble, scientifically speaking, the famous 'black holes'.

This darkness will not escape the Power of Christ who will someday subjugate it to the service of the perfection of the "Economy of Salvation" called *Mdabrânûtâ* (ܡܕܒܪܢܘܬܐ) in Syriac.

Symbolically, this issue of taming the unlimited Light, inconceivable by what is limited, was used by God, on Mount Sinai, to satisfy Moses by allowing him to see only His 'back', covering him with His 'hand', him who so insisted that God show him His 'face'.[452]

Reading this scene through Ephrem's 'mirror' makes clear that, by covering Moses with His hand, God did not close his eyes, but rather, He concealed Himself to Moses' eyes so that he would not see God's 'face' and die. In other words, we say: God limited Himself; He admitted a self-kenosis. He hid Himself *(kasyâ)* under

---

and the initial substance of the visible (*Ët Êrêtz*)." Once one considers things from their substance, all numbers, accidents, and quantities become irrelevant.

451 Gal 4:4. Some Bibles consider *Tohu*, emptiness, and *Bohu*, formlessness, and others the opposite. We wonder what pushed the exegetes to consider these two terms related only to the lack of light and of physical shapes, and unrelated to the order of things, time and sense of history. After all, the text assures us that since day one (*Imâmâ*) the conception of history was introduced by the following words: "And there was evening and there was morning, the first day." The enactment of the Light generated by the primordial Fire and its camouflage in the expression "day-light", and then the reduction of the dreaded darkness in the expression "day-darkness", based on a harmonic sequence according to the rhythm of "evening and morning", is what proves the creation of chronological time. It is thanks to the light of the first day that the chronological order has started the order of the universe and in the universe; otherwise, there would never have been the 'fullness of time'. Should we find it strange to see physicists measuring time by Light-years? It remains to raise the issue of the beginning of time, before the timer starts counting.

452 Ex 33:18-23.

an element that can be easily compared to ashes.[453] He did almost the same in His all-night fight with Jacob at Penuel (פְּנוּאֵל).[454] He hid Himself behind the shape of an angelic fighter. In other situations, we find Him concealed in the breeze.[455] Like a Father-Friend, it does not matter for Him what form He limits Himself to. What matters to Him is rather to enter into an intimate dialogue, face to face, with those who have been 'devoured' by zeal for His House.[456] It becomes evident from the foregoing that while any elimination of distance between creature and Creator seems to be unreasonable for humans, it is not so for God.

Classically speaking, any absence of distance and separation is followed by the loss of any possibility of freedom and free encounters in physical time and space; the situation is not the same in the *râzâtic* case.[457] The state of 'ember' (*gmurtâ*) seen from the *râzâtic* angle (the *gmurtic* one) reveals this phenomenon of distance and non-distance at the same time.[458] Accordingly, both the sacred author and the mythical one dared to describe the foundations of such a meeting between creature and Creator, but followed two different paths. The first wove his style around the anthropomorphization of God and the second around the deification of the human prototype. Both ways led to the same results. Both built on the same starting point, the strong desire for a warm encounter, or the sacred 'Heat' of an intimate meeting, until this 'Heat' produces a much-desired unifying 'Spark' that carries both Creator and creature in the same 'Flame'.[459]

In the case of the Sinaic encounter, to say that God has a hand is much more in keeping with the culture of Israel than to say that He has concealed Himself under a thick layer of ashes, or worse, that Moses was transmuted into a god to be able to see God without dying. Not only is this 'Fire' directly unapproachable, but even the Son who

---

453　*H Nat* 11, 5; *Cf. CSCO* vol. 186, p. 70.

454　Gen 32:24-32.

455　Dan 3:50; 1Kgs 19:12.

456　Jn 2:18; *Cf. TLE*, p. 108; Brock comments on stanzas 2 and 3 of the Hymn on Faith (*H Fide* 19, 2-3). He says: "Elsewhere, just as Christ's human body is 'the garment of His divinity' (*H Fid* 19, 2), so too the Eucharist Bread is another garment: 'The Bread has hidden the Fire that resides within It' (*H Fid* 19, 3)."

457　Zeno of Elea's paradoxes offer the best perception of either the distance between Man and God, or the unity that could be joined between the 'one' and the 'multiple'.

458　We signify by '*gmurtic* state', the state of embers where the fire veils itself under the ashes. The passage behind the Wall of the Tomb means the recovery of that distance. After proving to humans that He loves them as Himself, out of respect for their freedom God repelled them by establishing anew the distance once abolished, to play the role of a 'vital space'.

459　Michelangelo's fresco reproduced above also refers to this.

claimed to be the Face of His Father would have made Himself 'ember' on purpose so that one could see His Father's Face in Him without running the risk of dying.[460] This same *Gmurtâ,* with a capital G, is what the Maronite Liturgy puts into action to accomplish the Eucharistic oblation by the sealing of the Cup of the New Testament.

All that we have tried to prove by what preceded is that in the core of each being resides a *râzâtic* fire whose presence is felt thanks to the emanation of its heat. We are aware of its presence once its light is perceived glittering or radiant behind the ashes that veil it. Our Creed affirms this truth by pointing out that God-created beings are visible and invisible and the Lord Jesus confirms that nothing that is invisible will remain hidden.[461] All that is required from Man is to clear the ashes, or to possess the Luminous Eye capable of piercing any kind of contingencies, to reach the substance and to bring the Light out of it.

Ephrem sang with great wonder of all the layers of ash piling up around the Incarnation. What remains to be emphasized at this level of symbolism, and which is of great importance to achieve the Passover (*ma'bartâ* ܡܥܒܪܬܐ) from the *râzâtic* reality to the quantic one, and vice versa, to settle the foundations of the 'Luminous Bridge', is the layer of the Tomb.[462]

Here, despite all its symbolism, it is no longer the ash common to the other six layers that can be traced even in other religions and civilizations. It is not even the ash of the famous Phoenix that regenerates from its own ashes to its previous reality. Instead, at this level, the layer of 'ashes' separates two parallel realities, as mentioned above, and only the chiasmus of the Incarnation, which gave to the Cross all its meaning, is able to join them. How are we to conceive of this layer of the Tomb which veils the Light from our eyes?

---

460 Jn 14:8-11.

461 Lk 8:17; Mt 10:26.

462 In the Maronite Liturgy of Easter, as well as in popular Maronite traditions, the Tomb plays a large-scale role. During the last three days of Holy Week, the Maronite liturgy did not linger on the Via Crucis, a practice developed by the Latin Church to substitute the pilgrimage to the Holy Land, but rather on the burial of the Crucified. Even today, in every parish a 'tomb' is prepared by the faithful of all ages. Everyone goes with sacred seriousness, because they prepare a 'tomb' in white for an event in 'black', but not without sacred joy, because the 'tomb' is tamed. Gradually, with time, 'death' itself becomes tamed, and the *râzâtic* transmutation of the two types 'tomb' and 'death' becomes much better conceived and conscientiously admitted.

2. The Light in the tomb

"Before the Resurrection, the tomb was an end; after it, it became a beginning." This is what Ephrem affirms:

> With You, I shall flee to acquire by You life in every place.
> With you, the pit would not be the pit,
> For with You one would ascend to heaven.
> With You, the grave would not be the grave,
> For You are also the Resurrection.[463]

Before the Resurrection, once their life's flame was extinguished, humans were placed in tombs.[464] Before, the Old Testament reminded them that to the dust (ʿafrâ ܥܰܦܪܐ) from which they were taken, they shall go back, and to dust from which they were moulded, they shall return.[465]

According to Ephrem, it is unfortunate that after the Resurrection this paradigm persists in the teaching of the Church, as well as being included in the rituals of the Eastern Ash Monday, and continues to be present without any transmutation.[466] Moreover, it is the Apostle who wished to highlight the necessity of this shift, one could say, by developing a theology of Hope with the very purpose of convincing the afflicted not to cry over their dead like those who have not known the Resurrection of Christ.[467] In order to be consistent with faith in the Resurrection, to satisfy the Apostle and consequently the believers entrusted to us, we propose a change to this formula of Ash Monday by saying: "*Insan,* remember that from God you came, and to God you shall return."[468]

---

463 *H Nat* 6, 6. Cf. *CSCO* vol. 186, p. 51. Cf. *Ephrem the Syrian, Hymns*; Translated and Introduced by Kathleen E. McVey, Paulist Press, New York. Mahwah, 1989, pp. 111-112.

ܥܡܟ ܐܪܘܥ ܐܬܪܐܣ ܝܠ ܕ ܗ ܥܡ ܘܗ

ܥܡܟ ܠܐ ܢܗܘܐ ܓܘܡܨܐ ܓܘܡܨܐ

ܒܟ ܓܝܪ ܢܣܩ ܠܫܡܝܐ

ܒܟ ܩܒܪܐ ܠܐ ܢܗܘܐ ܩܒܪܐ ܕܐܦ ܢܘܚܡܐ ܐܢܬ

464 According to Lebanese culture we say about a dead person: "The oil of his/her 'lamp' ran out" *(kulso zaytêto* خلصو زيتاتو*)*.

465 Gen 3:19. Ecclesiastes goes back to this teaching with a more precise scope, the one of giving to 'Caesar' what belongs to 'Caesar' and to God what belongs to God: "And the dust returns to the earth as it was, and the breath returns to God Who gave it." (*Eccl* 12:7)

466 For the Latin Church it is commemorated on Ash Wednesday.

467 1 Cor 15:20-34.

468 This term *"Insan"* comes from Hebrew *íš* – *íša,* and in Arabic, it was philologically developed to serve a quality of the Edenic person, non-umbilicated and asexual, or rather sexed but in a "differAnt" way. It allowed a more perfect interpretation of the story of the creation of humans in the image and likeness of God (Gen 1:27) Who is asexual Himself or, like His Edenic humans, sexed in a "differAnt" way, "differAnt" since

We are not sure that we will not find ourselves one day reprimanded by the ecclesiastical authorities for having often preached this opinion. In spite of this, we hope to be able to say one day, with the whole Church and never apart from the Church: "Remember, *Insan* (human being), that you are 'Light', and 'Light' you must become again."[469] What we are sure about, however, is the incompatibility of the two concepts of life and light with the tomb.[470] Are we supposed to *râzâtically* read Ephrem's aforementioned verses to discover the richness of their synergistic symbolism?

2.1. Incompatibility between 'light' and 'tomb'

Indeed, no Gospel speaks of a light coming out of the tomb or shining out of it at the dawn of that famous Sunday. The Magdalene confirmed it well, saying: "They have taken the Lord out of the tomb, and we don't know where they have put Him."[471]

It is especially strange that our Syriac liturgical books take advantage of the few 'lights' and 'flashes' introduced on 'stage' by the evangelists (we do not know at which period) to praise the 'light' that broke forth from the tomb, that illuminated Sheol and destroyed it, and that blinded the guards.[472] However, nothing of this exists in the official texts of the New Testament. The references that testify to this idealization of the light shining from the tomb are innumerable. We borrow from Ephrem one of them that seems to be an apex in this field:

---

the Holy Spirit is feminine in Hebrew. Here too, the advantage of the use of capital letters by European languages is thwarted by the use of more distinct and clearer specialized Semitic terms. The translation of *Bar Nâsâ* becomes *Ebn - el- Insan*, and this includes the Woman, whereas in the majority of Romance languages, there is quite a confusion between Man and man, "Homme et homme" etc., which makes them more susceptible to patriarchy and machismo.

469   Jn 10:34-36; *Cf.* Gregory of Nazianzus in his homily entitled *"Let us give ourselves"* (Breviary of the Antonine Maronite Order; Easter time, 2009, p. 259), where he says: "Let us become gods for Him as He became Man for us." Here, God and Light go the same way. The problematic here comes from the perspective from which we consider humankind: the body's perspective, the *Noûs*'s or the Soul's, the latter being the Breath of God, the divine Spark, as Saint John Paul II called it.

470   Lk 8:16; Lk 11:33-35.

471   Jn 20:2. John is, in fact, the most objective in his description of the scene of the resurrection. In spite of all his objectivity, we see that the tomb was not at all 'empty' as the common denominator between the various titles used in editions of the Gospel suggests. This point is crucial.

472   *Cf. Hymn to the Light*: "The Light that has appeared to the Righteous..." (*Nuhrà Dadnah Lzadikê...*).

> Light rising (*denḥâ* ܕܢܚܐ) in the river,
>
> Brilliance (*ṣemḥâ* ܨܡܚܐ) in the tomb.[473]

We already recognize two verbs from this verse, *dnaḥ* (to rise) and *ṣmaḥ* (to shine), that both relate to the realm of light. With these two verbs, Ephrem describes the state of the 'womb' of the tomb at the time of the resurrection (*qiâmtâ* ܩܝܡܬܐ). He supports this analogy by recalling at once the various '-phanies' described in the Gospels. However, to say that a man 'rises' or that light 'rises' does not mean the same thing nor cover the same dimensions. Indeed, when the sun 'rises', especially since the discovery of the famous Italian astronomer Galileo, it leaves behind neither darkness nor emptiness, whereas when a human body is analogically said to be 'risen', that is when it mysteriously disappears or is removed by someone, its place is materially said to be empty. That is, at least, what is apparent. But, effectively speaking, especially in the case of the use of the Luminous Eye that both religious and scientific people may have, can it be said to be exhaustively empty? We invite a more delicate approach to the concept of 'emptiness', or better, 'void', especially in its absolute dimension. It is high time that theology, in general, reviews the concept 'void' because this discipline can no longer remain nonchalant to the quantum conceptualization of 'void' that seems to be non-existent. The French philosopher Jean Guitton must have discussed this with Pope John Paul II.[474]

Ephrem, whose eye and heart were fond of the divine Light and who is supposed to be behind the famous *Hymn to the Light*, could not say less than what the verses mentioned above allow us to perceive. The *râzâtic* 'Redemptive Light' and His evangelical transmission require that.[475] Ephrem had become so sensitive to the presence of Christ in the womb of Mary that he had projected it on every scene starting with the one of the Epiphany:

> Here is the Fire and the Spirit in the womb of Your Mother;
> Here is the Fire and the Spirit in the Jordan in which You
> are baptized.[476]

---

[473] *H Eccl* 36, 5-6; *Cf. CSCO* vol. 198, p. 91; *Cf. TLE*, pp. 91-92.

[474] Guitton, Jean, Grichka Bogdanov, Igor Bogdanov. *Dieu et la Science, Vers le métaréalisme;* collection directed by Jean-Paul Einthoven; GRASSET, 1991. p.46. From now on, we will refer to this book by *Dieu et la Science.*
Commentary by the author: Compared to the dizzying development of Science during these last three decades, this work could be considered chronologically limited, especially since the TV program behind it was, from the scientific side, a popularization to satisfy the curiosity of the masses. But, at the same time, it is Jean Guitton (1901-1999) who is in it, at a certain age, with the best of his philosophy of which Pope John Paul II has enriched himself. This is where the work draws its authority.

[475] *Cf. TLE*, Chapter 1.

[476] *H Fid* 10, 17. *Cf. CSCO* vol. 158, p. 51.

> [...]
> From above, the Almighty came down to us and from the
> bottom of the Womb, hope has shone for us (*ṣmaḥ*).
> The Word of the Father came from His Womb and put on a
> body in another womb:
> The Word proceeded from one womb to another and
> chaste wombs are now filled with the Word:
> Blessed is He Who has resided in us.[477]

"From baptism", what Ephrem described as the state of the Jordan takes place in the baptized. Everyone baptized becomes an incandescent lamp even if they are contained in 'vessels of clay'.[478] Everyone baptized becomes "light from Light", "*râzâ* from *Râzâ*", able to have the 'Luminous Eye' and to perceive the Light, by affinity, behind any obstacle, especially behind the 'ashes' of the tomb.

Of all that has preceded, especially from the burial of Life in death and the descent of Christ into Sheol, it is ascertainable that 'light', even though it is conceptually the antipode of darkness, has a certain compatible coexistence with it. That was the case between Creator and Creation at the Incarnation. The relationship is no longer a matter of rejection in the absolute, and much less of absorption, but rather a matter of non-annihilation, non-overlapping, non-mixing or blending. The hypostatic union between God and Man in Jesus Christ, as the Council of Ephesus teaches, gives a clear idea of it.

This union whose proper object is the old Adam passes beyond him to the very names that he gave to all creatures. Therefore, ontologically speaking, it surpasses him until it reaches all creatures, each in itself; hence the universality of Salvation. Ephrem expresses this idea as follows:

> Loving is the Lord
> Who Himself put on our names.
> Right down to the mustard seed
>   He abased and compared Himself.
> He gave us His names,
> He took from us our names;
> He magnified our names,
>   whereas our names have belittled Him.
> Blessed is the one who has spread
> Your blessed name over his own name,

---

477   *H Res* 1, 5-7; *Cf. CSCO* vol. 248, p. 79. *Cf. TLE*, p. 171.

478   2 Cor 4:7.

and adorned with Your name his own names.[479]

Furthermore, to speak of the universality of Salvation in the divine design, Ephrem adds:

> The Lord mingles with His creatures
>> that from far and from near, implore Him;
>> He holds those who gaze at Him as the One who made them.
> It would be like they are in His
>> hand no matter the distance that
>> separates them from Him.[480]

With Jesus and *râzâtically* speaking, there should no longer be any incompatibility between the 'light' and the 'tomb'. The 'light' can eternally reside in the tomb without any alteration in its nature or action, 'waiting' for someone to open the 'door' and let 'it' spring forth. Meanwhile, the tomb undergoes an essential transmutation without it being substantial or contingent. It means that even if the 'light' leaves it, it will not return to its original state, the one before its transmutation. The 'light' that passed through it actually impregnated it indelibly, hence the sacredness of the tomb in Jerusalem which continues to be respected today.[481] Therefore, the Light that came out of Saint Charbel's oil lamp and that which shone from his tomb are only a hint of this phenomenon. Ephrem so admired the womb of Mary in this sense – for nine months, by natural pregnancy, it welcomed the divine Son.

It would be absurd to speak of emptiness in the womb of Mary following the birth of Jesus. In this truth lies the enigma of the tomb that can never be empty,

---

479 H Fid 5, 7; *Cf. TLE,* p. 66. *La Pensée,* p. 210; *Cf. CSCO* vol. 154, p. 18. (The translation in *TLE,* has been modified by the author.)

[Syriac text, four lines in two columns]

[Syriac text, two lines]

480 H Fid 72, 23-24; *Cf. CSCO* vol. 154, p. 222. (Translated by the author)

[Syriac text, three lines]

481 The tradition of the Flame of the Sacred Tomb, which shines every year at Easter night, according to the Greek Orthodox calendar, at the unique imploring of the Greek Orthodox Patriarch, is but a proof. The lightning which springs from the tombstone lights all the candles of the temple, as well as the ones held by the people, without causing harm or accidents. This is a phenomenon to take into consideration.

according to Ephrem's description. He thus proclaims, as from a prophetic mouth, the basis of any hypostatic-*râzâtic* union between creature and Creator. Let us discover the basis of this possibility.

2.2. The enigma of the 'empty tomb'

Once the 'Light' was buried, humans thought they were finished with 'Him' Who bothered their eyes, their plans and their system. In his famous prologue, John aptly described the rejection of this 'Light' by humans. Plato, likewise, foresaw this refusal in his *Myth of the Cave*. In spite of this, it is also human to have exceptions in each case. In the case of rejection, the Marys, the disciples and the Fathers of the Church opened wide their ears and hearts to this Hero Who managed to return from the world of 'lights', as Plato foresaw, or Who is the very Light, as the prophets of Israel had taught. For the Marys, the Disciples and Evangelizers of the first decades, the difficulty was to convince oneself and others (crucifiers, Romans and all human beings) that this Man Who was crucified is resurrected and that His tomb, apparently left empty, has instead been transmuted into an everlasting source of life-giving Light. They had to do it from two perspectives: either He has resurrected, as He said, having received from His Father the power to have life in Himself,[482] or it is God who has resurrected Him, as it is written of Him in the prophets.[483] In both cases, the only sign must be a 'light' similar to that of the Transfiguration, which defies any emptiness. They were supposed to do this, being witnesses *par excellence* of the fact that this 'Man' Who was crucified, Who died and was buried, is indeed He, the awaited immortal Messiah[484] Who has risen and Who promised to be with them until the end of time.[485] But, as the doubt of the Eleven persisted, Jesus appeared to them and convinced them of His resurrection, then blew on them the Holy Spirit and entrusted them with the mission of forgiving sins.[486] Since then, the disciples have had neither the chance nor the idea to return to the tomb for a closer investigation and a meticulous inspection of it, as would happen nowadays after a crime. They have had less chance to examine Jesus' DNA[487] and to make deductions that would serve both Science and knowledge.

It is from their sensory observations, based on a reaction directed by the Holy Spirit, that they will launch the process of evangelization. They did so, being inflamed above all by an education built on the experience of miracles, that is,

---

482 Jn 5:26; 13:32.

483 Acts 2:32.

484 Jn 12:34.

485 Mt 28:20.

486 Mk 16:14; Jn 20:19-24.

487 As a matter of fact, can we not bring out the DNA of Jesus from the Shroud of Turin?

by faith in an aprioristic metaphysical Power that precedes all merit on their part and that allows them, by the unique merit of faith, to impress and convince.[488] Even Paul, the Pharisee of Greco-Latin philosophical education, will refuse to philosophize the things of faith. Ephrem will also do the same later, warning of the danger of Greek logic.[489] Unfortunately, all this will not take away the enigma of the empty tomb whose bait even the Areopagites will not bite. Ephrem then proposes to explore this enigma according to his Nisibian way and to decipher it by using any asset in which he is expert, especially his relationship with the Holy Spirit and the Virgin Mary. What was the key that allowed Ephrem to break through the walls of this puzzle?

2.3. From 'ember' to 'sun': the *yât* (ܝܬ)

ܐܝܬܝܐ ܕܡܢ ܐܒܐ ܓܢܝܙܐ
ܕܩܡ ܡܢ ܩܒܪܐ ܒܬܫܒܘܚܬܐ

Ityâ dmen Abâ gnizâ
dqâm men qabrâ bteshbuḥtâ

You, elusive Substance of the Father
Who rose from the tomb in glory.

St. Ephrem

As it turns out from the Ephremian contribution, the enigma of the tomb, illuminated by the One Who was buried in it in a sign of the resurrection of humans from death, lies mainly in the overlapping between contingency and substance. It goes back, as said above, to the 'angle of view' through which the perceptive eye seeks their union. Thus, while the disciples, and after them, the Evangelists and the first Fathers of the Church, did not discuss the emptiness of the tomb of the body of Jesus, Ephrem, as a pedagogue, took a more dialectical position.

Being a master in paradoxes, he was also master in fleeing any stalemate that may be suggested by the literal interpretation (*suʿrânâyit* ܣܘܥܪܢܐܝܬ) of the scene of the resurrection.[490] The positioning of the problematic in the form of a contrast between either "the tomb is full" or "the tomb is empty" does not seem to suitably

---

488 Acts 3:6.

489 Col 2:8; Rom 9:20; *Cf. Hc Haer 6; 9; 55.* With this point of view we join the one of Brock. (*Cf. TLE*, p. 17.)

490 He fled definitions because they aim to define the indefinable. *Cf. TLE*, p. 24, "The paradoxical theology".

serve his pastoral cause *vis-à-vis* the divine Light and the Economy of Salvation. In other words, saying that the tomb was found absolutely empty, without any trace of the Light that was buried in it, would mean a total loss of the Economy of Salvation. And, saying that it was absolutely full, and therefore that the Light was absolutely stifled, would mean as well that the world must have lost all trace of this Light.

However, what would the tomb be without the Light of Christ, and likewise, what would the world be without It? It is indeed indisputable that without the tomb and without the world, Jesus Christ would remain God, but in this case He would not have a Mother, He would not be Christ, and there would be no Kingdom. With Ephrem, we could continue – to infinity – with this dialectic of the nuptial synergistic union, to the point of asking: what would paradise be without this Light?

Actually, in which paradise is the good thief going to be with Jesus on that day, if not in His Light, no matter where Jesus will have to visit before returning to His Father and however long this day might be? With Jesus, the "today" that will start as soon as He puts His soul back into His Father's hands will no longer be a temporal "today", but rather 'God's Today' or 'God-day', the *râzâtic* Day that knows no sequence like the created ones.

Consequently, to be able to affirm that, "With Him, the Tomb would not be the Tomb," Ephrem decided to shift the very language from the level of logic to the *râzâtic* one and to address the *Râz* Who is in the Tomb by the substantive *Ityâ*.[491] The latter designates a being whose substance is called *yât* (ܐ) in Syriac. However, to abstract the substance from the matter, – its spirit, – and from the body of Jesus, His Person, becomes a linguistic necessity since the description of Jesus' DNA depends on it.[492] By these two concepts, *yât* and *ityâ*, Ephrem opens wide the spectrum to the symbolism which aims at the universal salvation, not only of humans, but also of every being coming from 'Being' and which is subjected to the verb 'to Be':

> God's Majesty that had clothed Itself in all sorts
>   of similitudes saw that humanity did not want to
>   find salvation through this assistance,

> So He sent His Beloved One Who, instead of the
>   borrowed similitude with which God's Majesty
>   had previously clothed Itself,

---

491   *Cf.* Payne Smith dictionary.

492   We will see later that Bergson and Teilhard will in turn bend before the 'Spirit of Matter'.

Clothed Himself with real limbs, as the First-born,
and was mingled with humanity:

He gave what belonged to Him and took what
belonged to us, so that this mingling of His
might give life to our dead state.[493]

Ephrem's approach to tackling these indescribable moments, especially when he speaks of the substance *Yât*, Whose role we discussed in the first verse of Genesis, pushes us to admit that we must look at the event of the Resurrection from a "differAnt" angle. It is the substance *Yât* that arose (*qâm* ܩܡ) from the tomb and at Whose resurrection (*nuḥâmêh* ܢܘܚܡܗ) the creatures on 'high', as well as those 'below', rejoiced.

With Ephrem, we admit that we avoid building our analysis on the contingent, this 'citizen' of the space and time of the universe, but rather to build it based on the 'substance' which from the other side of the ashes of the tomb frees itself from all kinds of gravity and belongs, henceforth, to what is universal, timeless andnon-spatial.[494]

It is the transmutation (*šaḥlef* ܫܚܠܦ) of the grey-matter of the brain and not of the ashes that should be realized to affect the eye of the observer and make it 'luminous'(*safyâ*), just as happened to Saint Paul on the road of Damascus:

The 'Sun' clothed Itself in a body that made us unable to
recognize Him.
He unveiled His radiation, in a limited way, on the
Mountain;

---

493 (*Hc Haer 32, 9*); *Cf. TLE*, pp. 42-3 and 156-7; *Cf. CSCO* vol. 169, p. 129

ܚܙܬ ܓܝܪ ܪܒܘܬܐ ܕܠܒܫܬ ܠܟܠ ܕܡܘܢ
Ḥzât gher rabutâ dlebšêt lkul demwon

ܘܠܐ ܨܒܬ ܢܐܫܘܬܐ ܕܬܚܬ ܒܥܘܕܪܢܗ
W lâ ṣâbêt nâšutâ dtaḥêt bᶜudrânêh

ܫܠܚܬ ܠܚܒܝܒܗ ܘܐ ܚܠܦ ܕܡܘܬܐ ܕܠܒܫܬ ܫܝܠܬܐ
Šelḥêt lḥabibâh wa ḥlâf demwâtâ dlébšêt šîlâtâ

ܗܕܡܐ ܕܫܪܪܐ ܠܒܫ ܒܘܟܪܐ
Hadâmê dašrârâ lbêš buḳrâ

ܘܐܬܡܙܓ ܥܡܗ ܕܢܫܘܬܐ
Wetmazag ᶜamâh dnošutâ

ܝܗܒ ܕܝܠܗ ܘܢܣܒ ܡܢ ܕܝܠܢ
Yahb dileh wansab men dilan

ܕܡܘܙܓܗ ܢܚܝܗ ܡܝܬܘܬܢܐ
Dmuzâgueh neḥyê mitutane

494 *Cf. Les Origines*, p. 71.

The apparent eye was unable to see Him.
Let us go to Him with the eye of the 'spirit' (mind).[495]

It is neither to the mystification nor to the mythicizing of Light that Ephrem invites us. He invites us rather to the *râzâtic* art of seeing and reading the Light especially in both Its poetic and musical dimensions. In short, once the 'ash' of the tomb veils the substance, the ashes of the 'cooked concepts', as well as those of the 'undesirable definitions', also veil it. Conventional reality, in its materiality, ceases to play the role of subject. Verbs now run away from any conventionality and no longer apply to situations limited by the context of the universe. From this new perspective, they yield everything to the new substance, the *râzâtic* one. And from now on, the new rules, those that Jesus Christ, the Wisdom-God, has laid out during the Last Supper, the night of His Passion, will apply.

Speaking of the *Ityâ* – Christ Who excludes all ashes and Who resurrects Himself (*etnaḥam* ܐܬܢܚܡ) – and of the revitalization of matter which follows the resurrection in various forms, each kind according to its germ, allows us to approach the sun and the photon from the same angle. Speaking of the *Ityâ* redeemer allows the acceptance of the idea of the assumption of Mary, as her Son ascended, with a body which has undergone a proper transmutation.[496] Mary's 'passage' with her body that knew neither death nor corruption (*lâ ḥbâlâ* ܠܐ ܚܒܠܐ) beyond the sensible world where she continues to live, in 'God-day', becomes a prototype of perfect holiness. This 'passage' finds support either in the 'fire' or in the 'light' that accompanied key persons in the Economy of Salvation. We mean the 'fire' that answered the prophet Elijah's prayer and the 'light' that baptized St. Paul's eyes.[497]

---

[495]

ܫܡܫܐ ܠܒܫ ܦܓܪܐ ܒܬܚܦܝܬܗ ܬܥܐ ܣܟܠܐ
Šémšâ lbêš faghrâ btaḥfiteh t{c}ao saklê

ܐܣܠܦ ܗܘܝ ܙܗܪܝܪܐ ܙܥܘܪܐ ܒܛܘܪܐ
Aṣlêf ḥaoui zahrirâ z{c}urâ bṭurâ

ܚܒܛ ܡܢܗ ܥܝܢܐ ܓܠܝܬܐ
ḥobâṭ meneh {c}aynâ glytâ

ܬܥܐ ܢܨܘܕ ܒܗ ܒܥܝܢܐ ܕܬܪܥܝܬܐ
tao neṣud beh b{c}aynâ dtar{c}itâ

Bou Mansour explains this by saying: "Moreover, this veil can constitute the cause of the wandering fools (*saklé*), who are unable to move from the vision of the visible eye to that of the mind's eye. (ܥܝܢܐ ܕܬܪܥܝܬܐ {c}aynâ d-tar{c}itâ). (*H Eccl* 29,12). *Cf. La Pensée*, p. 236; *Cf. CSCO* vol. 198, p. 72.

[496] *H Nat* 3, 7 and 9; The Old Testament tells also of two cases of crossing without natural death: Enoch and Elijah, who passed through this "differAnt" reality a few centuries earlier, thanks to the same phenomenon. *Cf.* also 1 Cor 15:35-54.

[497] For the Maronite Church's 'Day', the sign is the 'light' that sprang from Saint Charbel Makhlouf's tomb. He was canonized in 1977 by Pope Paul VI. *Cf.* http://www.

But this visit of Ephrem to the tomb does not cover only the resurrection from death. Due to the union between the Light of the primordial Fire on the one hand and between the Light and the matter represented by this tomb on the other hand, it also covers any regeneration that comes from the miracles that form the strength of the Christian religion. It mainly covers the regeneration of the 'Ember of the Incarnation' whose layers of 'ash' hide the 'Triune Sun'. For greater clarity and precision, let us discuss this hypothesis in the light of the substantial unit *'yât'* itself, instead of the light glowing.

2.3.1. The *yât* (ܝܐ) and the void of the 'tomb'

As a result of the command, "Let there be Light," the initiator of the Biblical whole, and of the 'light' that resulted, the entrance of the divine Light in the contingent darkness of Creation no longer allowed 'darkness' to be written with the same meaning it had before.

In fact, as we have said, the sacred author exchanged this concept for that of 'night'. If the day could be said to be existing ontologically, since it takes its existence from the *yât* of the 'Light' existing in itself, the night should only be an expression which designates the absence of the day. Without the daylight, the night loses all reason to exist in this world of dichotomies. It is from this analysis that God can be called "God of days and nights". The concept 'darkness', since it had to exist, will continue to serve as a contrast to the luminous *yât* and as a description of all that the collective intellect would like to consider as absurd, nothingness, anti-life, anti-name, antimatter and anti-memory.[498]

Given the causes and effects that accompanied the 'choreography' between darkness and light, it is no longer enough, in our opinion, to speak of divine Light or the Epiphany of God. Rather, it becomes necessary to clarify what the Luminous Essence is and then to focus on Its behavior. It is indeed the secret of the Ephremian theology which allowed its author to see beyond the heresies of his time and then pushed all the neighboring churches to translate his works during his lifetime, in order to transmit them to their faithful. We did point to this ontological 'light', taken in the sense of an epiphany of God, in the first chapter. It cannot be considered as a physical light submitted to our

---

saintcharbel-annaya.com; [Accessed August 2019].

498   Antimatter refers us to the famous cosmic black holes, and anti-memory refers us to the Bergsonian theory of *Matter and Memories* mentioned in *Dieu et la Science*, p. 88. However, a popular Lebanese expression, *"Diar el bele"* [lit. "dwellings of nothingness"], which is common to the Semitic languages, deserves to be quoted there, because it conveys a wisdom coming from the collective sub-consciousness of the inhabitants of this region. It describes absurdity as a place of annihilation.

laboratories. Then we further emphasized it in the third chapter, dealing with the *râzâtic* reality. Referring henceforth to this 'light' under its *râzâtic* quality would be like referring to the divine 'light' in a state of kenosis,[499] a *gmurtic* state we say, based on the 'ember' theory. It is a state appropriate to our universe, to our senses, to our intellect and to our chronology, tameable by 'Writing' (the Derridean *Écriture*) without the risk of burning either the pen or the paper.

> God has made small His Majesty
> by means of these borrowed names.
>
> For we should not imagine
> that He has completely disclosed His majesty:
> this is not what His majesty actually is,
>
> But it represents only what we are capable of:
>
> What we perceive as His majesty is but a tiny part,
> for He has shown us a single spark from it;
>
> He has accorded to us only what our eyes can take
> of the multitude of His powerful rays.[500]

So Ephrem begins writing his account of the resurrection with, "A tomb with You is no longer a tomb since You are the Resurrection." What does he mean by this statement?

According to this last assertion, we infer that it matters little to Ephrem that Jesus is already bodily out of the grave, no matter in what kind of body He has appeared. The tomb with Jesus is no longer a tomb because Jesus is there and not because He has left it. Is this possible?

Only one explanation could be given to calm our anxiety. Ephrem focused on Christ's substance (*itutâ* ܐܝܬܘܬܐ), not on what is contingent (*gušmâ* ܓܘܫܡܐ), the body that the Marys had decided to embalm. And since it is no longer appropriate to consider a contingency in Jesus Christ, the Son-God,

---

499 Would this be the first kenosis that the Father and the Son have put into action?

500 *Hc Haer* 30, 4; *Cf. TLE*, p. 65; *Cf. CSCO* vol. 169, p. 121.

ܐܙܥܪ ܠܪܒܘܬܗ ܒܡܠܝ ܫܐܠܬܐ
Azʿar lrabutêh bmêlay šêlâtâ
ܕܠܐ ܕܝܢ ܢܣܒܪ ܠܢ ܕܗܢܘ ܟܠ ܟܠܗ
Dlâ dên nsabar lan dhonâo kul kulêh
ܦܪܣܗ ܠܪܒܘܬܗ ܠܐ ܗܘܐ ܐܝܟ ܡܢ ܕܐܝܬܗ
Farsâh lrabutêh lâ hwâ ak mên d'Itêh
ܐܠܐ ܠܦܘܬ ܚܝܠܢ
Èlâ lfut ḥaylân
ܐܦ ܗܝ ܟܝܬ ܪܒܘܬܗ ܙܥܘܪܬܝ
âf hî kît rabuteh zʿurutoy

especially behind the tombstone, this puts us all in front of one of the oldest Platonic issues of the body-soul relationship. If for Plato an antagonism reigns between mater and spirit, body and soul, something that was approved by Jesus the night of His Crucifixion,[501] in the tomb, through Him and in Him, all antagonism disappears to give way to a simple unity where neither weakness nor force wields any influence. Body and soul give way to a unique Light which is neither physical, blockable by stones and walls, nor divine, invisible and unapproachable, but rather a 'light' of a *râzâtic* substance similar to that of the first moment of the first day of Creation.[502] This is the Light in question. Here, the phenomenon of synergy comes to our aid. The common denominator that intuition invites us to recognize between created light and Creator Light, sun and Sun, ember and Ember, is that of the *yât* (ܝܵܬ) of 'light', its ontological singularity, whatever the metamorphosis of this 'light' is.

Let us deepen a little more our meditation on this *yât* and its relation with the emptiness of the tomb.

2.3.2. The unveiling of the *râzâtic yât*

The problematic raised above is, therefore, to specify first what light is and then to study its behavior in order to fill the emptiness of the tomb.

The Aramaic epigraph of this chapter says that, "In the [*ït*] of the Beginning God created [*yât*] heaven and [*yât*] earth: *b-riš ït brâ alâhâ yât šmayâ w-yât arᶜâ.*" This epigraph leads us to say, without hesitation, that these two *yât* of *šmayâ* and *arᶜâ* are indeed what met in the tomb. Furthermore, it is the *Yât* of the divine Light Who was at the origin of both created *yât*, Who is now in the tomb where 'Heaven' and 'Earth' meet with their Origin.[503]

Indeed, 'Heaven' and 'Earth' have their origin in the Light without which nothing could have been, and 'Light' they should become again. Consequently, the substance of the Light in question seems to be, at the same time, creator on the one hand and creature on the other. How does one distinguish between the two? How how does one explain the behavior of this substance?

---

501 Mt 26:41.

502 The Maronite Church has a Christmas Arabic Hymn of unknown source that is common to various Syriac churches; it seems to be of Ephremian inspiration. This hymn wants Jesus to come out of Mary's womb "like a light" (*šibha daoʾén laḥ*).

503 This supports our wish to see the Church reminding the faithful the morning of Ash Monday/Wednesday that they are made out of 'Light'and that to 'Light' they must return.

These last lines, together with the two questions that flow from them, represent for us the peak of the taming process between Science and Religion. To be able to provide adequate answers to these questions, we have to first clarify what pushed Ephrem to deal only with the 'substance of things' and to avoid any kind of contingent. What interest had he to see in Jesus, his dear *Râz*, only a perfect Substance, the 'primordial Light' (Light from Light) of the 'primordial Fire' (God from God, Fire from Fire),[504] Creator of all that is visible and invisible?

Being the forerunner of the Chalcedonian doctrine, Ephrem refuses any separation between human and divine natures in Jesus Christ. He described their unity by the following attributes: no confusion, no compulsion, no bondage between the human and the divine natures in the Person (Hypostasis) of Jesus Christ. This is what the Council of Chalcedon will later assert. Therefore, the Christological Substance, *ityâ* or *itutâ*, is neither that of the *Ousia* of the Father without the *ousia* of the Human nor the opposite, but it is a *râzâtic* substance, an *'Ooussia' (sic)* with two 'o's, capital and small, "differAnt" from both and at the same time consubstantial with both.[505]

Ephrem's hymns *Contra Hæresis* and *De Fide* prepare the ground for the dogma of the hypostatic union between God and Man in Jesus Christ. Here are some excerpts:

> A: One in the other lives in an equal way without envy
> (They are) mixed (*ḥliṭin* ܤܠܝܛܝܢ) but not confused (*lâ blilin* ܠܐ ܒܠܝܠ),
> Involved (*mzighin* ܡܙܝܓܝܢ) but not enchained (*lâ asirin* ܠܐ ܐܤܝܪ)
> Gathered (*kniš̌in* ܟܢܝܫܝܢ) and not compelled (*lâ ališin* ܠܐ ܐܠܝܨ),
> Free (*šârên* ܫܪܝܢ) and not confused (*pehyân* ܗܗܝܢ)[506]

---

504   *Cf. La Pensée*, p. 364: "But the Christ Who is 'clothed' in Baptism is not only called divine Power. He can borrow a qualification belonging chiefly to the Father, that of devouring Fire which is not only 'clothed' in Baptism but is also approached by the senses in the other sacraments (*DeEp 8, 21*) [*H Epiph*] ... Water and oil attenuate the strength of this Fire so that humans, of weak nature, can bear it. (*DeEp 8, 3*)" (Translated by the author)

505   This formula rehabilitates Jacob Bardaisan's approach and helps to understand where Nestorius could no longer see. To say that in the Person of Christ there are two natures and two full and perfect wills, '*tartên itutâ - ousia –* and *tartên awyutâ'* (ܬ̈ܪܬܝܢ ܐܘܝܘܬܐ ܘ – ܬ̈ܪܬܝܢ ܐܝܬܘܬܐ), does not prevent us from saying that Jesus of Nazareth, as Christ, represented a Person with an *itutâ- Substance* and an *Awyutâ- Will*, but "differAnt" from those of humans as well as from those of God the Father. We recall here the phenomenon of Adam's navel which is supported by the absence of sin in Jesus. *Cf.* Chapter III, 3.2.4.

506   *H Fid 40, 3 (*Translated by Bou Mansour); *Cf. La Pensée*, p. 210.

ܚܕ ܒܚܕ ܫܪܝܢ ܘܐܘܝܢ ܕܠܐ ܚܤܡ

ḥad bḥad šâryên w'awyân dlâ ḥâssem

B: ... Separate (*psiqin* ܦܣܝܩܝܢ), not confused (*lâ blilîn* ܠܐ ܒܠܝܠܝܢ);
Separate, mixed, bound and unbound (*asirin w-šârên* ܐܣܝܪܝܢ ܘ ܫܪܝܢ)
Huge astonishment (ܬܡܗܐ ܪܒܐ).[507]

This description, which was made possible thanks to the kenosis of the Son, provides us with the perfectly necessary material to understand the hypostatic union between the Son of God (Word, Verb, Light) and the human being, the rational animal, who enjoys as a personal substance, a 'Breath' of the same God. This Breath represents the basis of their relationality. The synergistic result of this hypostatic relationality has become "the Son of Man".

In short, Ephrem, respecting the integrity of his perception vis-à-vis the reality of the Incarnation and Salvation, could not treat the **womb** of Mary, the **womb** of the Jordan and the **womb** of the tomb as different entities. To do so, avoiding any risk of pantheism, he has resorted to the *yât* as *minima naturalia* discernable behind every being who, or that, shares his/its being with the Being as a Word or as a Name. What Ephrem avoided most, considering the phenomenon of the Incarnation in its various phases from the same perspective, was falling into contradictions as others before him had done. Through the Holy Spirit, he understood that we could make Christ lose His salvific purpose if death and resurrection were treated differently from birth and baptism, whether to please the Jews or to please the philosophers. We support this necessity with the following analysis:

> Since from the first *râzâtic* moment of His conception
> Jesus was perfect God and perfect Man, He should be
> perfect God and perfect Man in the Jordan as well as in
> the tomb. Therefore, if leaving the tomb where He was
> buried, He left it empty, He should also have left void the
> womb of His Mother and the water of the Jordan.[508]

ܫܠܝܡܝܢ ܘܠܐ ܒܠܝܠܝܢ، ܡܙܝܓܝܢ ܘܠܐ ܐܣܝܪܝܢ
ḥliṭin w lâ blilin, mzighin w lâ asirin
ܟܢܝܫܝܢ ܘܠܐ ܐܠܝܫܝܢ ܐܦ ܫܪܝܢ ܘܠܐ ܦܗܝܢ
Knišin wlâ ališin of šorên w lâ pohên

507　*Cf. La Pensée*, p. 49 (*H Nat* 8,2) and p. 210 (*H Fid* 73,8). *Cf. CSCO* vol. 186, p. 59, and vol.
　　154, p. 224.

ܟܕ ܠܐ ܦܣܝܩܝܢ ܘܐܦ ܠܐ ܒܠܝܠܝܢ
Kad lâ fṣiqin w'of lâ blilin
ܦܪܝܫܐ ܚܠܝܛܐ ܐܣܝܪܝܢ ܘ ܫܪܝܢ
prišê ḥliṭê asirin w šorên
ܬܡܗܐ ܪܒܐ
Temhâ rabâ

508　The wood of the Cross should not be forgotten either.

From this quasi-syllogistic analysis, it becomes clear that, for Ephrem, it was no longer appropriate to preach an empty tomb, and if it were, we should specify what it is empty of and how it is so. We will get back to this later. For the moment we will say that with his recourse to the substantial *yât*, Saint Ephrem has not only discovered a theory of a *râzâtic* **fullness**, but also a theory of a *râzâtic* **void** which will affect any theory belonging to space and gravity. Even time will lose its precision because of the effects of this discovery. How can one specify when the tomb began to be void and when it was absolutely void? Furthermore, all that is said of **void** already applies to **full**: when did the tomb begin to be full? What was it filled with? Was it a corruptible body?

For Ephrem, the problem is solved. Since everything is based on 'light', everything, therefore, should for him be based on the substantial unity of the same 'light', the *'râzâtic* photon', the *Yât* that the tomb contained as the ocean contains a drop of water. At this level of analysis, could it not be said that Ephrem has developed a 'constant' which could well be named after him, that of the *'râzâtic yât*' (Ephrem's constant)?

For the sake of clarity, we draw below a representative diagram of this 'constant' to be used in understanding the upcoming scientific part.

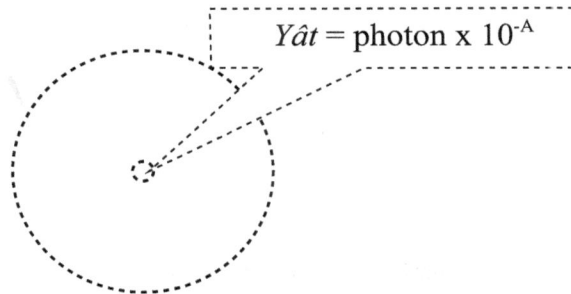

$$Yât = \text{photon} \times 10^{-A}$$

Figure 13: Ephrem's constant
(The A is that of the Derridean "differAnce")[509]

---

509 Since the beginning of the twentieth century, Einstein submitted the light of the world to the rules of quantum physics. Its substantial particle is a detectable quantum of light (photon) existing potentially in a quantum of energy as discovered by Max Planck in 1900. This potential photon passes into existence in act each time the particle of energy reaches a certain rate of heating, as the diagram of quantified energy describes. We will develop this phenomenon in Chapter V.

For Ephrem, the very substance of this particular 'photon' belongs to the category of the *yât*.[510]

We conclude this part by repeating the following: "Without Christ, the tomb would have kept its quality of 'door to Sheol', the place of darkness and annihilation, but with Him it becomes the catalyst, the place of transmutation at the substantial level of *yât*, the 'place' of the Resurrection."[511]

Therefore, we invite theologians and physicists to join us at this tomb because we may recognize in it the much-sought Tent of Meeting (*Maškanzabnâ*), the Place of His habitat with us, of our scientific and theological encounters with Him:

> See, Fire and Spirit in the womb that bore You,
> See, Fire and Spirit in the river in which You were
> baptized.
> Fire and Spirit in our Baptism,
> In the Bread and the Cup, Fire and Holy Spirit.[512]

And we add with Ephrem: Fire and Spirit in the 'womb' of the tomb to wait for us when we will be able to resurrect light from matter. Is this possible?

---

510 It is thanks to this scientific data that we have been able to detect, in the interpretation of the text of the Resurrection, a new dimension that did not exist before Ephrem's Marian teaching: we mean the consolation due to Mary. As it turns out, there is no mention in the New Testament of an apparition of Jesus to His Mother to console her. Should He not have done so, since she is a priority, at least to restore her soul after the sufferings He caused her? Or, is it to be supposed that Mary's suffering was "differAnt", *râzâtic*, and that she did not need to review what is contingent to believe in the Living Substance of her Son, this Substance that as a matter of fact never left her?

511 Ps 6:5.

512 *H Fid* 10, 17. *Cf. CSCO* vol. 158, p. 51. *Cf. TLE*, p. 94.

ܗܐ ܢܘܪܐ ܘܪܘܚܐ ܒܥܘܒܐ ܕܝܠܕܬܟܝ
ܗܐ ܢܘܪܐ ܘܪܘܚܐ ܒܢܗܪܐ ܕܒܗ ܥܡܕܬ ܗܘܝܬ
ܢܘܪܐ ܘܪܘܚܐ ܒܡܥܡܘܕܝܬܢ
ܒܠܚܡܐ ܘܟܣܐ ܢܘܪܐ ܘܪܘܚܐ ܩܕܝܫܐ

3. Resurrecting 'light' from matter

> The 'body' You wear, Lord, is a fountain of medicines:
> In Your visible 'dress' dwells Your Hidden Power.
> Again, some saliva from Your mouth became light
> Great amazement, the 'light' in the clay He made![513]

Since nothing exists that is not created by God, whenever light or energy or a certain life emerges from matter, whatever this matter is, it is one of the epiphanies similar to that of the resurrection of the elusive Substance called in Syriac ***Ityâ Ghnizâ*** (ܐܝܬܝܐ ܓܢܝܙܐ). And since 'light' can also find its place in 'mud', this resurrection proved that the incompatibility between Light and the tomb is not existential, but rather conceptual. Furthermore, taking into account the cause-effect relation, this incompatibility may well also be called etymological because between its two poles signs still exist that can be grasped by human understanding. One of the two poles represents the cause which exists ontologically and the other, the effect which exists only contingently at the expense of the first. For example, the aforementioned blind man was blind only because of the 'light' that was missing in him. This is why Ephrem interpreted this scene as if Jesus introduced this 'light' again, through the eyes, evoking the gesture by which His Father introduced life into Adam's 'mud'. According to Genesis, life does not exist properly without the Light (*lâ ït* ܠܐ ܐܝܬ); thus, it is with the sight that simply cannot be without the 'rays' of Christ by Whom all things have been made. This is how Ephrem interpreted the difference between the 'oil' of the wise Virgins and that of the foolish Virgins, especially the 'oil' that the latter went to buy:

> Here in the 'olive oil' are His rays:
> Anoint yourself, shine forth with knowledge,
> As shone the 'Thief' thanks to the ray Who in front of him
> was raised...
> For all you who believe in the Cross,
> Double light is there for you.
> Illuminated by the anointing, you will be enlightened by
> Baptism too.[514]

---

513 *H Fid* 10, 7; *Cf.* Feghali, *Bayna Ma'ida wa Ma'ida*, p. 21. *Cf. CSCO* vol. 154, p. 50. (Translated by the author)

ܢܚܬܟ ܡܪܝ ܐܝܬܘܗܝ ܡܒܘܥܐ ܕܐܣܘܬܐ
ܒܠܒܘܫܟ ܓܠܝܐ ܥܡܪܐ ܗܘ ܚܝܠܟ ܟܣܝܐ
ܪܘܩܐ ܙܥܘܪܐ ܕܡܢ ܦܘܡܟ ܬܘܒ ܢܘܗܪܐ
ܗܘ ܪܒܐ ܕܢܘܗܪܐ ܒܛܝܢܐ ܓܒܠܗ

514 *Cf. La Pensée*, p. 80 and specifically p. 351; *H Epiph* 3, 30-31; *Cf. CSCO* vol. 186, p. 153. (Translated by the author)

In other words, if the 'cave' carved in the rock and the body of Jesus lying there represent two signifiers, the two relative signifieds must be created matter and incarnate Light. In their turn, the latter refer both to a physical '*yât*' on one of the two sides of the tomb and to a *râzâtic* '*yât*' on the other side. Therefore, it can be deduced that, at the level of the *yât*, these incompatibilities no longer count because the *yât* in itself is not subject to our categories. So to the question, "Why is he blind?" Jesus, according to the analysis that follows, makes this truth very clear, giving an answer that involves none of the categories mentioned by His questioners.

If the Bible informs us that in the Beginning, God created the *Êt-Shamayim* and the *Êt-Êrêtz*, it also assures us that on both sides, God created just the *Êt*, the substance. That is all that He created. Then, in the DNA or RNA of each being (*êt* with a small e) of both sides, He introduced all the necessary potential (*ḥaylâ* ܚܝܠܐ) to help each become what it should be, according to a given design He as Creator has foreseen. But are these respective DNA or RNA substantially different? Nothing says this is so.

It is true that 'heaven' and 'earth' will be different in their accidental Matter, but taking into consideration the axiom of the co-Creator Light, in their substantial Form they must not be so. Without the Light (the Son), nothing could have been, so they are both based on the same Light. All creatures are creatures, whether physical or

---

This reading of Ephrem makes us discover the whole tradition of the sacramental use of oil and even its medical use at the level of popular piety. First, we confess that we had never heard or read about the dimensions of oil as Ephrem writes, but then, all of a sudden, we heard of the miraculous profusion of oil that the Middle East has known, from the image of Saint Charbel in a small village in Lebanon (ᶜAin-ᶜAlaq) in 1979 to the Soufanieh in Damascus, a phenomenon in front of which we have stood speechless. We wonder today if it was not due to the *yât* of the Light since olive oil has always been a perfect combustible that leaves no residue. These profusions could have happened by the power of the Holy Spirit Creator, that the Light transforms Himself into Oil. Is it the same light that springs from the burning oil which finds again its origin? The answer resides in the equation: oil → fire → light vs Light → Fire (of faith) → Oil, where between the lowercase letters and the uppercase letters matters are only "differAnt". Otherwise, how can one understand that after many tests all this oil proved to be pure olive oil? *Cf.* http://www.soufanieh.com/FRANCAIS/2014.syr.fra.soufanieh.en.syrie. et.dans.le.monde.web.pdf [Accessed August, 2019]
If we admit this reading, all the miracles of Jesus, as well as Mary's apparitions, the Light appearing to Paul, and the Gospel itself, would be grasped in a better way.

metaphysical, visible to the naked eye or to the perceptive eye (ᶜaynâ d're̊yân ܟܢܝ ܕܪ̈ܢܝ) or invisible to any kind of eye and instrument, and therefore must be based on the same Light.[515] This explains why, for Ephrem, there is no way to confuse creature and Creator even if they sometimes share the same "name". This also explains the conflict between Jesus and the Pharisees who accused Him of casting out demons in the name of Satan (Beelzebub), the chief of the devils (ܡܠܟܐ).[516]

Man has been slow to grasp the role of the signifier, of the signified, and to what they both relatively refer. From birth, to believe the grace of the vocation which according to Isaiah takes root in the maternal womb,[517] Ephrem grasped more referents than the three dimensions allowed by the reality that surrounded him. He asks us nowadays, thanks to the great progress that humanity has reached, to go to additional referents beyond the fourth and fifth dimension, all of which are united in Christ. However, Jesus Christ Who is from neither of the two 'milieus' taken separately, neither from the 'Earth' nor from the 'Heaven', gathered in His Person exhaustively, by His incarnation from Mary, the one and the other. The aim was to assume them both in Him, in a *râzâtic* synergy of perfection, as His Father desired on the day of His *Shabbat*. This is why Paul insisted on saying that in Christ, there is no longer any difference between 'all' and 'all', between 'everything' and 'everything', space, time and directions included.[518]

Making 'light' rise from matter is henceforth possible through the *yât*. For every *yât* comes from Him, the divine Light, without ever being exclusively His own *Yât* and without the latter being reduced to any other or being affected by the creation of the other *yâts*. However, where could one imagine this osmosis if not within the 'womb' of the Sun phenomenon, in the core of which Ephrem was able to trace three Signifiers for three Referents at the same time: Fire, Light and Heat for the Father, the Son and the Holy Spirit. For the purpose of a wise evangelization, Ephrem invites us to reason exhaustively and multi-dimensionally through the substance *yât* and no longer through accidents.

This Ephremian device, based on Saint Paul's words, "The letter kills but the Spirit gives life",[519] will require a distinction between the Resurrection (*qiâmtâ* ܩܝܡܬܐ), based on the verb "to get up" (*qâm* ܩܡ) (in Arabic قَام), and the Resurrection (*nuḥâmâ* ܢܘܚܡܐ), based on the verb "arise" (*naḥêm* ܢܚܡ; Arabic

---

515 This allows saying that the Light could be included *a priori*, even if in an apophatic sense, in the raw material.

516 Rev 12:7; 2 Cor 11:14.

517 Is 49:1.

518 Eph 4:9; Col 3:11; Gal 3:28;

519 Jn 6:63; 2 Cor 3:6.

انبعـث).[520] It is necessary to do this in order to be able to join a solid evangelization free from any reductionistic syllogism.[521] What does this distinction consist in?

3.1. Distinction between *qiâmtâ* (ܩܝܡܬܐ) and *nuḥâmâ* (ܢܘܚܡܐ)

The angel said: "Do not be afraid: I know that you are looking for Jesus, the Crucified One. He is not here because He is resurrected as He said (*qâm leh ghêr aikanâ dêmar*)... He is resurrected (*qâm leh* ܩܡ ܠܗ), and behold, He goes before you into Galilee; that's where you will see Him."[522]

What exactly did the angel mean by the following words: *qâm leh*. Is it "resurrected" as the verb "to arise" means or simply "woke up" as someone gets up from his sleep?

In literary terms, *qâm* (ܩܡ) means "to get up", "to wake up", as Jesus suggested in the gospel passage about Lazarus' revival, while *naḥem* (ܢܚܡ) instead means "to resurrect". It is a phlogistic verb, *i.e.*, related to fire and inflammability. To shed light on this luminescent distinction, we must meditate again on the nature of God. Is it not Saint Peter who invites us with insistence in his letters to do so?[523]

In fact, Holy Week began with Jesus' glorious entry into Jerusalem while the resurrection of Lazarus was still fresh in the minds of the urban community. Some praised this miracle; others feared it. Let us return, in a flashback, to the scene of this miracle which marked the 'beginning of the end' of Jesus' days in this world, in other terms, the 'beginning of the end' of the stay of the Light in our universe.

At Lazarus' tomb, things seem to happen as in a 'Mirror' that reflects Jesus'

---

520   After examining the dictionaries and the texts, we emphasize that it is not enough to take the verb "to emerge; resurrect" in its meaning "come out of the tomb" or "return to life". It would be too limited to express the reality of the Christological event *par excellence* as wanted by the verb *qâm* ("to rise") or by *naḥem* ("to resume igniting or radiating"). The verb "arise; resurgence" should rather be taken under the phlogistic or luminous dimension of its meaning "explode", "spurt", "flow", "reverberate", "spring", or else "emerge", "pierce", "dawn", "come out", "burst" to meet the demands of the resurrection of the 'Light of the World' as Syriac Christology would want. The issue as Ephrem grasped it is that the same phenomenon must be respected at the birth of this 'Light of the World', at His declaration as Christ at the Jordan River, and at His Resurrection, "Like a Ray which springs." It is the Luminous field of interpretation, not just any field.

521   With this opinion, we agree with Sebastian Brock's distinguished position. *Cf. TLE*, p. 17.

522   Mt 28:5-7.

523   2Pet 1:4-5.

tomb.[524] Women are there, their doubts too, but above all, their persistence. Even the Light is present, visible to some and not visible to others. Jesus had already foreseen this happening by assuring his disciples, before turning to Bethany, that, "Are there not twelve hours of daylight? Those who walk during the day do not stumble, because they see the light of this world. But those who walk at night stumble, because the light is not in them."[525] Therefore, the whole problem lies in the fact of already possessing or not yet possessing the Light in one's own self.

By these words, Jesus clearly pointed to Lazarus and not to Himself since He told His disciples that first He would return to Judea to see His friend Lazarus who "is resting" and that He would wake him up. Analogically speaking, and taking into account the tradition of the oil lamp as well as that of the wise virgins mentioned above, we can put in Jesus' mouth the following words: Lazarus apparently died ... his 'light' or even his *râz* "rests" in the sense that the 'oil' of his 'lamp' has run out ... and I will, for your lack of faith, 'resurrect' his 'light' from its rest.[526] If the poor disciples did not understand what Jesus meant, it is because, in our opinion, unlike Lazarus, they had not yet had the grace to have the same Light in them, but only with them: *Emmanu-El*.

Lazarus and his sisters are witnesses that during the lifetime of Jesus there had been some people who had already loved this Messiah and were able, in a homeopathic way, by a grace of the Father, to have the *râzâtic* light in them. Neither the Canaanite woman nor the Centurion whose servant Jesus resurrected were excluded from receiving this Christological grace. All those to whom Jesus granted a resurrection (*nuḥâmâ*) or even a healing, without making too much distinction between miracles, represent people of the Kingdom, as Jesus wanted us to know it. He later proved this to Peter in his vision on the terrace of a certain Simon in Joppa: "Get up, Peter; kill and eat... What God has made clean, you must not call profane."[527] Therefore, all miracles, without exception, especially those which aimed at the eyes and the sight, were epiphanies of a resurrected 'light' veiled by the 'ashes' of a humanity exhausted by the Law, or rather by the thousand and one laws of every kind and color imposed by the Sanhedrin.

---

524 The *râz* of Lazarus' tomb is like a road sign that points the way to the Tomb of the *Râz* Who is the Way.

525 Jn 11:9-10.

526 *Cf. La Pensée* p. 351: "Not entirely absent from the *Hymns of the Epiphany* [H Epiph 3, 30-31], the symbolism of the oil in relation to the light is expressed by the affirmation of the presence of the rays of Christ-Light in the oil, the effective presence which endows the oil the ability to illuminate." (Translated by the author)
From this symbolic dimension of oil, the lamp of the Lebanese Saint Charbel takes on more importance, especially since Ephrem also links the baptismal water to oil (*mešḥâ*).

527 Acts 10:15.

Eventually, Jesus was resurrecting His own Light, even Himself, in each of these miracles, no matter what circumstances accompanied them. The most prominent example is the case of Lazarus, His friend. The latter would always have the chance to rise again on the "last day", and this according to the tri-dimensional Jewish-Pharisaic faith mentioned in Martha's answer, "I know that he will rise again in the resurrection on the last day."[528] However, it is precisely at the level of the *râzâtic* dimension, that 'added value' which dominated the dialogue between Martha and Jesus, that Lazarus will immediately be resurrected, as the rays of the sun arise. This dimension is underlined by the Lord's declaration: "I am the resurrection and the life. Those who believe in Me, even though they die, will live."[529]

Jesus mirrored on two occasions the founding order of Genesis, "Let there be light," once commanding, "Take away the stone," followed by the command, "Lazarus, come out."

Following Jesus' resounding words that had followed His communication with His Father, the dead man came out, his feet and hands tied with bandages, and his face wrapped in a shroud. He came out like a beam of light rather than like a man! Jesus then ordered them to free him and let him go. Is it not as if He were asking them to put aside the 'vessel' that blocked His own Light which had been hidden at a certain moment within Lazarus' *yât?*[530]

Ephrem pushes the eye-light relationship toward its extreme application on the *yât* of beings, *kyânê*. He suggests that the Incarnation has even made the 'eyes' of the universe, represented by Lazarus' tomb, luminous and able to testify to the glory of the Incarnate.

"Nothing in Creation is isolated from the rest of nature, and the world is, beside Scripture and with it, a Bible of God."[531] This is what Louis Leloir writes introducing Brock's French version of *The Luminous Eye* (*Œil de Lumière*) which highlights Ephrem's famous verses that follow:

> So that both the natural world and His Book
> Might testify to the Creator:
> The natural world, through humanity's use of it,
> The Book, through humanity's reading of it.[532]

---

528   Jn 11:24.

529   Jn 11:25.

530   Lk 8:16.

531   *TLE*, French Copy (*Œil de Lumière*), Preface. (Translated by the author)

532   *Cf. H Par* 5, 2-end, *CSCO* vol. 174, pp. 16*ff.*

ܟܝܢܐ ܐܟ ܗܘ ܚܠܝ   ܘܢܫܡܗ ܠܟܘܢܐ
dnashêd lboruyâ   kyânâ of sefrâ

From all this apology, we deduce that *qâm* and *etnaḥam* refer to the same phenomenon of the springing of 'light' from the core of 'matter' but with the *râzâtic* dimension of "differAnce" which requires, to be joined, a *râzification* of hearts, minds, and eyes, the cleansing of all sin by the glowing Ember (*Gmurtâ mḥasyânitâ*). It is in just such a case that the exchange between *qâm* and *etnaḥam* will be acceptable. It is at the level of paradigms and various dimensions that the *râzification* should be accomplished in order to make the comprehension of the translations of the two Syriac verbs accessible in modern languages that use verbs such as resuscitate - resurrection, to rise - risen, *Auferstehung*, ανεστι, *risorgere - resurrezione*, قـام – القيامـة, بعـث, الانبعـاث etc.

This is all well and good but it still does not sufficiently satisfy our hypothesis that the light resurges from matter. Having emphasized the distinction between *qiâmtâ* and *nuḥâmâ* in the texts that describe the Resurrection in its various dimensions, let us further consolidate our position on the union between 'light' and 'matter' in treating the Light-Tomb relation before going on to the various considerations of the most renowned physicists *vis-à-vis* this phenomenon.

3.2. Resurrecting the 'light' from the 'tomb'

During the last years of his life, the philosopher Jean Guitton began, after his master Henry Bergson, to seek the spirit of matter. Previously, his master was able to discover, almost at the end of his days too, that there is memory in matter, the memory of a harmony and a symmetry which existed at a certain moment, and were later broken. Both insights are a proof that what Ephrem was able to reach in the full spring of his age was not a concern of his own, but rather of divine inspiration due to his close relation with the Holy Spirit.

Thereupon, we continue to say, with Ephrem, that if only one 'portion' of the created universe is away from the divine Light, it represents an infraction against the concept of the omnipresence of God as well as against the universality of Redemption. Even *Gehenna*, Hell, Sheol, etc., that we previously considered as types or symbols of the primordial darkness remaining from the first day of creation must not remain strangers to this axiom. From the tomb, Jesus visited

---

ܟܝܢܐ ܒܚܘܫܘܚܗ   ܟܬܒܐ ܒܩܪܝܢܗ
Kyânâ bḥušoḥêh   *ktâbâ bqeryâneh*

The following references (*H Virg.* 20, 12; *Cf. CSCO* vol. 223, p. 70) also serve as strong support and bring out the concept *râzâ*:

ܒܟܠ ܕܘܟ ܐܢ ܬܗܘܪ ܪܐܙܗ ܬܡܢ   ܘܐܝܟܐ ܕܬܩܪܐ ܛܘܦܣܘ ܬܫܒܚ
Bḵul duḵ en teḥur râzêh tamân   *waykâ dteqrê tufsao tšabaḥ.*

Which can be translated by the following: "Wherever you turn your eyes, there is God's *râzê* (symbols); whatever you read, His types glorify."

them.[533] He visited all that is considered contingent, non-existent in itself, according to our dimensions, so that nothing remains outside the dimensions of the divine Salvation. Thus, there is nothing meaningless and senseless. Since that visitation, then, these supposed non-existents (*lâ it* ܠܐ ܐܝܬ) could no longer be the same. They will enjoy the verb "to be". They will emerge from the absolute darkness to the *râzâtic* completeness.

Saying that Jesus brought out from the darkness of Sheol those who are supposed to be waiting for Him, since Adam and Eve's day, is like saying that He brought out the 'spirit' from the absurd 'matter'. This is confirmed by the fact that the essence of every prayer of the three theist religions is to remind God of the 'works' of His hands, or to commemorate the role of the 'Prophets', 'Just ones' and 'Saints' who have pleased Him since day one 'until today'. The utmost prominent role is given to Mary, the *Genetrix* of God on "earth". This means that the divine Memory is what holds the Creation in union as in evolutional synergy.[534] This Memory also includes all those who have rested, the so-called dead, as well as the angels and the archangels in the same unity. The prayers end up asking God, the Almighty Father, to join us to our beloved ones and to allow no one to get lost by reminding Him that, of those who have known the earthly globe, "only His Son is without sin". [535] Since the 'memory' of matter cannot be anything but God's Memory and His Archive, it gives reason to the Bergsonian theory. Guitton dared not to mention it or to give it a definition. He seemed to apply Ephrem's warning: not to define the undefinable.

In our opinion, the issue 'God–spirit' and 'matter–memory' regains its fullness once conceived from the *yât* perspective. The *râzâtic* theory makes it possible to say that in every *yât*, be it that of the 'Heavens' or that of the 'Earth', a **Sound Stem** of the Voice that has given them the order to be, is registered. 'Heavens', 'Earth', and their inhabitants remember that Voice by anamnesis, by revivification, at each moment of time, at the risk of losing their being and their existence.

Likewise, an **Information of the Light** from which all three have taken their being must be recorded in each of their respective *yât*. 'It' orients the direction of their course in the same sense of the *yât* of Time which moves toward its Source, its Omega (*Tao* in Aramaic), which is nothing other than its Alpha (*Olaf* in Aramaic), to prevent them from re-entering the primitive chaos.[536]

---

533   *Cf. Hymn to the Light.*

534   According to Saint Ephrem's theology, our Lady Mary, Mother of Jesus Christ, thanks to her fullness of Grace and divine Motherhood, holds the most prominent role in bridging human memory and that of the Triune God.

535   From the Maronite liturgy, an application of the teaching of the Nicene Council.

536   The sunflower, which turns to the beams of the sun, is a good example of this

Bergson would be correct to think that matter is only the result of breaking a certain symmetry that reigned somewhere at one time. But did he speak of matter under its contingency or under its substance? This matter, whatever it may be, should be made of a substance (*yât*) which due to a certain Bang, no matter whether Big or Small, has detached itself from its 'root' and entered a chaotic state. Then, thanks to the "Sound Stem" of the Voice that carries the Creator's Word as well as to the indelible information of the Light that are immanent in each of the *yâts*, this matter was pushed to form, according to a certain design, a Creation as beautiful and as orderly as the present one. This was possible due to a certain governing Natural Law in a certain field of Action[537] given to the *yâts* to adhere to each other, either by concomitance or on a basis of sympathy and antipathy specific to their nature. This Natural Law, that is, the information the *yâts* keep in themselves, allows this possibility. This phenomenon can be said to be, in computer language, a simulation of the 'Primordial Reality' based on the triunity "*unum-verum-bonum*" from which they departed and, at the same time, to which they can only lead. On the anthropological side, St. Augustine joins us in perfectly expressing this relationship between the 'heart' of Man and of his Creator: "Our heart is restless until it rests in You," he wrote.[538] At this point, we would also like to mention Khalil Gibran who captured with his perceptive mind the same inspiration as Bergson. In the early twentieth century, he wrote in his book entitled *The Prophet*, addressing married couples: "You were born together, and together you shall be forever more. You shall be together when the white wings of death scatter your days. Ay, you shall be together even in the silent memory of God." Is there any term more beautiful than "Love" and "God's Memory" to speak of 'spirit' and 'memory' specific to a matter made of an oriented *yât*?

And, to the famous questions asked by Philosophy, "Toward what is matter oriented, *i.e.,* the whole cosmos, be it the macrocosm or the microcosm?" or even, "What will be the limits of the expansion of the cosmos as described by astrophysicists?" Ephrem's answer is clear: it is toward the creative *Yât*, the Alpha-Omega, the *Unum-Verum-Bonum*.

Our own answer must wait for the final synthesis. It will reveal our opinion, which will certainly be *râzâtic*. In the meantime, we invite ourselves to explore a very delicate dimension that the *râzâtic* theory should respect. It is that of clarity

---

phenomenon. Computer software programmers understand this figure as being an application of what is called "directed objects".

537   Recently called the Higgs Boson Field.

538   St. Augustine. *Confessions,* Lib 1,1-2, 2.5,5; *CSEL* 33, 1-5. *Cf.* https://www.crossroads initiative.com/media/articles/ourheartisrestlessuntilitrestsinyou/ [Accessed Aug 22, 2019]

and distinction that René Descartes, father of modern philosophy, demands out of respect for human reason. Does Ephrem's *râzâtic* theory satisfy this requirement?

3.3. Descartes and Saint Ephrem (Rationalism vs Exhaustivism)

Different philosophies and their interpretations caused some Christological problems during the first centuries after Christ. At the level of modern philosophy, some elements have made Christianity more enigmatic, especially with regard to the Incarnation, the Resurrection, Life after death, and all that relates to these "mysteries". It is of these elements that philosophies, such as the existentialist and materialist ones, could well have taken advantage to counter the phenomenon of the Christian faith and therefore of Religion in general. It is Descartes' Method that we will target with what follows, hoping to confront it summarily with Ephrem's approach.

Descartes based his intellectual activity on the principle of "clarity and distinction" in order to overcome, in general, every doubt and to reach reassuring rational certainties, thinking this way to serve God and, above all, the Christian faith. His Method urged him to proceed by separation, categorization and elimination to the point of separating the cogitative 'I am' (*sum*) from the comprehensive 'I am' belonging to the human collectivity. Who, after all, are those who could see 'clearly' and 'distinctly' at the level of God, whether at the level of His Oneness, of His Trinity, or especially at the level of the Incarnation? Are they not the persons who enjoy a heart and an eye illuminated by His Light? We would have liked to see Descartes, the Catholic philosopher, read Ephrem's works before writing his philosophy.[539]

In fact, since this father of modern philosophy called for rationalization, the positive sciences began little by little to replace philosophy. After Descartes, the celestial mechanics of Laplace and then psychology, as mentioned above, were enthroned as the saviors of humanity. They made God and His precepts a burden that Religion had to defend at all times. Many philosophers, especially Maurice Blondel, criticized Descartes, having foreseen where this Cartesian method would lead the world of the spirit (religion) and conscience.

At the level of the 'Spirit', whether it is the Spirit of Wisdom ($\Sigma o\varphi\iota\alpha$), the Spirit of matter, or the Holy Spirit, any method that aims to separate it from what incorporates and expresses it could only submit to the model of the parable of Wheat and Tares.[540] To separate one from the other, one must wait for the end times. Moreover, it is clear that only the One who created them can separate them.

---

539   "I remained a follower of Descartes until the day I became a follower of Ephrem." (Quote from the author)

540   Mt 13:21-30.

Meanwhile, we must leave them together, substance and contingency, understand them, especially their behavior, reveal the wisdom behind their 'synesseration',[541] and finally understand how to behave in relation to their whole.

This is what Ephrem wanted us to get to by his dichotomy *kasyâ* and *galyâ* (veiled and unveiled), a basic principle of his own philosophy. It supports the Ephremian *râzâtic* theology that distinguishes the rational–gnostic mystery from the Christian one sprouting from the truth of the Incarnation and the Resurrection.

It is a philosophical 'method' that helps to clarify the obscure, something the Cartesian method could never accomplish, not by inability to do so, but – we allow ourselves to say so – because its dialectics was founded on purely rational dimensions. It does not include the *râzâtic* dimension. It is in fact this limit of reason which had prompted St. Augustine to write as a philosopher, *"Crede, ut intelligas,"*[542] making it clear that faith cannot be acquired by pure reasoning.

How has the twentieth century, paralytic in Philosophy yet considered the century of the victory of rationalism, of illuminism, and above all of technology, ended? We send all persons curious to know the answer to the reality of today's world. Does it present more predictability and certainty to humans? Do humans profess to be happier and healthier than they were a century ago?

Let us sincerely say that this twentieth century ended on a chaos of values at the global level with a drastic crescendo of materialism's unique value, the god Mammon, with a victory of the ethical and moral relativism that gave excuse to all possible and imaginable corruption, and facilitated its infiltration even into the Catholic Church. Furthermore, let us not talk about the global side of environment, nature, health and climate change caused by the industries that pollute.

With Ephrem and the famous Syriacist scholars like those quoted in this book,

---

541 The substantive 'synesseration' is a concept that we had to forge, in collaboration with Prof. Wahib Keyrouz (+2012), curator of the Museum of Khalil Gibran, to express the state of being together (*Mitsein*) and not simply of existing together. We have forged it for the Arabic language from Syriac *kayen, Kyânâ, yât* and *hwâ* to be able to correctly translate the Cartesian motto *"Cogito ergo sum"*, without causing any confusion at the ontological level between the Being and the being; the being of someone, the being of the other, and the Being of the "Wholly Other". We hope that one day this term, with its derivatives and its verb 'to synesserate', will be admitted by the English-speaking academics. *Cf. Taḥawol al Mafahim Fi Bina' Al Joumhouria*, Sader Publishing House, Beyrouth, 2006.

542 "Believe in order to understand." *Tract. Ev. Jo.*, 29.6. *Cf.* http://www.dictionaryof spiritualterms.com/public/Glossaries/terms.aspx?ID=431

a new philosophical era is now breaking through. It is that of openness to meta-rational dimensions, such as Ephrem knew them. Until a few decades ago, they were still hidden in the 'womb' of the Syriac language, which in turn has remained neglected for centuries.

We can start with Ephrem, the lyre of the Holy Spirit. Being provided with the 'Luminous Eye', he equally respected the three constituents of the human being: the body (*soma* σωμα), the spirit (*psyche* Ψυχη) and the soul (*pneuma* πνεύμα), from all perspectives. He was able to distinguish between them thanks to the multi-dimensionality with which his 'eye' was endowed. And, surprising him in meditation in front of his 'pearl', his 'mirror', the 'oil', the 'salt', the 'embers', the 'sun', and generally in front of the 'Book of Nature', we discover that he is in no way Cartesian. The same can be said of his *râzâtic* position *vis-à-vis* the relationship of Sun and sun. And as by his symbolism he would have pleased the philosopher Gadamer, his *râzâtic* philosophy would have been pleasing to Bergson, Guitton, Chardin, Derrida, as well as several other philosophers and linguists.

To see the 'light' in the tomb and to say that the tomb, with Christ, is no longer a tomb is not only to move away from seeking clarity and distinction between entities and concepts of a different nature, but above all, to ensure that what was revealed by the intermingling of what is human and what is divine, thanks to the Incarnation, is a *fait accompli*, irrevocable, as indicated by Pilate's words: "What has been written, has been written." It is the "differAnt" reality mentioned above, perceived only by 'luminous minds' to whom this possibility has been given, which must, henceforth, prepare for a "differAnt" philosophy. This "differAnt" reality proved to be impervious to rational distinction and clarity, but very open to a **comprehensively loving** conception of truth and life.

The relevant experts are supposed to admit this last conception as they admit those of the perceptive minds in Science. Similarly, the perceptive minds of both sides, Science and Religion would come, on the one hand, to respect and believe each other, and on the other hand, to seek together to overcome the riddles that are always causing them conflict and distress and which, at the same time, hinder the well-balanced progress of humanity.

Saint Paul says no less in his second Epistle to the Corinthians, dealing with Moses and his Veil, contrasting Law (Veil) and Spirit (Liberty):

> Since, then, we have such a hope, we act with great boldness, not like Moses, who put a veil over his face to keep the people of Israel from gazing at the end of the Glory that was being set aside. But their minds were hardened. Indeed, to this very day, when they hear the reading of the old covenant, that same veil is still there,

since only in Christ is it set aside. Indeed, to this very day whenever Moses is read, a veil lies over their minds; but when one turns to the Lord, the veil is removed. Now the Lord is the Spirit, and where the Spirit of the Lord is, there is freedom. And all of us, with unveiled faces, seeing the glory of the Lord as though reflected in a mirror, are being transformed into the same image from one degree of glory to another; for this comes from the Lord, the Spirit.[543]

The main enigma is, therefore, never to separate in Christ 'fire' and 'womb', 'fire' and 'water', 'fire' and 'oil', 'fire' and 'wood', and through Him, light and cave, light and pearl, light and face, light and mirror, light and eyes, light and faith, light and cross, light and Kingdom, light and Love, light and Tomb, light and Sheol, light and Resurrection, light and *Parousia*. The main enigma is especially never to separate Fire, Light, and Heat at the level of the 'Sun', *i.e.,* the Son, nor at the level of the sun, but rather to comprehend and interpret everything exhaustively in the light of the "differAnt" philosophy, the *râzâtic* one.

Once the different '*yâts*' come into coexistence, or better yet, in *synesseration*, a separation between them in order to make greater 'clarity and distinction' at the service of reason and its logical certainties would no longer be possible. And if a revelation of their union is still possible, which would require a contribution between Heaven and Earth similar to that of the Incarnation, it would surely be *râzâtic*. A proof of this kind of contribution comes out from the 'door of the Tomb'. In order to be opened at the dawn of that famous Sunday, a contribution was necessary between the women (or Eve) and an angel, just as at the Annunciation, under the aegis of the Holy Spirit and by order of the Father.[544] In other words, to whom would the Father open the door if no one could go to the Light except by Him?

With all due respect to Descartes, the father of modern philosophy, we conclude this comparison between him and Ephrem assuring that, if he had read Ephrem before developing his Method, he could have spared humanity much embarrassment. He could also have spared the Catholic Church much discomfort by giving greater value to the impact of the Incarnation on the *cogito* and on the *sum* of his axiom "I think, therefore I am" (*Cogito ergo sum*), and, exhaustively speaking, on human cogitation itself in its relationship with all the components of the universe.

---

543   2 Cor 3:12-18.

544   Many Syriac prayers compare the birth of Jesus from Mary's womb to His leaving the tomb as 'light', *i.e.,* without having to 'break the seals'.

Conclusion

Ephrem prayed, "You, elusive Substance of the Father Who rose from the tomb in glory... With You, the tomb would no more be the tomb for You are also the Resurrection." This key *râzâtic* angle of view helped Ephrem to propose his new constant, a *yât* similar to Planck's quantum. This is what fits with the Fullerian theory of "seeing exhaustively". Constants drawn out of the Book of Faith or out of theBook of Nature are necessary for Science's comprehensiveness as well as for the universal scope of Religion. For Ephrem, the very Planck's constant that, according to Einstein, became the particular 'photon' with its heat, light, and wave belongs to the category of the *yât*. It offers humans the unique instrument valid to reconcile Science and Religion, the 'light'.

Accordingly, we conclude Part II of this book by reaffirming with Ephrem that indeed, without Christ, the tomb would have kept its quality of 'door to Sheol, to chaos', the place of darkness and annihilation of every mindset. However, with Him it becomes the catalyst at the substantial level of *yât*, the place of transmutation of philosophical paradigms as well as of the scientific concepts, the 'place' of the Resurrection. Therefore, we invite theologians and physicists to join us at this tomb because we may recognize in it the much-sought Tent of Meeting (*Maškanzabnâ*), the Place of His habitat with us, of our scientific and theological encounters. The fact is that, according to Saint Ephrem, the Fire and Spirit that met in the womb, in the river, in baptism, in the bread and the cup, are the Fire and Holy Spirit mentioned by the divine Word. Furthermore, they also meet, undoubtedly, in Science and in Religion to form the 'Luminous Bridge' for their triune communication. In the 'womb' of the tomb they wait for us to resurrect 'light' from 'matter', Religion from Science and vice versa. Is this possible?

Once more, we repeat with Ephrem: "With You, I shall flee to acquire by You life in every place. With you, **the pit would no more be the pit** for with You one would ascend to heaven." We see the scientific 'pit' included here.

We apologize for having lingered so much around the tomb and in its 'womb'; however, a physicist in his laboratory would do no less, especially when his goal is to bring out light from matter without risking blindness. We consider ourselves excused especially as we approach our goal, which could be the desired *Maškanzabnâ* for Faith and Reason, as well as for Religion and Science.

For the moment, we are going to halt our investigation (*bsâ*) of the tomb with our *râzâtic* eye to do it with our quantic eye, while asking ourselves once again: "Would this venture help us obtain greater distinction and clarity to assist the reconciliation of Science and Religion?" It is the quantum unit, this *yât* of Nature by which Science is overwhelmed today, that will guide us to the answer, as the Star guided the Magi to Bethlehem. Let us move on to the 'quantum tomb'.

PART III

Quantum Mechanics:

From the Birth of the Photon to the Tomb of the Light

Chapter V
Planck and the *quanta*[545]

"After the experiment of modern physics, our attitude toward concepts like the human spirit, the soul, life, or God, will be different from that of the nineteenth century."

Werner Heisenberg
Physique et Philosophie[546]

The patristic era, the era of Saint Ephrem, which laid the foundations of the Christian religion, extends over the first seven centuries of the Christian calendar. The quantum era, which is ours, has been underway since the beginning of the 20th century. Max Planck inaugurated it in 1900, and the Copenhagen Interpretation in 1927 represents one of its landmarks.

Figure 14: Participants in the Solvay Congress [547]

---

[545] Quantum is the Latin word for *amount*. In modern understanding, it means the smallest possible discrete unit of any physical property, such as energy or matter. Quantum came into usage in 1900, when the physicist Max Planck used it in a presentation to the German Physical Society. Planck had sought to discover the reason why radiation from a glowing body changes in color from red, to orange, and, finally, to blue as its temperature rises. He found that by assuming that radiation existed in discrete units, called *quanta*, having a value represented by *hv*, where h is Planck's universal 'constant', whose value is Js, and *v* the frequency of the radiation. This theory is at the base of all modern physics, even postmodern physics.

[546] Morrannier, Jeanna. "La Totalité du Réel, La Matière et l'Esprit", V, Lanore, Paris, 1991, p. 10.
Cf. https://books.google.com.lb/books?id=4d2QBL5I5dgC&lpg=PP1&pg=PA9#v=onepage&q&f=false [Accessed August, 2019]
Cf. also, *Dieu et La Science*, pp. 168-169.

[547] A souvenir photo of the Solvay Congress (1927) held in Brussels, during which the

It was not just the beginning of an era of physical science, but rather the onset of transmutation at the level of the collective comprehension of reality, matter and universe alike. This was thanks to a group of scientists who were able to marry mathematics and physics. The entire positive cognitive phenomenon confronted two new challenges: on the one hand, Planck's quantum, and on the other, Einstein's relativity.[548] Many more discoveries marked by close collaboration with other geniuses followed, especially the Big Bang theory.

For a long period, the 'molecule' reigned as absolute master, and space and time imposed themselves as two distinctly separate and absolute blocks limiting all scientific experimentation.[549] Now it was time for Planck's subatomic quantum and Einstein's space-time theory to occupy the royal seat.[550] Energy revealed itself, the fourth dimension materialized, and cosmic gravity came into play at the epistemological horizon of humanity.[551]

We have two reasons behind our particular interest as 'mystics' in Einstein's theory of space-time (*Raumzeit*): first, to highlight Saint Ephrem's way of reading it and then to evaluate its impact on the theory of Creation according to the Bible;[552] second, to grasp the Kabbalist interpretation of space-time when it comes to answering the question, "Where did God create the universe?"

This *Raumzeit* theory also imposes itself as necessary for locating the 'meeting place' (the Tent of Meeting) for Science and Religion for which we are looking.

Copenhagen Interpretation, the foundation of quantum physics, was established. Some of those who appear in this photo "should" be considered as prophets of a quantum theology. (are consisered prophets at the level of the *râzâtic* theology)

[548] *Cf. E.U.*, "Espace-Temps". Article written by Jean-Pierre Provost, Marie-Antoinette Tonnelat.

[549] *Ibid.*

[550] *Cf. E.U.*, "Antimatière". Article written by Bernard Pire; Jean-Marc Richard.
The beginning of the twentieth century saw the birth of the theory of relativity, which modifies our conception of space and time, establishes the equivalence between mass and energy, and corrects classical mechanics when high speeds are at stake. The other great progress results from the apparition of quantum mechanics, which deals with microscopic processes. (Translated by the author)

[551] By analogy, we would say that such a dimension was concretized and entered into action in the Triune God's horizon at the very moment of the Incarnation. *Cf.* Mk 13:32 and Jn 10:30.

[552] Jesus, therefore, realized salvation by staying, consequently, in Mary's womb, in the Jordan River's 'womb' and in the Tomb's 'womb'. Moreover, this salvation is considered as having already been thoroughly realized in the time of the first event, because the past, the present, and the future meet in an eternal Present of the liturgical time. *Cf.* Brock, *La Harpe de l'Esprit, Florilège de poèmes de Saint Éphrem*, commentaire des *Hymnes sur la Nativité*, # 11) p. 244. From now on, we will refer to this work by *La Harpe de l'Esprit*.

However, our direct interest shall first be oriented toward quantum mechanics in general, for the same reason raised by Heisenberg in the above-mentioned quotation. We do not claim to be experts in modern physics. We will count, very humbly, on the great minds of this science, in all its branches, to lend us a hand so that we can draw the framework of that 'meeting place' with the best common understanding. To do this, we propose our theological contribution to those Nobel Prize-winning perceptive minds, the same way they propose their contribution to us. It is through the 'word', 'writing', and the different forms of symbolic and inspiring expressions that we do it. We eagerly desire that the information provided by modern Science, especially by quantum physics, be considered a source of inspiration for Religion, and vice versa. Indeed, this is what Saint Ephrem, *Doctor Ecclesiae,* meant by drawing attention to Nature as the second Sacred Book opened to humans by God. It is from this perspective that we approach the data of quantum physics since it represents one of the two bases of the 'Luminous Bridge' mentioned in the title of this book.

By his theory of Special Relativity, Einstein performed the incredible transmutation of 'space' and 'time' into 'space-time'. This new concept is consolidated by 'c', the symbol that represents the speed of light, which led to a new natural dimension untapped until then. Consequently, the traditional conception of two absolute and fatally distinct blocks of space and time, on which the metaphysics of theistic religion had carelessly rested since the waking of the human mind, will collapse. The strength of the human understanding and intellection will find itself unable to recapture the flat reality on which it rested and that laid the foundation of its habits, values, and especially its spiritual aspirations. Conversely, it will have to recognize and get used to a so-called quantum reality, multidimensional, uncertain and unpredictable, in which it will not know where to put its foot.[553] However, in the pragmatic sense of things, this strength of understanding continues to function.[554] But, at what cost?

---

553   Hawking Stephen, *A Brief History Of Time*, electronic book: https://archive.org/details/ABriefHistoryOfTimeByStephenHawking/page/n37 [Accessed Sep 2019]; From now on, we refer to this electronic book as eHawking. We will keep the French references for more accuracy: *Une Brève Histoire du Temps: Du Big Bang aux trous noirs,* Collection Champs science, Flammarion, 1991, 2004, pp. 29-30.

554   For more epistemological precision, it is necessary to distinguish, along with Stephen Hawking, between "the general relativity which describes the force of gravity and the large-scale structure of the universe, up to 1024 km", and "quantum mechanics, which is interested in phenomena with extremely reduced scales as at 10^{-24} of the centimeter". *Cf.* eHawking, p. 7; Hawking, pp. 30-31. The first adds nothing to Newtonian physics but develops it in four dimensions and creates a new unit of measure that is the speed of light, while the other is unheard of and deals well with what Guitton, in his book *Dieu et la Science*, took for a meta-reality that joins the infinitely small to the infinitely big.

To answer this question, we will tackle the major causes of this understanding under the following points:

1. Max Planck (1858-1947)
2. The realm of *quanta*
    2.1. The "blackbody"
    2.2. Light: definition and behavior
    2.3. Light: nature and constant
        2.3.1. The intervention of Einstein and Bohr
        2.3.2. Young's double-slit experiment and the dual nature of light
            2.3.2.1. The double-split experiment
            2.3.2.2. The dual nature of light
        2.3.3. Scientific symbolism
        2.3.4. The value of anti-light
        2.3.5. The wave dance
        2.3.6. The *gluon*
        2.3.7. The interception of the light beam by an opaque disk
3. De Broglie's wave-particle duality
    3.1. Again and again (*sic*), interpretation[555]
    3.2. The uncertainty principle of Heisenberg and "Schrödinger's cat"
    3.3. Reality according to Heisenberg's principle
    3.4. "Schrödinger's cat" in the tomb
4. A different approach to the paradox of "Schrödinger's cat"
    4.1. Moses at Horeb: strangeness
    4.2. Laplace's intervention
    4.3. Planck's intervention
    4.4. Einstein's intervention
    4.5. Schrödinger's intervention
    4.6. Guitton's intervention
    4.7. Ephrem's intervention

Conclusion

---

555 'Again and again' (*sic*). An expression taken from the Maronite Liturgy as a sign of accentuation. A similar emphasis was used by Jesus Christ when He said, "Amen, Amen".

1. Max Planck (1858-1947)

Max Planck, the son of a pastor, was known for his religiosity and seriousness. He was born in Kiel, capital of Schleswig-Holstein, on April 23, 1858. He belonged to a line of old bourgeoisie and was the sixth child of a law professor.[556]

In 1879, at the age of twenty-one, he presented his dissertation on the second principle of thermodynamics, which earned him the opposition of Kirchhoff, the father of the 'blackbody' theory. The following year, he passed his *habilitation* thesis on "The Equilibrium States of Isotropic Bodies at Different Temperatures".

The summit of Planck's scientific career was in 1899-1900, when he concentrated on the then-current scientific problems, especially radiation emitted by a 'blackbody'. As an independent spirit and deeply informed of the importance of "the notion of entropy for everything that concerns energetics", he discovered a possible formula for interpolation between the indications deduced from the experimental results. In this formula, the constant h, a new universal constant in physics, appeared for the first time.[557]

Planck's formula rapidly proved to be satisfactory, and since then, h, known as Planck's constant, has later been demonstrated several times to be 6.625×10^{-34} Js.[558]

---

556   *Cf. E.U.,* "Planck (Max) 1858-1947". Article written by Josef Smolka.

557

$$u_v = \frac{8\pi v^2}{c^3} \times \frac{hv}{e^{(hv/kT)} - 1}$$

558   We have already mentioned Max Planck in the notes of the previous chapter. By reading the work of Jean Guitton with the Bogdanov brothers, we were referred to his discovery that inaugurated the quantum era at the beginning of the 20th century (December 1900). The mention of a famous physicist like Planck by a philosopher of the stature of Jean Guitton has made us more convinced that the collaboration of philosophy with Religion and Science is always possible. Philosophy is believed to be able to take stock of the issues that continue to multiply like humans and their ideas. Accordingly, we call for the admission, at the epistemological level, that what is new, as we understand and conceive it today, is no longer located in the discovery in itself because once discovered, the discovery is no longer in question. Instead, what is new resides in the questions raised by any discovery. That Guitton, starting from modern physics and its virtuoso findings, has proposed a journey in the *meta-real* toward a *meta-reality* in the light of the wonders of quantum physics is only a sign of support for what we have proposed as a *râzâtic* reality. After all, what has preoccupied Guitton's mind and pushed him in search of the Bogdanov brothers, or vice versa, is the answer to a question never considered before this era. We formulate it as follows: "Is there any different reality between the physical world and the metaphysical one, a reality of which neither the 'Greeks' nor the Bible had any idea?" In any answer to such questions, Planck, his constant, and the quanta must now take part.

2. The realm of quanta

The importance of Planck's discovery, introduced on December 14, 1900 during a session of the German Physics Society in Berlin, does not lie in a formal operation or mathematical skill. It resides instead in the revolutionary interpretation he advanced to unveil the physical meaning of the constant *h*. From the beginning, Planck gave it the name of "elementary quantum of action". He did this because it has, at the same time, the two dimensions of an action (J) multiplied by a time (s) and, in short, it only occurs in integer multiples. This idea was indeed introducing a granular composition where all physicists thought that continuity reigned.[559]

This quantum of action, called Planck's constant since then, is used to measure the granular character of energy exchange. Although Planck was not immediately aware of it, this irruption of discontinuity in the problem of the blackbody sounded the death knell of classical physics and ushered in the quantum era.[560] But what is a blackbody?

2.1. The "blackbody" [561]

According to Merriam-Webster Dictionary, a blackbody is an ideal body or surface that completely absorbs all radiant energy falling upon it with no reflection and that radiates at all frequencies with a spectral energy distribution dependent on its absolute temperature. An example of a theoretical blackbody is an isothermal enclosure with a tiny opening. Thus, a spherical box with one small aperture is generally a good approximation. Each of our eyes presents a good example, even if it is not with the same perfection of absorption.[562]

---

[559] *Cf. E.U.*, "Planck (Max) 1858-1947". Article written by Josef Smolka.

[560] *Ibid.*

[561] We purposely put the name "blackbody" in quotation marks. It is to draw attention to the fact that Planck did not use a natural black physical body for his specific research on energy, from which he discovered the secret of its emission, radiation and propagation. It is instead a theoretical blackbody, without ever denying that the experiments made on the heating of a black physical body, such as an iron rod, were the precursors of that of the theoretical blackbody.

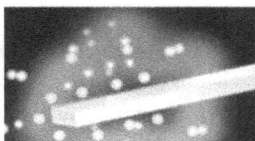

[562] Could we not compare the tomb of Christ, the ultimate end of His journey in this world, as well as the light described as springing from it, to this kind of blackbody?

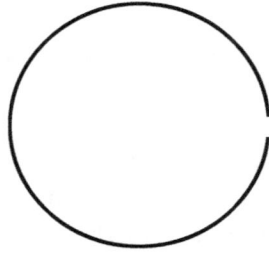

Figure 15: A theoretical blackbody

By definition, any black colored, metal or other body, which is subject to real objective experimentation and whose behavior seems to derogate from the laws of thermodynamics and electromagnetism is called a blackbody if it completely absorbs the radiation received and then emits it in the form of energy or light.[563] It has been proven that the emission of light by incandescence is the same, on an equal surface, for all blackbodies and superior to that of all non-blackbodies.

We limit ourselves to this amount of information on the theoretical blackbody. For the moment we prefer to continue acquiring a better knowledge of the radiations emitted by blackbodies and the very nature and behavior of these radiations, especially under the luminous state.

2.2. Light: definition and behavior

Visible light is what allows our eyes to see. It has the sun as the primary source for our terrestrial globe, but it is also emitted by any fire, as well as by the incandescence of specific bodies heated to high temperatures.

Before Planck, theories indicated that the spectrum of radiation emitted by a blackbody was such that at a given frequency, the amount of energy emitted was to increase indefinitely at such a rate that the total energy radiated would be infinite. Such a supposition proved to be in flagrant contradiction by experiment. By going beyond such paradoxes, Science modifies its theories and concepts. Max Planck supposed that the amounts of energy exchanged between matter and radiation are 'discrete' and not continuous. By that, Planck brought a solution to the blackbody paradox. This was the trigger for what would become quantum physics, revolutionary in many ways.[564]

---

563 *Cf. E.U.*, "Couleur". Article written by Pierre Fleury, Christian Imbert.

564 Discrete in Science is the opposite of continuous: something that is separate; distinct; individual. *Cf. E.U.*, "Paradoxe". Article written by Yannis Delmas-Rigoutos, Étienne Klein. Discrete, here, in the physical and mathematical sense, refers to a quantity consisting of distinct units (as opposed to continuous quantities) of a variation proceeding in whole quantities. In computing, the latter is equivalent to 'digital' and 'bits'.

Max Planck has proved that a blackbody re-emits in a discontinuous way the energy that it has absorbed in the form of electromagnetic radiation. The amount of energy re-emitted depends on the temperature of the blackbody. This is what has been known since as the 'Blackbody Radiation Law', which allows measuring the value of the energy emitted according to temperature.

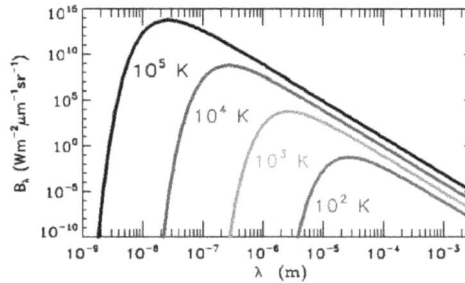

Figure 16: Curves showing a range of wavelength peaks
of blackbody radiations, under a variety of temperatures.

This demonstration that deals with the proper behavior of energy as well as light is vital to our contribution. Saying that a blackbody could emit light calls to mind the importance of the analysis done in previous chapters. This analysis tackled all that relates to the 'Incarnation of Light' and to the absurdity of the 'empty tomb' theory, once emptiness is taken in the absolute sense.[565] Given its importance, we will develop this point to be able to elucidate the nature of the light emitted by blackbodies, as well as the nature of light in general.

2.3. Light: nature and constant

From what preceded, we have learned that any real object, metallic or of another material that can be subjected to feasible experimentation, is a blackbody if it fully absorbs the radiations received and then emits them in the form of energy or light. Any piece of iron in a blacksmith's forge can become a source of light once energized (heated). What kind of light is this?

It is a light emitted in small quantities (quanta) which have the value of Planck's constant (h) and which are transmitted by waves. From now on, the description of the nature of light will be in question. This description is what should have reigned first in human understanding, from the day of Creation as soon as God commanded, "Let there be light," **for if it were not so from the beginning, light would never have been what it is today.** What do Einstein and Bohr have to say about this?

---

565   Does this not give rise to the question, for example, of whether or not the tomb of Christ continues to emit energy into the universe?

2.3.1. The intervention of Einstein and Bohr

With the development of the theory, thanks to the successive contributions of Planck, Einstein, then of Bohr, the graduation of the level of energy generated by heat is classified according to the following schematic:

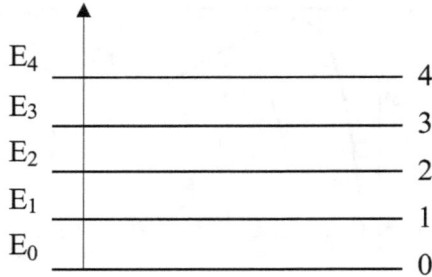

E_4 ———————————— 4

E_3 ———————————— 3

E_2 ———————————— 2

E_1 ———————————— 1

E_0 ———————————— 0

Figure 17:   Table of quantified energies

The energy levels at which the atom will be carried determine its behavior and that of its constituent particles (*i.e.,* electrons, neutrons, protons). In the fundamental state, the energy of the atom is E_0. The higher the state of energy reached by the atom, the more excited it becomes. It turned out, from these experiments, that an atom could remain in an excited state only for a quantum time, circa 10^{-8} s, after which it must regain its stability. This is why, we say, it de-energizes by emitting photons of energy ($h\nu$ where h is Planck's constant and ν *(nu)* the frequency of the emitted light) until it returns to its original state: thus, light is generated.

Therefore, it becomes evident that for light to come out of iron, *e.g.,* from an electrode, we need a source of energy, *e.g.,* fire (*nurâ*), embers (*gmurtâ*), oil (*naphtha*) or electricity, *etc*. These sources of energy make it possible to have a source of light which could reach quite high intensities, as in the case of the cathode arc of a film projector in a movie theater or the light emitted by arc welding.

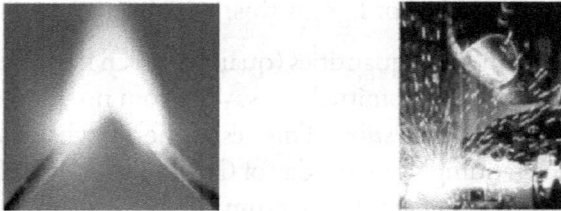

Figure 18:   1- Electrical Arc of 3000V between two nails;   2- Welding

We can say that the light emitted by the sun, or by methods similar to the two mentioned above, has a light intensity so high that the naked eye is incapable of discerning its graduation and propagation. This result was the basis of the impression humanity would have about light until Newton. It is sufficient to put oneself inside the 'theoretical blackbody', this enclosure equipped with a pupil-like opening, and to observe the beams of light which enter by this opening into the dark environment to have the impression of seeing a block of light wrapped in black.

Figure 19: Isaac Newton

The aspect of the luminous block that pierces the black block stands behind the impression of the incompatibility between light and 'darkness' reflected in the creation narrative of the Book of Genesis. Henceforth, the aforementioned physicists have abolished this impression. It would be interesting to consider rewriting the biblical text in a way that respects modern Science, or to think, at least, of elucidating the challenges of its symbolism and the dimension of the *Sitz im Leben*, that is, of the particular socio-historical context which often accompanies "writing" and "cultures".[566]

In the preceding chapter, we have already explored the various deductions that the Gospels have elicited from the emission of the Light from the tomb. Let us immediately explore the analysis that Planck, his predecessors, and his successors, have succeeded in making of their experiments on radiation. We do it to recognize the extremes before which Planck once stood.

---

[566] It is no longer permissible, as we wrote earlier, to teach children that the sun rises every morning, even if that is the impression that all children have.

2.3.2. Young's double-slit experiment and the dual nature of light

Almost one hundred years before Max Planck, another scientific genius, Thomas Young (1773-1829), had already demonstrated in 1801 the wave-like character of light in the so-called double-slit experiment. He deduced that light waves are capable of cancelling and amplifying each other. However, Young could not explain why. Let us review the experiment:

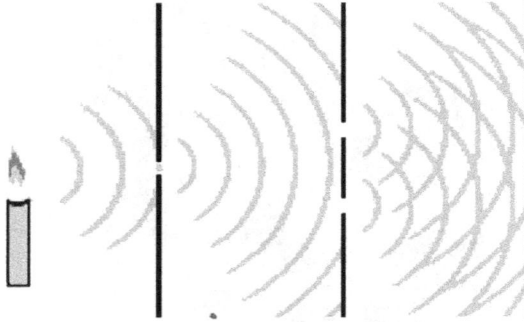

Figure 20: Diagram of Young's double-slit experiment. The orifices are in the λ range, and the distance between the interception walls must be much higher than the one between the slits.

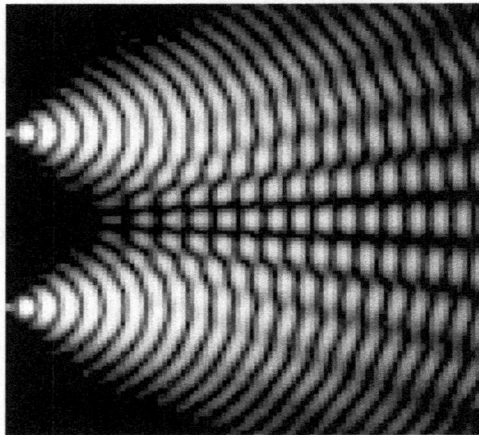

Figure 21: The propagation is produced by an electronic simulator.

Up to the threshold of the nineteenth century, the nature of light was still unknown. Once totally visible, light divides into visible and invisible as soon as it passes, under particular conditions, through a double or multiple slit opening. Even if we intercept a free ray of light with an opaque disk, the same phenomenon occurs.

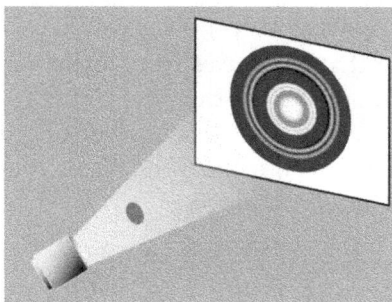

Figure 22: Rings caused by the interception of a ray by a disk

Out of scientific curiosity, let us follow the description of the double-slit experiment.

2.3.2.1. The double-slit experiment

Consider figure 21 and the following description given by Steven Hawking in Chapter 4 of his book *A Brief History of Time*:

> Consider a partition with two narrow parallel slits in it. On one side of the partition, one places a source of light of a particular color (that is, of a particular wavelength). Most of the light will hit the partition, but a small amount will go through the slits. Now suppose one places a screen on the far side of the partition from the light. Any point on the screen will receive waves from the two slits. However, in general, the distance the light has to travel from the source to the screen via the two slits will be different.

Figure 23: Numerical simulation of the interference fringes

This will mean that the waves from the slits will not be in phase with each other when they arrive at the screen: in some places, the waves will cancel each other out, and in others, they will reinforce each other. The result is a characteristic pattern of light and dark fringes. [567]

The following schema clarifies what we mean:

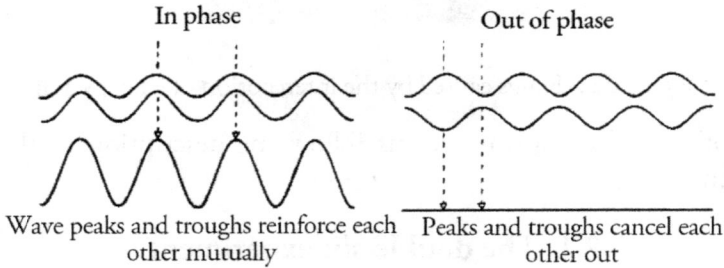

Figure 24: Constructive interference   Destructive interference

The   experiment was the starting point of transmutation at the level of understanding and intellection of the nature of light of whose wave-like characteristics we have just presented proof. Let us look deeper into this nature of light.

2.3.2.2. The dual nature of light

Let us go back to the revelatory figure that puts the two stages of Young's experiment in succession:

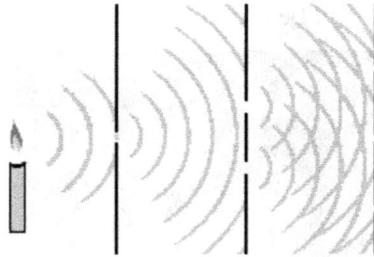

Figure 25: The revealing difference (more accessible schema)

According to this schema, a proportional brightness spreads gradually in the

---

567   *Cf.* eHawking, p. 32. [accessed Sept. 2019]. *Cf.* also *Dieu et la Science,* pp. 120-124. *Cf.* also the French copy: Hawking Stephen, *Une Brève Histoire du Temps: Du Big Bang aux Trous Noirs,* Collection Champs science, Flammarion, 1991, 2004, pp. 29-30. From now on, we will refer to this book by Hawking.

first chamber, as if a lamp lights it. One might say, "Light ousts darkness."

In contrast, for the second chamber, when the light passes through the two slits of the receiving wall, whether the slits are longitudinal or round, instead of illuminating the whole, the light produces on the receiving screen a pattern of bright fringes, the intercalation of light and darkness, a sign of resistance of some darkness to light.[568] How do physicists explain this? According to Huygens's discoveries in 1690, this phenomenon is due to the interference of the waves of light between themselves.[569]

Strange! At this point, a series of questions come to mind. This time they concern the nature of the 'blackness' dominating the room, 'darkness' itself. Could 'darknesses' have a specific nature in themselves? Would it be necessary, in order to identify the nature of light, to know beforehand the nature of 'black', the 'dark'? Could darkness be the 'mysterious milieu' existing before the *râzâ,* the Word of God, and which continues to contain the cosmos with all its lights and to provide it with gravity?[570]

We will return to this point in the following chapters. For the moment, let us continue our analysis concerning the nature of light.

Energy is the source of light. According to Einstein and Bohr, energy becomes visible only when under excitation of the atom, the frequency of the emitted wave, multiplied by Planck's constant, falls between the extremes of the optical spectrum of visibility. It is thanks to the equation $E = h\nu$ that it will be proved that the propagation of the energy is also the propagation of a physical entity of corpuscular aspect, later called the "photon". It is a discovery that we consider prophetic for the impact it has on the prologue of the Gospel according to Saint John, as well as on the concept of Light itself in the Gospel. The same impact will apply to the Creed of Christianity, as well as to everything related to the Christological Light and, in our case, to the *Hymns to Light* of Ephremian inspiration.[571]

$E=h\nu$ is, therefore, the equation necessary to calculate the energy contained in

---

568 We recall the very demanding conditions (mentioned above) for the success of this experiment.

569 Here we insist on honoring Huygens (1629-1695), who deserves the title of founder of wave physics by reproducing one of his most important discoveries in the field of light exploration. *Cf. E.U.,* "Onde (physique)". Article written by Mikhael Balabane and Françoise Balibar.

570 *Cf.* Gen 1:2.

571 Jn 1: 9-11. "The true Light [He; the Word] which [Who] enlightens everyone, was coming into the world. He was in the world, and the world came into being through Him, yet the world did not know Him. He came to what was His own, and His own people did not accept Him." If they did not do so, it was because they expected some customary light, at least intense enough to blind their enemies, but never of a "differAnt" nature and equally beneficial to everyone.

each photon. How mystical this equation seems!

It is from this subatomic level of measurement that the wave propagation is calculated, and the quanta of energy are carried by the respective wave and are detected under well-defined conditions. Thus, light made of photons would be detectable either as a set of particles or as a spreading wave, but never as both at the same time. Why is it impossible to detect the wave and the particle at the same time? Answering this question is also part of the analysis of the nature of light.

Igor Bogdanov describes this to philosopher Jean Guitton as follows:

> If I decide to verify experimentally that the photon is indeed a particle crossing a defined slit, then our photon behaves exactly like a particle passing through an orifice. On the contrary, if I do not strive to follow the trajectory of each photon during the experiment, then the particle distribution on the screen eventually forms a wave interference pattern.[572]

Étienne Klein, the author of the book *Petit Voyage Dans le Monde des Quanta,* details this point. He asserts first that it is not possible to explain this, other than with approximate words meaning that the particles are 'disturbed' or 'disrupted' by the measurement. Then Klein ends up pushing his deductions, which he has classified into five lessons. He ends up declaring that the error does not come from the wave phenomenon and its components, but rather from the fact of allowing oneself to assume that one can simultaneously observe the interferences and identify the slit used by each electron.[573]

Who would believe this! The enigma is still relevant, and something symbolic derives from it. Faced with this enigma, Jean Guitton, already an octogenarian, exclaims that the theory of his master Bergson (that of the "spirit of matter"), which is also his, is now supported and that matter proves well, once again, that it has a "spirit".[574]

For the moment, we will no longer venture into the descriptive details of this experiment. It seriously presents us with the difficulty of dealing with a *minima naturalia* which is not unity, but duality, as Einstein predicted. We draw attention to the terms and verbs used in this experiment and the detailed description of its

---

572 *Dieu et la Science,* p. 172. Klein, Étienne. *Petit Voyage dans le monde des Quanta,* pp. 29-35. Both references give a well-developed scientific explanation. From now on, we will refer to the latter work by Klein.

573 Klein, p. 34.

574 The word 'spirit', here, is taken for intelligence. Ephrem would say it differently. He would say that matter reveals the *râzâ* of its Creator, *i.e.,* His Wisdom, His Word. *Cf. Dieu et la Science,* p. 142.

activity to infer that this double-slit experiment puts us in front of two components of the same unit: the wave-like and the corpuscular ones of the photon. From the point of view of physics, and until proven otherwise, they are neither separable nor fusible in an absolute unit. At the same time, and under their two distinct essences, they are neither detectable nor observable at the same time. They are only revealed on the back of the Platonic cave by their effect, to leave humans in a state of uncertainty, with a relatively broad ambiguity of discernment between what is real and what is not, *i.e.*, what is the shadow.[575] How was the human intellect able to reduce this ambiguity and proceed toward what provides better predictability and certainty?

2.3.3. Scientific symbolism

Physics thus proves that the behavior of light in darkness and the reaction of the eyes with both are more complex than we think. This is why this phenomenon has been, since Plato, a source of symbolism.

Figure 26: Fringes produced by a monochromatic light

This figure shows fringes of light and darkness that coexist. It calls into question the first figure of Genesis when the light was commanded to be in our world.

Thus, since we have learned, according to Genesis, that obscurity – in other words, darkness – does not stand up to the light, two questions come to mind: what makes darkness, in Young's experiment, resist the light? and what causes this interference between light and shadow on the screen?

It seems caused by a hidden kind of anti-light value. As a result of Young's experiment and the wave-particle duality of the photon, the importance of the black appears, as well as of the dark called, in our opinion, allegorically, if not falsely, 'darkness'. A new entity also comes out and solicits our attention: anti-light. It comes from the light/darkness dichotomy. This entity refers us directly to the moment of biblical creation with its temporal rhythm translated by the day/night, evening/morning intercalations. In simple terms, in comparison with antimatter, it is anti-light.

---

575   Is it not a phenomenon that reflects Christological dimensions?

2.3.4. The value of anti-light

Even the Creator, by ordering the light to be, has respected the existence of its opposite – its anti – which, on the one hand, highlights it and, on the other, seems indispensable to reduce its impact on creatures who have eyes to see. The light has to adapt to the spectrum of human sight. This 'anti' also seems a prerequisite to ensure the necessary graduation between bright and dark, day and night, light and darkness and this is meant to protect humans from blindness.

But the challenge of the duality being/anti-being is dangerous and must respect very critical rules, otherwise the whole rationality risks annihilation and, consequently, absurdity. This is what quantum mechanics affirms through Stephen Hawking's words. The latter states that:

> All the known particles in the universe can be divided into two groups: particles of spin ½, which make up the matter in the universe, and particles of spin 0, 1, and 2, which, as we shall see, give rise to forces between the matter particles. [576]

In order to maintain the vital equilibrium within matter, between particles and acting forces, nature has furnished them with an 'exclusion principle' that was discovered by the Austrian physicist Wolfgang Pauli in 1925. This exclusion principle says that:

> ... two similar particles cannot exist in the same state; that is, they cannot have both the same position and the same velocity within the limits given by the uncertainty principle. This exclusion principle is crucial to understand why the particles of matter do not collapse in a state of very high density, under the influence of the forces produced by the particles of spin 0,1 and 2... [577]

Pauli also affirms that:

> ... If the world had been created without the exclusion principle, quarks would not form separate, well-defined protons and neutrons. Nor would these, together with electrons, form separate, well-defined atoms. They would all collapse to form a roughly uniform, dense "soup". [578]

And since all mechanics are imperfect, quantum mechanics also needed the addition of perfection. Thus arrived Paul Dirac, in 1928:

---

576   eHawking, p. 38. *Cf.* also Hawking, pp. 94-95.

577   *Ibid.*

578   *Ibid.*

> Dirac's theory was the first... to combine in a single theory, the quantum mechanics and the special theory of relativity predicting, for the first time, that the electron had to have an anti-partner: the anti-electron, or positron. The discovery of the positron in 1932 confirmed Dirac's theory... and quantum mechanics recognized the obvious existence of any kind of 'anti'. [579]

By popularizing these discoveries, Hawking pushes the analogy to warn that there might even be anti-worlds and anti-people made of anti-particles. [580]

This analysis, once applied in Religion, could well influence the basis of moral theology. Forgiveness would have a highly interesting epistemological scope which would be an additional benefit for classical theology and not an abstraction. Hawking adds, not without humor: "If you meet your antiself, don't shake hands! You would both vanish in a great flash of light."[581]

Let us concentrate for the moment on revealing the 'mystery' of fringes made of light and anti-light. We will do this under the title "The wave dance". Why does the same light beam passing through two slits, or intercepted by an opaque disk, change behavior under the effect of its free passage, and how do we understand the light/anti-light coexistence?

2.3.5. The wave dance [582]

The interweaving of light rays, originating at the same time from various sources and in multiple directions without any clash or hindrance, has preoccupied modern physics since Huygens. The exploration of light raised various enigmas that added to the mythical one of Prometheus. In 1800, the phenomenon of the

---

579 *Ibid.*

580 This theory may explain the psychological foundations of the dichotomies anima-animus and sympathy-antipathy.

581 Analogically, it can be quite true in the case of "love at first sight"!

582 In the Kerygma of the Church, the concept 'perichoresis' is well placed to explain the relationship of intrinsic trinitarian Cohabitation, Generation and Procession. It is a "differAnt" dance between the three divine Persons. It is also a characteristic of Johannine theology (Jn 16:12-15). The divine-human 'dance' is nothing but its reflection. It is also a particularity of the religion of the Incarnation. Through all the miracles it contains, the Gospel affirms that the Word builds His relationship with the 'ear' step by step. The same applies to the relation of Light-eye. (However, it can be vice versa – Word-eye or Light-ear – but always by the Holy Spirit.) If we humans do not take into account this relationship, we will never be able to understand either revelation or the prophets, and especially not the history of Salvation. We will risk, therefore, misunderstanding the relationship between God and the universe that is, in turn, taken by this 'Love dance' between Creator and creature.

fringes, which is the amplification and cancellation between the waves of the same light, was added to this list of enigmas. Let us examine where this figure of the 'wave dance' comes from.

When the rays propagate from a single source in a free space-time, they go to infinity at the speed of light, which is about 300,000 km/s. However, once these rays are channeled through two or more slits, under exact experimental conditions (not necessarily in a dark room), their behavior changes.[583] An interlacing between the waves forming at the exit of the two slits dominates and dark 'reappears' to form an interference pattern on a receiving screen, regardless of the intensity of light. What is objectively going on? Where do shadows come from? Can we imagine darkness residing in light that prevails over it or vice versa?[584] Would it not rather be a challenge similar to a 'dance', whose rhythm and choreography would be prepared somewhere in creation, according to hidden laws, difficult to decipher, and among which Planck's constant would be one of them? This conception is what inspired our idea of the 'wave dance'.

Figure 27: Rhythm and symmetry recall

Moreover, the English proverb "It takes two to tango" helps us figure it out. There is also the music that has to accompany it. Besides, when dancing, is it not rude to trample on one's partner's foot and/or not to respect the vital space of other dancers?[585]

Using the metaphor of the tango, we will approach more readily the reality of light and its behavior, a reality that has taken hundreds of years to be elucidated and has left its impact on Philosophy and human knowledge in general, except theology.[586]

Theologically speaking, in regard to the apparition of the fringes, we categorically oppose any assertion by virtue of which darkness would prevail

---

583   Measurement of slits, the shape of slits, thickness of the wall containing the slits, distances separating the slits, and the two walls of emission and reception.

584   Mt 12:25.

585   *Cf. E.U.*, "Lumière et Ténèbres". Very interesting article written by Alain Delaunay.

586   After Teilhard de Chardin, a Jesuit priest and anthropologist who was forbidden to teach theology because of his unprecedented ideas, no one dared to approach the Theology-Science 'marriage' again. Georges Lemaître, the real father of the Big Bang theory, was simultaneously a physicist, cosmologist and Catholic priest who affirmed that Science does not contradict Religion.

over the light. Darkness, this 'free existent' that is, according to Genesis, present before light, seems to us playing the role of background, or of blackboard, in any experimentation concerning light and sight. This 'free existent' will have its share of analysis, thanks to the sign of Yin-Yang explained in the final chapter dealing with Creation.[587]

Denying, as shown by the pattern, that one side can prevail over the other remains the second supposition, that of a choreography which surpasses all human expectations. This also supports Guitton's affirmation that matter has a 'spirit' in itself. Let us go back to the experiment: The following two figures show how monochromatic waves, like the perfect ones of a laser, propagate through two slits:

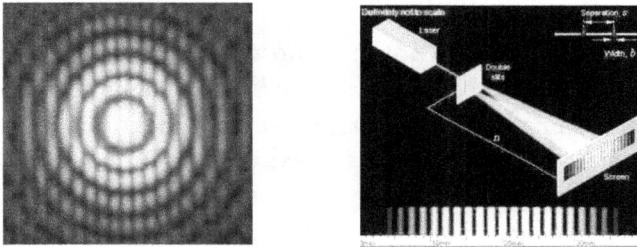

Figure 28 : Round slits=holes

Figure 29 : Laser Beam Ray and longitudinal slits

And the following figure shows the normal propagation of the waves following their interception by an opaque disk.

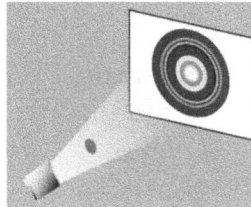

Figure 30 : Interception of a ray by an opaque disk [588]

Out of these figures, we deduce that just as there is rhythm in music, there are

---

587 *Cf. E.U.*, "Lumière et Ténèbres". Article written by Alain Delaunay. "All *gnoses* rest on this latent conflict. On the one hand, reigns the frightening realization of the darkness of the experience of the soul – 'Save me from matter and darkness', *Pistis Sophia* pleads. On the other hand, the glimmer of hope born from this very observation responds to it [the soul] universally, the star is the symbolic image of the saving light. In the night of the soul, only the guiding star shines (North Star, Shepherd Star, Magi, Alchemists' Sparkle, etc.)." (Translated by the author)

588 Courtesy of the *E.U.*

phases in waves. Allegorically speaking, the phenomenon that accompanies the absence of light in one area, then its reappearance in another, is seen as treading on the foot of the partner. In physical terms, this is expressed as "interfering in phase or out of phase".

When the peaks and hollows of two waves coincide, simultaneously, peak-to-peak and hollow-to-hollow (*cf.* Figure 24), they are in phase. They are added, they reinforce each other, and they give a maximum of light power. In the opposite case of perfect coincidence of the crest and the hollow, crest-to-hollow, they cancel each other out completely. Between the two extremes, they give way to a progression or regression of darkness relative to the degree of phase shift. The preceding figures of the fringes objectively show the results.

It would be truly shocking to admit, almost three thousand years after the writing of Genesis, that the dance of the waves of the same light may allow darkness to keep its vigor, according to a choreography whose hidden rules are previously established.[589] In other words, the light itself sets its own limits according to some hidden 'decrees'. Analogously, we allow ourselves to say that it self-applies a certain 'kenosis', a kind of voluntary submission of its perfection to binding conditions... Poor 'darkness'! It has nothing to do with it. It plays the role of the ribbon on which the oscilloscope inscribes the rhythm, the movement and even the vibrations of the waves, whether they are sound or light.[590]

The big question is: What would humans do if there were no black, night, or dark background? It has never been said that these last three concepts existed on their own. Despite all the analogy that the Bible emphasizes *vis-à-vis* black, night and darkness, Chaos and Sheol are in no way similar to them. Our eyes must be convinced that such analogies would belong to the realm of nomadic description. For Neanderthals, night and chaos would for sure be identical. Note that for people of the biblical era (c. 2500 BC), 'reality' could only be one of two possibilities: either the physical reality in two dimensions, as delivered by their eyes; or the reality advanced by dreams, underpinned by metaphysical sources and accepted as indisputable truth. They could not conceive of other types of 'reality'!

All this leads us to ask ourselves if darkness or 'darknesses' also enjoy a wave system and a choreography of their own which respects that of light. Would they be out of God's control? If they were, would this not jeopardize the omnipotence of God, an omnipotence that relates directly to the attributes of God the Father in the Creed?[591] The darknesses that form the two extremes of the spectrum of the

---

589   Sir 43:13; 45:5; Ps 148:6.

590   *Cf.* http://www.falstad.com/coupled/ [Accessed Oct. 5, 2019]

591   We wish to make known that at this point the Nicene Creed in Syriac or Arabic is much

color black, or the spectrum of color visibility as shown in the following figure, are made of waves, and are not so bad after all.

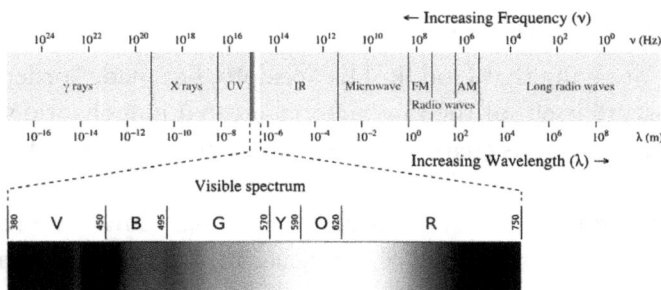

Figure 31: All is waves

In addition to their contrast which highlights both light and colors, they have proved to be useful to humanity, scientifically and spiritually, as much as the visibility waves, if not more. [592]

To our surprise, we have found somewhere, throughout the surface of the human imagination that now fills cyberspace and deals with quantum physics, an interpretation of the *Fiat Lux* command pronounced by God at the first moment of Genesis. This interpretation frees God from all linguistic determinism and intimates that God has made laws (formulas and constants) hidden in the universe and that it is up to humans to discover them. Instead of saying in Hebrew *Yehieh Or* or in Latin *Fiat Lux*, God would have somewhere written the laws and elementary constants discovered by Newton, for example, by Avogrado, Boltzmann, Maxwell, Planck, Einstein or Bohr – not just these laws and constants, but all those found by the ancient Greeks as well as the ones that will continue to be discovered until the end. He would also have developed the laws that manage the map of the DNA

more successful than European languages in expressing the 'control' that God the Father exercises over creation. While the *Catechism of the Catholic Church* teaches in #268, "Of all the divine attributes, only God's omnipotence is named in the Creed: to confess this power has great bearing on our lives. We believe that his might is universal, for God who created everything also rules everything and can do everything." Indeed, we believe. But, still, the Syriac and Arabic versions of the Creed emphasize much more the Controlling Creator using the attributes *aḥid ḳûl* ('holding all') in Syriac and *dabet al kull* ('controlling all') in Arabic, very well-oriented attributes. The *omnipotens* attribute, instead, is used in the opening Trisagion for the readings: Holy is God, Holy is the Almighty, Holy is the One Who never dies *(qadišat Alâhâ, qadišat ḥayeltânâ, qadišat lâ moyutâ)*.

592 From a 'non-nocturnal night' to another 'nocturnal night' *(Dark Night)* as it pleased Saint John of the Cross to call it.

that has lately been considered the imprint of God in the universe. [593]

Would there be a law for the dark? In other words, was the darkness of Genesis subjected to a specific divine law?

It is a way of saying that God, in His Wisdom, has made "order" an intrinsic principle of nature itself, of the raw material created in a chaotic state. The *yât* mentioned in the previous chapter is the nucleus of that 'order'. He will likewise make His Word, Whom He will send us later, a Principle of the same 'order' for all information that forms human knowledge which has remained in disarray until His Incarnation.[594] Those quanta of information, especially the ones that orbit around the wave-particle duality, play a contrasting role. The same can be said of the quanta of information which revolve around the impossibility of giving reassuring solutions to the enigma of the relationship and behavior of the two components of this duality. On the one hand, according to epistemological awareness, they make the dimension of disorder much more oppressive; hence, the sense of collective worry that reigns in today's world.[595] On the other hand, they reassure the collective consciousness, since they make more promising the horizons of taming this disorder. The fact is that Man, God's only partner, has already grasped the simplest of the divine Laws put into effect to 'rule' the universe. It implies greater responsibility on the part of humans to respect these Laws.

Indeed, it is worth noting that from now on the geniuses of quantum physics are able to preach a more precise and more distinct 'God', by answering the question "What God is it about?" Is He a God Who inspires certainty and

---

593   *Cf.* Lewandowski M.D., Ray. *The Imprint of God: Secrets in Our Genetic Code,* Morgan James Publishing, NY, 2008.

594   *Cf. Dieu et la Science,* pp. 114-115. We feel invited, in this case, to consider that all the wave-particles of which the universe is made are quanta of information, among others the divine laws. These laws, constants, axioms and rules are part of the *yâts* of the origins, since the *minima naturalia,* be it a particle or wave, cannot be admitted under the concept 'being' without an intrinsic law (an address) compatible with the Being as a Name as well as a Verb. All of them, known and unknown, form exhaustively the framework of knowledge, of Science, of conscience, as well as of wisdom and faith. There can be no substance (*itûtâ, Dasein*) without law. The problematic that opens up is to be able to guess whether God created a single law that undergoes the same expansion as the single 'created' particle that inaugurated the Big Bang, or whether He created none. In the latter case, the organizing *yâts* responsible for symmetry and harmony, being not created, must be seen as emanating from Him as the expansion of the material universe progressed. The Church is clear in this point: no 'emanation' occurs in the sense of overflow, without a deliberate and voluntary act. It contradicts the doctrine of Creation.

595   *The Quantum World,* p. 247.

predictability in a world that is becoming more and more constraining? They can answer this, inasmuch as they probe deeper into the nature of light in correlation with the nature of the human eye, perception, imagination and the world of signs, signifiers, signified and referent. In addition to all this, they must not fail to develop the only method of correct and sound interpretation.

So far, it can be deduced, thanks to Young's experiment and all the work about particles, photons, and above all, the wave-particle duality initiated by Einstein, that we must speak semantically and epistemologically of a 'Being', the creator of laws, principles, rules and constants, and moreover, that He does this in the service of initiating the existence of the creation and the process of its expansion in order, symmetry, and harmony from the chaotic state called 'dark' to the 'reasonable' state called 'luminous'.[596]

This phenomenon of being in order, symmetry and harmony – at the level of all dimensions – is revealed, as much as the "Information" of the Economy of Salvation, which began with the apparition of the duality Man-God, reveals Himself too. Physicists unveil it as much as theologians, with the simple difference that physicists act positively, by experimental deduction, with 'certainty', while theologians have continued to do it almost the same way for four thousand years, by induction. They base their hypotheses on the *a priori* of prophecies supported only by miracles and hierophanies, said to be supernatural, – among others the Incarnation, – as if they build *a priori* on previous *a priori*.

Thereupon, we have the right to wonder about what will become of us tomorrow, since we are quite sure nowadays that even the information that forms the fabric of reality, such as that unveiled to us by quantum physics, will not remain the same. At such questions, people resorted – if not by conviction, at least by habit – to Religion, to an 'all-purpose' God, to find reassuring answers for the heart. Today, taking into account the New Age movement as well as the

---

596   *Cf. Dieu et la Science,* pp. 50-51. Starting this moment, we find it convenient to introduce the reader to the role of the theory of emanation to which every approach to theology under the light-perspective phenomenon refers. The natural theosophies, which took the sun for God, were no farther than the Bible, we would say, from recognizing the true God – the Trinity – as Ephrem has explained it to them. Heat and rays of light emanate from every source of light. It is the crucial point on which quantum physics weighs in most. On this problematic, the Kabbalist theory does not deny a God Who deliberately changes form (*zimzum*) to make space, within Himself, for the creation (*Cf. E.U.*, "Luria"). The new point that we will explain in the following chapters wants the product of emanation to remain chaotic until the One Who allowed it to be enables it to be in a specific order that reflects His Spirit, His Reasoning, in forms that will emerge from chaos at His order.

awakening of the large Protestant Churches, we find that they make good use of the splendid discoveries of Science. They do so to assure their believers that a God (creator, architect, omnipotent, omniscient) controls the universe, and whatever changes there may be, a better future awaits the coming generations.[597] Thus Science, which sows anxiety and dread, becomes itself the reason for the growth of this assurance. Apophatic theology did the same thing in the past. The laws of order, symmetry and harmony established by the Creator guarantee that the information will no longer be the same, – as has been the case since the Incarnation of the Word, – without a "pilot wave"[598], as discovered by De Broglie in 1926, or without a "guiding wave", as David Böhm postulated almost thirty years later by his theory which established instead a field of information that imposes a trajectory onanyparticle.[599]

So, nothing is random. Speaking of symmetry and harmony on a large scale, as in general relativity, Jean Guitton raised a kind of "gauge field" to support the philosophical theory of his master Henry Bergson. After the breaking of the perfect symmetry, which triggered the imperfect and imperfection (which means the creation), something had to accompany this imperfect existence to maintain it in a specific order and balance. This imperfect existence had to be protected from collapsing, from crumbling, and especially from losing its purpose. The forces of interaction mentioned by Lemaître and their mediators (photons, bosons w and z, gluon, all of which are called gauge bosons) affirm it. [600]

Is there still indispensable information left to 'delimit' the Tent of Meeting (*Maškanzabnâ*) of a quantum or *râzâtic* nature?

Yes, indeed. It remains that all those theories which disclose the enigmas have to affirm the existence of a force, not subjected to the categories of this same universe, which undoubtedly manages it by imposing, scientifically speaking, certainty and predictability in spite of every imperfection and entropy.

Theologically speaking, it remains to assert that certainty and predictability are

---

[597] Klein, p. 144.

[598] De Broglie's theory through which he contributed to better seeing what stands behind the wave phenomenon in Young's experiment. *Cf.* Klein, pp. 199-200.

[599] Klein, p. 194.

[600] A gauge field theory is a specific type of quantum field theory, in which matter fields (like electrons and quarks, which make up protons and neutrons) interact with each other via forces that are mediated by the exchange of vector bosons (like photons and gluons, which bind quarks together in nucleons). The Standard Model provides a consistent theoretical description of all of the known forces except gravity. To know more about this theory, *cf.* https://www.hep.phy.cam.ac.uk/~gripaios/gft_lecture_notes.pdf [Accessed, Oct. 2019]

present at the level of the universal Redemption despite every persisting presence of evil and suffering.

It is like saying that the 'dance of the waves' leads to the discovery that a 'dance', following a specific 'rhythm' and whose choreography is determined by some "differAnt" gauge field, must also exist between God and Man, between Religion and Science. The Church's gauge field, in general, is the Gospel as She has lived and understood it before any interpretation, that is, Jesus Christ Himself, Informative Light and Luminous Information. The diffraction of the Informative Light in Luminous Information forms the chiasmus mentioned in the previous chapters. [601] This latter chiasmus belongs, according to Ephrem, to a 'mystery' as it were "differAnt" from the mystery evoked by the successors of the fathers of the Copenhagen Interpretation. It is indeed the famous *râzâ*. We may well say that since Ephrem, a new kind of quantification of Religion has come into play. [602]

On both sides of the 'gauge fields' deemed to maintain symmetry and harmony, whether quantic or Christological, especially the primordial symmetry between the forces of the universe, there appears a 'corrective gum' once hidden in the essence of the underlying order on which nature rests. [603] From a choreographic point of view, we can say that the dance and its ethics count intensely on its 'magnet'. Indeed, in subatomic physics, this is the gluon element that appeared in the fourth phase of the breaking of the Big Bang, as described by Lemaître. The gluon, as its name indicates, serves as a 'sign' or 'symbol' for a glue that maintains the relationship between subatomic particles and makes them unperturbable and unobtrusive. Given the importance of the gluon concept for the following chapters, we will allow ourselves to develop it a bit more, with the help of Stephen Hawking.

2.3.6. The gluon

Hawking also informs us that the formation of the gluon began with the Big Bang. It is one of the four elementary components that Science now sees as

---

601 Saint Ephrem by joining Nature to Scriptures in his poems and by casting the elements considered sacred, such as light, sun, oil, pearl, etc., only transmuted the 'deaf' reality of the first or second dimension into a meta-quantum reality, *i.e.,* into information of a "differAnt" nature than the quantum one. According to our contribution, we called it *râzâtic*.

602 Ephrem's quantum *par excellence* is the Host to which he will refer either by the 'Ember' or by the 'Pearl'. Moreover, the 'Pearl' in its perfection is the typical quantum, because it emits Information: "This pearl has satiated me, replacing all books, all their interpretations and all their readings." *(H Fid* 81, 8). *Cf.* Bou Mansour, *La Pensée,* p. 68; *TLE,* p. 106.

603 *Dieu et la Science,* p. 110.

building blocks of which everything is comprised.[604] He places the apparition of the gluon, without which there could not have been a universe, at the level of the fourth category of primordial elements that came out of the Big Bang according to the process specified by Lemaître.[605] They are:

- The first category: gravitational force

- The second category: electromagnetic force

- The third category: weak nuclear interaction

- The fourth and last category: strong nuclear interaction,

This latter force holds the quarks together in the proton and the neutron, and protons and neutrons together in the nucleus of the atom. It is believed that this force is carried by another spin 1 particle, called the gluon, which interacts only with itself and with quarks.[606]

Physically speaking, the gluon, which is of neutral charge and which is supposed to convey the strong nuclear interaction, has the role of keeping the particles of the nucleus of the atom attracted to one another in a certain symmetry, despite any distance that separates them.[607] One of its most astonishing

---

604   eHawking, Chapter 5; Hawking, p. 92.

605   *Dieu et la Science*, p. 169. Georges Lemaître, (born July 17, 1894, Charleroi, Belgium— died June 20, 1966, Leuven), Belgian astronomer and cosmologist who formulated the modern big-bang theory, which holds that the universe began in a cataclysmic explosion of a small, primeval "super-atom." In 1927, he proposed his big-bang theory, which explained the recession of the galaxies within the framework of Albert Einstein's theory of general relativity. Lemaitre's approach, as modified by George Gamow, has become the leading theory of cosmology. (*Cf. Encyclopedia Britannica Online*) [Accessed Sept. 2019] We wish here to pay tribute to a man devoted to the Catholic Church, Mgr. Lemaître, whose famous theory was sarcastically described as "Big Bang" by Fred Hoyle. (*Cf. E.U.*, "Hoyle F." Article written by Marek Abramowicz.) Pope Pius XI appointed Mgr. Lemaître in 1935 to the Pontifical Academy of Sciences. The following photo of the learned priest with Albert Einstein appeared in an American newspaper with the annotation, "They have profound respect and admiration for each other."

Lemaître and Einstein
Source (Internet): *New York Times Magazine*, 19 February, 1933.

606   eHawking, p. 38; Hawking, pp. 97-101.

607   eHawking, p. 40; Hawking, p. 102 and *Dieu et la Science*, p. 111.

characteristics is that the more spaced out the particles of which it is in charge, the more the attraction increases between them, and not the opposite.[608] The gluon is said to 'hold hard' behind the phenomenon of symmetry, the basis of synchronicity between all the components of the atom.[609]

Would the following deduction, considered as the nucleus of the quantum 'mystery' from Young's experiment, be understandable without the qualities of this gluon? Hawking himself describes this experiment:

> If electrons are sent through the slits one at a time, one would expect each to pass through one slit or the other, and so behave just as if the slit it passed through were the only one there – giving a uniform distribution on the screen. In reality, however, even when the electrons are sent one at a time, the fringes still appear. Each electron, therefore, must be passing through both slits at the same time![610]

Moreover, referring to string theory, Hawking informs us of the very presence of a "... collection of gluons whose colors add up to white. Such a collection forms an unstable particle called a glueball." [611]

Concluding his discussion of the curious property of the strong nuclear interaction that physicists have named "confinement", Hawking surprises us with his sincerity when he claims:

> "The fact that confinement prevents one from observing an isolated quark or gluon might seem to make the whole notion of quarks and gluons as particles somewhat metaphysical." [612]

We stop for the time being at this "metaphysical" nuance, at the gluon as well as at this comparison that for a "Christian mystic", even a *râzâtic* one, does not seem absurd at all. However, we will return to this in the following chapters, especially in the final one.

---

608 *The Quantum World,* pp. 80-81. (*e.g.,* By counter-analogy with the elasticity of the spring.)

609 *Dieu et la Science,* p. 84.

610 eHawking, pp. 32-33; Hawking, p. 85.

611 eHawking, p. 39; Hawking, p. 102.

612 *Ibid.* We note that Ephrem, given his categorical refusal to submit what is metaphysical, especially God, to human definitions, called for us to be satisfied with confinement. "Confinement", as a concept, would fit well this level of metaphysics, *i.e.,* to recognize God's realm from the confines that stand between the physical realm and the metaphysical one; the same separates the quantum reality from the *râzâtic* one. By this, we reduce the separating confinements to the maximum.

In our opinion, and taking into consideration the second Ephremian source of revelation, Nature, we can substitute the *Fiat Lux* by a series of God's laws and constants.[613] That series of God's laws and constants would allow the Energy produced by the divine Fire to heat the primordial *yât* and cause the Big Bang. Then, this same Energy would confine all wave-particles produced in a chaotic state into a 'milieu' of order and symmetry called, nowadays, a 'boson field', in which the gluon plays the aforementioned role. This series of laws and constants that manages gauge bosons, and specifically gluons, pushes all kinds of created particles and waves to join together and convert into various shapes and materials, starting with the four primordial elements: air, earth, fire and water (and ether in case it exists). As for the formation of energy and light at human scale, it is thanks to the same series of laws and constants that these two were able to be, and that Planck and Einstein were able to tame them. Consequently, it also became possible through the collective effort of all scientists to probe their secrets and begin the organization of their waves in quanta and ranges. Other laws would have been set by God to form hydrogen from the helium that preceded it at the Big Bang, then the rest of the creatures that the Bible witnesses. Ephrem, for example, states that God created, *"En arkhêi"* (*b-riš-it*), only the *yât* of 'Heaven' and that of 'Earth',

---

613   The Bogdanov brothers list the 'mysterious beings' whose unveiling seems to be preserved to God: gravity, time, space and the electromagnetic forces that manage the movement of the particles of the atom. These elements do not seem to have been created. The Bogdanovs say, "And it is perhaps there, at the heart of quantum strangeness, that our human minds (*esprit*) and the One of this transcendent Being whom we call God are brought to meet."(p. 131)

Furthermore, if we consider the Arabic saying "knowledge is a light" (*Al ʿilmu nuron*) and we contrast between the two parts of the light-darkness dichotomy, the light would be nothing but the *Noûs*, the "universal ability to think and act". Jean Guitton, following his master Bergson, asserts its belonging even to Matter: "the spirit of matter". (*Cf. Dieu et la Science*, pp. 25-26.)

That God said *Fiat Lux* would mean that He introduced sense into nonsense, order into the 'hustle and bustle' and the possibility of distinction and clarity where it had been impossible. All this remained in 'potency' (source of symbolism), in the form of tacit laws and constants, until it became in 'act' through the 'Rational Animal' that God is not going to 'create', but rather to model, from all the 'soup' of *yâts* already present. He made the human be the image and likeness of the creation (small cosmos) in order to be the image and likeness of the Creator of the universe.

As we said in the previous chapters, God modeled a partner for Himself, "... in the image of God created He him; male and female created He them" (Gen 1:27), so that this partner leads to perfection what one day was created in rough draft, with passion (receiver) and with action (transmitter or executor) in the best sense possible. Expressing this reality with a feminist joke would add the dimension of humor that could remain absent when it should not. Q: Why did God create Adam before Eve? A: God had to do the rough draft first. It makes sense, doesn't it?

and that all the rest was only organization and shaping.[614]

Figure 32: Refraction of light through a prism.
One of the laws that stands behind the rainbow, the sign of
the Covenant established after the Flood. [615]

Finding, discovering or revealing the rest of the constants and laws existing within the various parts of what exists, like DNA for example, and that stand behind any dance is for scientists the equivalent of wishing to come, one day, to what they consider to be the Grand Unified Theory (G.U.T.). It is the theory supposed to explain by itself all the forces that govern the universe.[616]

Thanks to Young's double-slit experiment, we have explained the phenomenon of light which became the basis of the early stages of the quantum era thanks to Planck and Einstein. We have unveiled the kind of energy and/or light we have to deal with, henceforth, and the spirit of matter that is confronting us. We have unveiled the cause of the apparition of the fringes and the implicit 'dance' between the dualities (and not just the dichotomies) in all that exists. We finished with the role of the last two elementary categories according to Lemaître, the weak nuclear interaction and strong nuclear interaction, represented on the one hand by the w and z bosons, and on the other hand, by the gluon.

At this stage, we deem it necessary to prove that what is true in the case of the interception of light and its channeling through two or more slits is also true if an

---

614 *Cf.* Ps 148:6. What is interesting in this verse of Psalm 148, which affirms the existence of all scientific discoveries in the Law (*nâmûsâ*) that God used to organize creation, is that it can have two different interpretations. The Jerusalem Bible and the majority of other translations, including the Syriac one, suggest that there is *"a decree which shall not pass away"* (*nâmûsâ yahb w lâ ᶜobar*). Louis Segond's version alone distinguishes itself by writing: "He has given laws, and He will not violate them." The "He" here refers obviously to God, and this is what physicists should have known for a long time.

615 Gen 9:12-17.

616 So we suggest that the G.U.T. must be applicable to the whole as well as to the parts of the Creation with its Creator, since even semantics requires it. (*Cf.* eHawking, p. 40). That is on the one hand. On the other hand, the physicist David Böhm thinks that matter, consciousness, time, space and the universe represent only a tiny 'ripple' in comparison with the immense activity of the underlying plane which comes, itself, from an eternally creative source that is beyond space and time. (*Cf. Dieu et la Science,* p. 49; Hawking, p. 103.)

opaque disk, free of any slit, intercepts the light beam. This experiment highlights much better the circular as well as the spherical behavior of light waves, a crucial phenomenon for our theory of creation.

2.3.7. The interception of the light beam by an opaque disk

Figure 31 shows a new way of seeing light diffraction according to Young's double-slit experiment. In this experiment, the rays prove that, whatever the obstruction which prevents them from continuing their course, they "enjoy" the feature of flexibility. The latter is based on the principle of reinforcement and cancellation, intrinsic to their waves. It allows them to overcome any restrictive obstacle.[617] What is interesting in this experiment with a disk which intercepts a light beam is that the circular and spherical base of the waves of the light is strongly apparent. The earlier description of this experiment, referred to as "the diffraction of a light beam by a disk on a screen", provides a characteristic figure in concentric rings. It is the same with quantum particles. It is said that either the waves of the first case or the particles of the second case are delocalized between the transmitter and the screen. As for us, concerning the first case, we prefer to say that it is the supposed 'choreography' of the dance of the light waves. It involves the reorganization of the photons to create shapes and to illuminate as much space as possible, even if with different intensities. Indeed, everything proves that our human eyes, so weak, can never bear the light in its full intensity and purity. It is preventive that they enjoy little instead of nothing. The wave and the corpuscular quality of the light remain the same before and after the shutter, whether in the double-slit or disk experiment. They allow the light to change shape, to take different layouts although everything stays in the box of the laboratory just as in that of the universe.

Let us see. The universe! Can it be said to be included in a box phenomenon, a theoretical 'blackbody'? If so, where could this spherical box exist in relation to the omnipresent God of the Bible? And then, while respecting the principle of quantum discreteness, we ask ourselves: why does light move in concentric waves following a sinusoidal path and not in cubic waves, for example, following an angular path such as that the painter Picasso, co-creator of the art of cubism, would have allowed himself to imagine? We will try to answer these questions as and when they arise, directly or indirectly, by reinterpreting the scene of Creation.

With this last experiment, we have reached all that we wanted to highlight *vis-à-vis* this point which deals with light, 'waves', 'particles' and their 'behavior' as quantum physics has revealed them to humanity. However, we are sure that the reader must have noticed that a new wave-particle duality has been established

---

617 Flexibility comes from the four qualities the wave enjoys: reflection, refraction, diffraction, and interference.

where we sought, under the principle of *râzâtic* triunity, "unity in plurality".

Let us see how the principle of wave-particle duality, initiated by Einstein for light and completed by De Broglie for matter, will put the 'train' of quantum interpretation back on a single track with two rails. It is just as the single particle in the experiment of the double slit was able to go through both simultaneously without losing its unity.

3. De Broglie's wave-particle duality [618]

For a physicist of the classical age, that is, before 1900, the distinction between matter and light was clear and sharp. This partition of the world into two categories of objects will already be shaken by Einstein's article published in 1905 under the title, "On a Heuristic Point of View about the Creation and Conversion of Light". In this article, Einstein shows that for the explanation of the thermal properties of light, it is necessary to consider light energy as generated in the form of grains $E = h\nu$. By this, Einstein introduced the notion of a quantum of light, later named the photon. In 1909, he postulated, at the level of luminous radiation, the theory of the wave-particle duality.[619] Would this duality also be applicable at the level of matter (atom, electron), and could it reach the universality necessary for a scientific principle? This is what Louis De Broglie succeeded in proving. Consequently, the honor of the principle of this duality is linked to his name. His starting point was the hypothesis that the wave-particle duality, established for radiation since the conception of the photon by Einstein, had to be extended to matter.

In collaboration with Schrödinger, De Broglie was able to develop parallelism between classical mechanics and optics, and thus achieve the design of wave mechanics. In this theory, the study of the motion of an electron – or any atomic system – must no longer be based on classical trajectories, Newton's solutions and equations. This study should instead be based on a wave associated with the electron, the De Broglie's wave, which defines a complex function of the coordinates of space and time, Q (x,t).

This duality, some dimensions of which remained to be unveiled, was considered subjective and deemed not to suit scientific progress. It had to wait until the 1920s to be transformed, thanks to De Broglie himself, into objective duality. This transformation could associate what he called a wavelength with each particle, according to the relation λ, known as De Broglie's relation, where ρ

---

618   *Cf. E.U.*, "De Broglie L.". Article written by Marie-Antoinette Tonnelat. "Electron". Article written by Jean-Eudes Augustin.

619   *Cf.* https://physics.ucf.edu/~ishigami/Teaching/Phys4083L/lab%20descriptions/NETD/blackbody%20theory.pdf [Accessed June 22, 2021]

symbolizes the quantity of vibrations also called momentum.[620]

The quantum theory of the electron began with the hypothesis that he made in 1923; it suggests that the wave-particle association observed for the photon is universal.

In 1924, he made a revolutionary proposition: that of also associating a wave with particles of matter. The frequency of the wave associated with each particle will be proportional to its energy, according to Planck-Einstein's relation $E = h\nu$.

In 1925, he developed wave mechanics by combining waves with the various particles (electrons, protons, etc.). We are, therefore, in the presence of a double aspect, presented by both light and matter. The connection between these two qualities (wave-like and corpuscular) led to a statistical conception which states that in the case of light as in that of matter, it is the wave that determines the probabilities of the presence of particles, photons, protons, neutrons, etc.[621]

The results obtained in diffraction experiments of electrons or neutrons, on the one hand, and in collisions of photons on nuclei, on the other hand, have amply demonstrated the veracity of this liaison or unification.

In 1927 and 1928, this hypothesis was spectacularly confirmed. By channeling electrons through a crystal, De Broglie obtained a diffraction pattern similar to that which light gives. As with optical waves, we can superimpose waves linearly (that is, by addition or subtraction) with different phases or frequencies. Doing this, we manage to consider more generally either "wave packets" or "wave functions" associated with each particle. The link between wave and particle is ensured by the following postulate: the probability of finding the particle at a given point is measured by the intensity of the wave.

Before De Broglie and the transition from dualism to duality, the interpretation of quantum theory became bogged down in false questions of the kind: are electrons (or photons) waves or particles, waves and particles, sometimes waves and sometimes particles, or waves that transform into particles or particles that transform into waves? [622]

---

620   *Cf. E.U.*, "De Broglie L."

621   Statistics: The science that deals with the collection, classification, analysis, and interpretation of numerical facts or data, and that, by use of mathematical theories of probability, imposes order and regularity on aggregates of disparate elements. Moreover, their passing into the laws of phenomena and theoretical models likely to represent them.
Cf. https://www.dictionary.com/browse/statistics and *E.U.*, "Statistique". Article written by Georges Morlat. (Translated by the author)

622   Can we not say, analogically, that the Christology of the first centuries suffered from such a situation?

The controversy ended when it was decided to admit that the classical concepts of waves and particles, with their own determinations, were no longer valid in what is known as the quantum field, numerically delimited by "*h*", Planck's constant value. Electrons, photons and, more generally, all quantum elements are neither waves nor particles but objects of a different nature, linguistically undeterminable.[623]

Judging this indeterminability as extremely significant for our work, it comes to our mind to exclaim with Jacques Derrida that this should be rather said to be of "differAnt nature", and to consider the concept "differAnce" as a marker between quantum and *râzâtic*.

This conclusion is just one more proof that the problem of semantics in Science is strictly linguistic. It gets closer to that of theology, and specifically of Christology, and supports our *râzâtic* theory.[624] It was necessary for physicists, from the beginning of the twentieth century, to start putting a new vocabulary into effect.[625] It would at least have reduced the problem of interpretation we have raised above. Is it true that even in a positive discipline like physics, experts can suffer from an issue of interpretation? Before moving on to the third point, Heisenberg's Principle of Uncertainty, let us say two words about this point that is crucial for our work.

3.1. Again and again *(sic)*, interpretation

In his book *Petit Voyage Dans Le Monde Des Quanta*, Klein raises this problem of interpretation at the quantum level. He devotes an entire chapter to discussing it and defines the problem as follows:

> As early as 1927, its founding fathers had to specify the rules according to which it [quantum mechanics] should be used. They were led, little by little, so to speak, to debate the type of discourse concerning physical reality that it authorizes or forbids. This new physics had something special in their eyes. By the fact that it questioned the "subject-object" pair, it was literally said to be "unprecedented" (as everyone wrote, each in his/her own way). Indeed, never before, at this point, had a science required another specific discipline, interpretation,

---

623   *Cf. E.U.*, "Ondes (Physique)". Article written by Mikhael Balabane and Françoise Balibar.

624   This truth is becoming more and more unanimous. *Cf. Petit Voyage*, p. 57. To escape it, the Bogdanov brothers and Jean Guitton had to push their interpretation of things a bit towards what is considered classical anthropomorphism, recognized and well criticized by biblical exegeses, such as by talking about particles having knowledge. *Cf. Dieu et la Science*, p. 127.

625   Klein, p. 144.

in order to be understood and applied. By its very structure, quantum physics questions the relationship between the world and its representation.[626]

It has been rightly said that the whole of the quantum phenomenon finds almost all its coordinates in Young's double-slit experiment, about which the American physicist Richard Feynman wrote:

It highlights a phenomenon that is impossible to explain in a classic way. It houses the heart of quantum mechanics. In reality, it contains the one mystery of quantum mechanics.[627]

It is just like in the Christian religion where one unique 'mystery' feeds all kinds of interpretation. Is quantum reality developing as Christianity has developed, beyond any definition, as Ephrem says, since its 'raw material' is indefinable? Moreover, could the Ephremian *râzâtic* contribution save it so that it can always be favorable to the creature-creator relationship, especially where the subject-object pair is inseparable as in the case of the Incarnation?

At this point, and thanks to the phenomenon of the synergistic interpretation that accompanies imagination and analysis, we can affirm with Klein that there is something new under the sun, even if it might displease the author of Ecclesiastes.[628] This is the principle of indeterminacy, more commonly known as the 'Heisenberg Uncertainty Principle'. The allegory of Schrödinger's cat will highlight it.[629] Let us examine together, while holding our breath, how far this 'cat' leads us.

---

626   *Ibid.* p. 141. The author adds a commentary on p. 144, saying: "Should the interpretation be a sub-theory within the quantum theory itself, as some physicists think today, or should it consist of a commentary inspired by previous philosophical theses...? These debates, often passionate, sometimes bitter, generally confused, have lasted for more than seventy years. Some discussions have even gone beyond the epistemological framework to venture into areas where metaphysics becomes indeed smoky." Klein whispers in one word "Fortunately", to then criticize the abuse made by "the New Age scribes who have seized quantum physics to build, with great support of syncretic plaster, a kind of holism that is both hybrid and unbridled, based on quantum non-separability, uncertainty, matter-energy, wave-particle duality, space-time, planetary cybernetics..." Then, Klein adds, "But I'd rather prefer to stop here the austere list."

628   *The Character of Physical Law*, Chapter 6. *Cf.* https://i-phi.org/2017/12/13/ richardfeynman-the-two-slit-experiment-contains-the-one-mystery-in-quantum-mechanics/ [Accessed 28 Oct 2019]. Reference mentioned also by Jean Guitton in the French book *Dieu et la Science*, p. 121.

628   Eccl 1:9; *Cf.* Klein, p. 142.

629   Klein, p. 51; specifically note 1, pp. 57-58.

3.2. The uncertainty principle of Heisenberg and "Schrödinger's cat"

The language must adapt to the facts and not the other way round. Trying to accommodate the interpretation of a phenomenon with a language already formed and filled with a priori can only lead to false conclusions about the nature of things.[630]

Ludwig Wittgenstein

In other words, this epigraph of Wittgenstein means that, unless we decide to initiate a linguistic and semantic transmutation such as that raised by Ephrem in the patristic era, we will be forced to admit our inability to meet the various demands of progress. Only this 'stubbornness of the gods' to which we hold, unfortunately, and the abuse of confidence in our superiority over all that is created (visible, invisible and meta-invisible) makes us capitulate. It is that, instinctively, we rejoice in dropping our scientific humility and the 'Mold' in whose form we were formed. Confidence in a proper idiom and a specific language sufficient by itself to tame all that is strange (from the 'fox' of Saint-Exupéry's Little Prince to Schrödinger's 'cat') any time scientific progress surprises us, is a pathetic and confusing confidence.

Given the delicate aspect of the principle of uncertainty or indeterminacy, we have chosen to scrutinize its applications on our *râzâtic* theory and to highlight the deductions that can be drawn from it at both the theological and Christological levels. However, readers will profit from researching Heisenberg's principle of uncertainty in the countless references in libraries and on electronic sites before criticizing our input. We content ourselves with reporting the key statements necessary to assimilate our angle of view.

At first glance, we draw attention to the fact that once De Broglie identified the *minima naturalia* with a quantum that enjoys the wave-particle nature, what is "new under the sun" in regards to Science began to exist. This new *minima naturalia* has also proved to be matter and light, neither in a complementary way nor in a consecutive way, and even less in an alternative way. It is so at once and not at once, as proved by Young's experiment when making sure which slit the quantum of light (the photon) passes through. The problem, as Heisenberg described it, is neither in the *minima naturalia* nor in the human mind. Rather, it comes from three major unpredictabilities in the whole system that accompany

---

630 *Ibid.* p. 51.

the experiment: the unpredictability of the space-time in which the measured element is located, the unpredictability in the perfection of machines and the circumstances in which the experiment takes place, and the unpredictability in the limits of the human being who conducts the experiment in respect to both physical and mental abilities. In other words, it is as if we hear Heisenberg say, "How can one claim to reach perfection in our positive knowledge when this world is full of imperfections, among others those of the human being?" Hence, his principle of uncertainty or indeterminacy, which was aimed at deterministic physicists, was aimed at all the physics of the eighteenth century. It was targeted above all at Laplace, representative of the century, who so praised the certainty and predictability of all the physical laws of nature, and pretended presumptuously, before Napoleon Bonaparte, to no longer need the 'God hypothesis'.[631]

Hawking backs Heisenberg's position expertly, saying exactly that, "The principle of uncertainty is an ineluctable fundamental property of the world."[632] He proved that in the process of measurement mentioned above, calculations are made on particles that no longer have well-defined or precise positions or speeds that can be observed. Instead, they have a quantum state, which is a combination of their situation and their speed. Consequently, it becomes absurd to promise oneself precise results that inspire certainty from calculations based on imprecise or uncertain data. In the best case, the result could be equivalent to Planck's constant but never inferior to it. Until further notice, in the quantum field only Planck's constant, and it alone, represents the sole determinant and certainty-inspiring datum.

The uncertainty principle has had a profound impact on how the world is viewed. Even after almost a century, its implications have not been fully accepted by several philosophers and are still the object of much controversy. An infinity of questions is generated by these new discoveries, as we mentioned above, one of which now haunts the human mind, that of how to predict future events accurately if it is not even possible to measure the present state of the universe with precision. Our guide Jean Guitton joins Heisenberg to make a relevant observation. He sees that, indeed, according to the discoveries of modern physics, our attitude toward concepts like the human spirit, the soul, life, or God will be different from that which existed in the nineteenth century. How much and in what way will it be different? Guitton leaves the answer to the care of the generations of 'quantum linguistics'. As for the question "What is it about?" which arouses embarrassment in the face of elementary particles that seem to behave like abstract entities, Guitton asserts that trying to find an answer means, "We have to abandon our world, its laws and certainties, as we must admit that the universe is not only stranger than

---

631   *Cf.* eHawking pp. 30 and 92; Hawking, pp. 79 and 216-17.
632   *Ibid.*

we think it is, but even stranger than we can think it is."[633]

Despite these difficulties at the philosophical level, including the problem of the "dice of God" raised by Einstein, most scholars were willing to admit quantum mechanics as determinative because it fits perfectly with the experiment. It underpins almost all modern science and technology. The only areas of physical science in which quantum mechanics has not yet been truly integrated are the understanding of gravitation and the large-scale structure of the universe. We will come back to the phenomenon of gravitation at the same time as the two phenomena of space and time that the Book of Genesis does not mention among created things, as was the case with the phenomenon of 'fire'.[634]

Before saying anything about the famous 'cat' discussed above, we close this point on Heisenberg's principle with a fundamental question. Now we have some idea of the repercussions of this principle of uncertainty on the way the world is viewed. Would this principle have an impact on 'reality', this concept so dear to our thesis, that is to say, on the way 'reality' can be perceived?

3.3. Reality according to Heisenberg's principle

The controversy between Philosophy and Science, as it turned out, has prompted some critics to address seriously the implications of this uncertainty principle on the way reality is considered from now on. Indeed, speaking of these repercussions, the story of Plato's Allegory of the Cave becomes the clue. It points to the famous question that we raised at the beginning of our work, dealing with the confusion between 'illusions' and 'reality'. According to a translation made by Washington University, we read it as follows:

> "And if they could talk to one another, don't you think they'd suppose that the names they used applied to the things they see passing before them?"

---

633 *Dieu et la Science*, p. 86.

634 It is true that Einstein's relativity is precisely the relativity of space-time, but there are two critical points which support our doubt:
1- In his 1905 paper, Einstein makes a critical analysis of the methods of measuring length and time. As a result, he draws his theory of special relativity which causes time and space to lose their previously undisputed status: no absolute space exists. *Cf. E.U.,* "Espace-Temps". Article written by Jean-Pierre Provost and Marie-Antoinette Tonnelat. (Translated by the author)
2- "All that we believe about space and time, all that we imagine about the location of objects and the causality of events, what we may think about the separability of things existing in the universe, all this is only an immense and perpetual hallucination that covers the reality of an opaque veil..." *Cf. Dieu et la Science,* pp. 108-109. (Translated by the author)

Plato's point here is that the prisoners are mistaken. They are using terms in their language to refer to the shadows that pass before their eyes, rather than to the real things that cast the shadows. [635]

It is evident that every time someone refers us to this myth, he/she aims to contrast the two philosophical currents that have dominated the subject for almost twenty-five centuries: the idealist and the materialist.

On the one hand, says Jean Hamburger, author of a comprehensive article in *Encyclopedia Universalis*: "The idealists, among others, solipsists ... declare that reality is only an illusion; that nothing exists except thought, that everything is subjective." On the other hand: "The empiricists ... judge not only that the world exists, external to us, but that it is appropriate to bear on it the most objective look possible, by twisting the neck to any subjectivity." [636] The author adds that between these two extremes, almost all other philosophers agree on a mixed reality, made of objective and subjective, born of a dialogue between Man and Nature, descriptive not of the world in itself, but of the world as observed by humans. [637]

With Heisenberg, a new era will begin because he will mention and prove that even what humans see through the eyes, as well as what might be observed from experiment through our analytical mind, is not exactly as it appears to be. The more we become aware of what we know and of the methods we use to reach knowledge, the more we shift toward uncertainty or indeterminacy. [638] Even the semantics we use have become archaic. It is semantics that must adapt to the facts and not the other way around, as Wittgenstein says. [639]

In this sense, Igor Bogdanov, who was during his television program trying to compete with Jean Guitton, does not hesitate to state that: "According to Heisenberg's principle of uncertainty, we do not observe the physical world: we participate in it. Our senses are not separated from what exists in itself, but they are intimately involved in a complex process of feedback, whose final result is, in fact, to create what 'is' in itself." [640] And a Christian "mystic" would add that, if it were not creating what 'is' in itself, it would at least be making everything that is in itself new, even language, by introducing a new meaning to it.

How significant this finding would be for people who, like us, are in front

---

635   https://faculty.washington.edu/smcohen/320/cave.htm - #6 - [Accessed Sep. 15, 2019] *E.U.*, "Réalité (Concept de)". Article written by Jean Hamburger.

636   *Ibid.*

637   There, we recall, lies Ephrem's strong point.

638   There lies Ephrem's second strong point.

639   There lies Ephrem's third strong point.

640   *Dieu et la Science*, p. 174.

of the tomb from which we have just come out! It urges us to ask: what would the resurrection be, as deduced from the empty tomb, since according to Heisenberg's principle, all the categories of 'real' and 'unreal' vanish? Would it be real or illusory? Could Schrödinger, who comes with his 'cat' to help Heisenberg, calm the worries of the Magdalene who cried out in front of Peter and John: "The Lord has been removed from the grave, and we do not know where He was put!"[641] Could he answer the angel's question to the Marys, "Why are you looking for the living among the dead?"[642] The height of this critical situation that shakes the human intellect is that just at the moment when the latter believes that it has its hand on certainty, or rather on the sole 'Certainty', it finds itself again uncertain, embarrassed.

On the scientific side, it is the situation of any deterministic mind that continues to be disappointed by reality in its unveiling process. On the religious side, the scene lived by three main characters of the Gospel highlights it. First, it was lived by the Magdalene; relieved to have reached the eagerly desired certainty, she exclaims "*Rabbouni*" and moves toward Jesus to kiss Him, to never let Him disappear. But He stops her in her impulse, saying, "Do not touch me, for I have not yet ascended to the Father.[643] Second, it was lived by Thomas, who when the Lord asked him to put his hand in the wound to become sure of what he heard and to be fully convinced it was He Who came out of death, he no longer dared to touch Him. He preferred to declare a 'faith' which grows freely in the probabilities of incomplete knowledge of his Master, rather than entering into a positive one which will remain frozen in its *hic et nunc* and will belong to the past, from the first moment of its existence.[644] Thirdly, like the last two in regard to disappointment with control over certainties, Peter, James and John experienced it following the Transfiguration.

In the end, and in front of the Tomb said to be empty, Jesus still wanted to reassure the Magdalene and entrusted her with a mission: "Go, find my brothers and tell them: I am ascending to my Father and your Father, to my God and your God."[645] Mary Magdalene comes to announce to the disciples that she has seen the Lord and to tell them what He has said to her. Will they believe her? Will they accept this new reality?

---

641   Jn 20:2.
642   Lk 24:5.
643   Jn 20:17.
644   Jn 20:27-29.
645   Jn 20:17.

3.4. "Schrödinger's cat" in the tomb

So with De Broglie and Heisenberg, all determinism and all formalism have fallen, and the centuries that were considered 'lights' in contrast with the Middle Ages have themselves fallen into the same situation they mocked. Their light has become dubious. In contrast, the light of the centuries of the wise Phoenicians and Greeks, these two peoples with a vast imagination and extreme intellectual capacity for abstraction who were able to establish the foundations of language, Science and Philosophy, has returned to strength. These wise men enjoyed besides, by nature, a logical rigidity in which they cast their reading of reality with a linguistic flexibility that made them, on both informational and methodical sides, the founders of universal knowledge.[646]

Schrödinger, a contemporary of the other two founders of quantum mechanics, but himself of mathematical formation, sharpened to the extreme the theory of indetermination (principle of uncertainty), thanks to the paradox of the 'cat'. For clarity and scientific precision, we will describe immediately, as accurately as possible, the bases of this paradox:

Imagine that a cat is locked in a box that contains a bottle of cyanide. Above the bottle, there is a hammer whose fall is caused by the disintegration of radioactive material. As soon as the first atom disintegrates, the hammer falls, breaks the bottle and releases the poison: the cat is dead. At this, the experiment is not surprising.

However, everything gets complicated as soon as, without opening the box, we try to predict what is happening inside. Indeed, according to the laws of quantum physics, there is no way of knowing when the radioactive decay that will trigger the deadly device will take place. At most, we can say, in terms of probabilities, that there is, for example, a 50% chance that disintegration will occur after one hour. Therefore, if we do not look inside the famous box, our power of prediction will be very limited: we will have a one in two chance to be deceived by affirming, for example, that the cat is alive. In fact, inside the box is a strange mix of quantum realities, composed of 50% live cat and 50% dead cat.[647]

---

646 *Cf.* Julian the Apostate, *Hymn to the Sun*. The inescapable contribution of the Greeks began with their Phoenician masters known as the Ancients who invented the alphabet and gave every human the chance to write. We name them: Melqart, Mochus, and especially Sanchuniathon, the first atomist, and Zeno of Elea, the father of paradoxes and dialectics. *Cf.* http://remacle.org/bloodwolf/philosophes/julien/soleil.htm [Accessed Oct 2019].

647 *Dieu et la Science*, p. 140; *Cf.* Klein, pp. 77-80. Klein takes up the mathematical description of this experiment where, instead of a 'cat', he speaks of state vectors of the quantum. These state vectors specify the internal state of the particle, for example, its spin state (the spin is an internal property of the particles, similar but not identical to the concept of self-revolving). He concludes his analysis by saying that quantum physics

Several theories have been developed to remedy this paradox, one of which talks about superimposed realities where the 'cat' must be considered alive and dead at the same time but not under the same state vector, that is, under the same experimental condition and the same space-time.[648] This theory was rejected because the experiment demonstrates that the quantum, once observed, cannot be particle and wave at the same time. And, if the wave and the particle do not belong to the same reality, the experiment will be absurd. Another theory then called for 'parallel universes'.[649] According to this theory, at the moment of disintegration, the universe would divide into two to give rise to two distinct realities. In the first universe, the cat would be alive; in the second, it would be dead. These two universes, being equally real, would have been duplicated, in such a way as to never meet again. The majority of physicists rejected this theory, too, because it postulates the existence of an infinity of universes that would be forbidden to them, forever.[650]

Igor Bogdanov explained to Guitton why this last theory is unbearable to quantum physics. We summarize his opinion immediately because it is very significant for Christology.

Let us suppose that one admits, before observation, – and because of the probability death/life, – the idea of two or more universes (or realities) in some way superimposed or adjacent to each other. Once the Copenhagen Interpretation is applied to the letter, the wave function, carrying the two 'cats' simultaneously, collapses at the time of observation, causing one of the two cats to fall. The disappearance of this 'feline' instantly causes the elimination of the second universe and of any other imagined universe and physical reality.

We are not going to dwell on Bogdanov's analysis and commentary and on his deduction that supports the existence of a Supreme Being, outside the box of our universe, observing it in a way that allows it alone to exist.[651] We will continue to focus on the findings of this strange experiment of Schrödinger's cat, which wants the wave function to have no physical meaning of its own and to be considered as

---

offers nothing better than probabilities. (Translated by the author)

648 *Ibid.* p. 79; Klein comments: "We, therefore, have no right to represent to ourselves, by thought, the state of the quantum superposition (a + b) as a mixture or coexistence of states a and b. What is it then? No one knows, and this is what makes quantum physics so delicate to interpret..." (Translated by the author)

649 Like the American physicist Hugh Everett's theory; *Cf. Dieu et la Science*, p. 141.

650 *Ibid.*

651 *Ibid.* p. 147. "On this basis, nothing prevents me from advancing the hypothesis that this complex network of interacting wave functions collapses into a single world when it is observed. Now, the whole question left is: who observes the universe?"

an abstract operative system that represents the physical state of a system and not of a particle. This wave function contains all the information about this system, but only allows the calculation of:

a- The probability that a physical variable (like the coordinates x, y, and z) which is written as a vector (ρ) could have a given value in a measure;

b- The transition probability of a physical system between two possible states (physical state, quantum mechanics state);

c- The average values of the physical quantities of a system in a given state.

All this probability situation is reflected in what is now considered a "statistical description", hence the increased questioning around the results in atomic and nuclear measurements. Should we regard these results as a temporary expedient that should, in principle, be replaced by a deterministic description? Or should they be considered as a consequence of the inevitable interaction between objects and measuring devices, in which case determinism does not make sense, especially when the observed action is of the order of magnitude of the quantum?

These questions become alarming for deterministic physicists and, consequently, for humans who are almost instinctively avid for predictability and certainty. Following Einstein's expression, "God does not play dice", said as soon as he was informed of these quantum mechanics findings made by his colleagues, we are all eager to know who, after all, plays dice, God or us?[652] With all the extra subtlety that accompanies the quantum process, if God is the One Who plays dice, its implications and their repercussions on human beings, as well as on human understanding of the reality that surrounds them, will be catastrophic for Religion. Neither freedom nor responsibility could any longer be taken into consideration. The universe, reality, Science and Religion, to believe Descartes, necessarily need that Supreme Principle in Whom all doubt ends and all certainty begins. In other words, and perceiving things in Fuller's exhaustive way, the micro-systems would constitute, according to the Copenhagen Interpretation, an indivisible whole with the measuring instruments, with the observer who shares these instruments, and with God Who is there, somewhere, at the heart of quantum strangeness itself.[653]

Thereupon, and as we said above referring to the English proverb "It takes two to tango", we testify here as Middle Easterners, experts in trictrac (backgammon),

---

652 *Ibid.* pp. 131-132.

653 *Ibid.* p. 131. Jean Guitton says, "And it may be there, at the heart of quantum strangeness, that our human minds and that of this transcendent being that we call God, are led to meet." Would Guitton have discovered the *maškanzabnâ* we are looking for?

that analogously "It takes two to play dice." (Trictrac is a social game of one-to-one which highlights the role of the dice.) Thus, the problematic reaches completely other dimensions. There is no longer any question about whether God is playing dice or not. The issue opens to encompass both questions, at once: who plays dice with God and who was the first to throw them?

Although Guitton is ignorant of trictrac, he interprets its rules: "This is why, he says, the elementary particles are not fragments of matter but, simply, the 'dice' of God."[654] Igor Bogdanov agrees with this statement and adds: "In fact, as the theory in question confirms, the dice do exist; however, from Einstein's point of view, it is not God who plays with His dice, but Man himself."[655] Jean Guitton has the last word to conclude an entire chapter entitled *L'esprit dans la matière* when he says, "And it is up to us to know how to steer them [the dice] in the right direction every moment."[656]

Hawking, Klein, the Bogdanovs and the majority of supporters of quantum mechanics assert that, although imperfect, for the human mind the act of wanting to know, going from the eye to the last cell of the brain, is also part of this indivisible whole. Ephrem, furthermore, says that only those who have acquired by grace the *râzified* luminous eye (*ᶜaynâ šafyâ râzânitâ*) may have a more in-depth knowledge of a meta-reality, a meta-metaphysics, or even, according to our thesis, a *râzâtic* reality. The observer, whoever he/she may be, Jacob son of Isaac, Mary Magdalene or Heisenberg, is not exempt from being subjected, him/herself, to the same conditions of the 'waves' of the matter which he/she observes since he/she is made biologically of the same kind of particles. Therefore, he/she must be subject to the same ratio of the probability of being or not being, which makes him/her in need of defending his/her "being" first, and then his/her "existence". Thus, at the moment of observation, unique as it is in space-time, if the human being does not "be" for what he/she observes, what is observed will not "be" for him/her either. If he/she "is", what is observed will "be". Based on this, the experiment of the 'cat' becomes crucial, since being dead or alive will no longer be particular to its state, but exceeds it to become the very state of the observer. This is, indeed, the basis of the trictrac rules. Let us now see how this reflects on the tomb scene.

---

654   *Ibid.* pp. 131-132.

655   *Ibid.*

656   *Ibid.*

4. A different approach to the paradox of "Schrödinger's cat"

In front of the closed tomb stand for the moment Moses, Laplace, Planck, Einstein, Schrödinger, Guitton and Ephrem. What could be the dialogue going on between them?

4.1. Moses at Horeb: strangeness.

Moses (*c.* XV century BC), the Patriarch of the monotheistic religion of Abraham, opens the dialogue: "Will you convince me, reverend fathers of the Copenhagen Interpretation, almost thirty-five centuries after my experience on Horeb, that what I took for a Burning Bush was only an Elementary Particle or rather the Elementary Particle of creation *par excellence*? ... that It was only the Wave-Particle *par excellence* that was transmuted into energy, light, and wave at the height of my sight and my hearing so as to pass to me a piece of Information that does not mean anything other than Itself and refers only to Itself: Yahweh, I am Who I am; I am the Being?

Would it not be the same Information that the Person who lies behind this wall, Whose story resembles that of my bush, wanted to transmit to you or to remind you of?"

4.2. Laplace's intervention

Laplace (Pierre-Simon, 1749-1827), the patriarch of determinists of the eighteenth century, wounded in his secular liberal spirit and intellectually humiliated by the hypothesis 'Yahweh' that he eagerly neglects, intervenes: "What you say, Moses, is absurd. My celestial mechanics exhaustively includes all the information needed for the universe and no more room ever for any new information. The present state of the universe rejects any supposed additional information. Nothing should remain uncertain to our intelligence, and the future, like the past, must be present to our eyes.[657] Any information, represented by this Person Who is behind the 'wall', being unable to find a place in our universe, is wiped out with Him, behind this 'wall'. This Person was rejected, as well as the Information He was carrying, and here we stand before His tomb, the symbol of fatal determinism that I speak of in my theories. All that is outside of my celestial

---

657 "...Laplace's determinism is scientific/causal determinism. According to this, if someone (Laplace's Demon) knows the precise location and momentum of every atom in the universe, their past and future values for any given time are entailed; they can be calculated from the laws of classical mechanics, thus eliminating free will... Quantum mechanics will eradicate Scientific determinism, *i.e.*, Laplace's vision, of a complete prediction of the future." *Cf.* https://www.quora.com/Whats-Laplaces-determinism [Accessed Oct 2019]

mechanics belongs to the 'hustle and bustle' of your Bible. Any other reading of this reality is but weakness. In the best of cases, for those who support a 50% probability of the existence of a mechanics different from mine or of different information that can take place and affect our reality, these must belong to another reality. Here I ask you to excuse my arrogance, but that would be the reality of madmen."

4.3. Planck's intervention

Humiliated in turn by the speech of Pierre-Simon de Laplace, Max Planck (1858 - 1947), profoundly Christian, intervenes in defense of Science much more than in defense of Moses and says: "Does it not seem to you that what you are saying here, Mr. Laplace, is a little presumptuous for a man of Science? Your presumption sins against scientific humility and therefore against all of us. What the prophet speaks of fits well in my scientific field, and I personally support the analogy that came to his mind. It is, seriously speaking, what made my heart burn a bit while progressing in my discoveries. Imagine, gentlemen, that for some reason, I have not been able to grasp all the dimensions of my constant h which is just one among myriads of information circulating in the universe. It took hard teamwork and deep collaboration until my friend Albert (Einstein) put his finger in the heart of the light and spoke out. My constant has thus become the cause of a significant turning point in the understanding of what a blackbody is, as well as the nature of light.

"This Man, Who is behind the 'Wall', once declared Himself to be the Light of the world and presented Himself as the One Who was signified by the sign of the bush of which Moses speaks. So, all that was left for us humans to do in order to understand Him was to know what "light" really is.

"The blackbody radiation helped me discover the constant h and helped me better understand the light that forms the *sine qua non* condition of all Newtonian physics, as well as of the celestial mechanics that you boast about, Mr. Laplace. Would it not help to understand the Information that was transmitted to Moses? Would it not also help to conceive what this Man would like to communicate to us from the "isothermal enclosure" in which He is locked up? Would it help us understand more objectively the "I Am", the Being He has often named Father and with Whom He is totally one? Did not He want to make us understand that They are one just as light, energy, and fire are? And, as knowledge, information, and the informant are?

"I leave the stage to my friend Einstein who had the favor of revealing the breath-taking dimensions of my constant h as well as those of the quanta of light, called photons. To conclude, I say to the Prophet that his deductions inflamed my heart and are part of the sublime common sense toward which we must all now turn our eyes, our thoughts, and our research."

4.4. Einstein's intervention

Einstein, circumcised son of the Bible, does not take long to intervene by addressing, first of all, his gallant friend Planck, saying: "You are so right, dear Max, to express your support for this point, especially since this 'phenomenon' has often been repeated in the Bible. It was so in the case of the Mount Sinai scene, in the Tent of Meeting, in the divine Fire of Elijah, different *hierophanical* cases where this phenomenon is taken for the Glory of God. So, one can well deduce that even the Glory of God depends on this luminous Information and your constant *h* which has helped us better decipher it.[658]

"With you, dear Max, I support Moses' deduction, and I affirm that this is precisely what we wanted to suggest to him without ever pretending that we know today more than he himself knew in his time. We want him to know that all that has changed is just the way of interpreting it and that, thanks to your discovery of quantum.

"I share with you the reproaches made to Laplace, especially since, according to his assertions, he who abuses the adjective 'celestial' for his 'mechanics', he would have been the first to be in the 'hustle-bustle', whereas this is not the case because he has a "name" that makes him "be" and makes him always present. "Being" by having a name is what allowed him to be with us in front of this 'Wall'. By the way, this is the case for all of us and forever. He does quite well to return to the Prophet Moses to learn from him the importance of a 'Name' and what it means to have one's name erased from the register of the 'living beings' to enter the 'hustle-bustle' state.[659]

"In fact, dear colleagues, to have an idea of the phenomenon that influenced the particles of the bush and made them luminous and sonorous according to a specific need, we must take into consideration the message that was transmitted to Moses. This message, rather this Information, the "I am the One Who is", I would say, was the Light wave that will forge the collective consciousness of my people, just as I explained in my theory of photoelectricity. This luminous Information was of crucial importance to two agents: the Transmitter and the final receiver that is the collective consciousness of the people to whom it was transmitted. The bush, the Prophet's intellect, his "twisted" tongue, and his stick have in this case only the role of the particle. Is it not the case that after all the *"-phanies"* we meet in the Bible, all that remains is the Name "I am He Who is; I am the Being"? So, the same for us,

---

658   Ex 24:16-17; 40:34-35.
659   Ex 32:30-35; Deut 9:14.

dear Planck. Of all the scientific "*-phanies*" you have experienced, all that matters is your constant *h*, and of all of mine, the equation of energy $e = mc^2$. We are lucky that these two symbols have not been substituted for our own names. Accordingly, I sincerely do not know what to say about this "Friend" Who is supposed to be behind this 'Wall'. Out of respect for the principle of non-separability, I cannot deny a specific correlation between His Being and our 'being'; otherwise, why would there be 'behind' and 'before'? And, respecting my general relativity, I will say that He is as much behind the 'Wall' for us as we are for Him. The same would be said for the concept 'in front', and therefore, these two terms, 'behind' and 'before' (or 'in front'), are transmuted at the level of my space-time theory and become a unit that could be called 'behind-before' or, quite simply, a relative 'here': He is here; we are here; etc. So, I dare not say that something separates us from Him, despite the 'Wall' that falls under our observation, just as nothing could separate you from the source of radiation emitted by the blackbody that you know better than I. My weakness has always been my inability to go beyond what is observable, and for the same reason I fail to admit that quantum mechanics is a positive science and that it allows reaching certainty and predictability.

"This has been the cause of the dispute between me and Heisenberg's team who contend that even the unobservable, according to Young's experiment, is part of quantum reality and must be taken into account in the calculations of uncertainty and probabilities. I will let Schrödinger, here present, a strong supporter of Heisenberg's uncertainty principle, clarify the state of being of this "Friend" Who is supposed to be behind the 'Wall'."

4.5. Schrödinger's intervention

"You are right, dear Albert, to say "supposed to be behind the 'Wall'" because, once the 'door' is closed, who can say with certainty if the One Who is locked behind it is still locked or not?[660] It is just like the energy, the formula for which you just mentioned. For me, energy is either mathematical vectors or a 'cat'.[661]

"In accordance with my famous paradox of the 'cat', and on behalf of my colleagues who support wave-particle duality, as well as the quantum unit of action revealed at each phase of the correlation phenomenon, I do not deny your principle of non-separability. Indeed, what makes the crowning achievement is the deduction you have made:

This correlation is not understandable unless, from the

---

660 Klein gives an example of the relationship between the fridge door and its internal lamp. *Cf.* p. 38.

661 *Ibid.* p. 80.

beginning, from the source, the pair of photons had the same property determining the result of the measurement. Otherwise, the existence of these correlations (entangle-ments) implies that the pair of photons constitutes a non-separable whole.[662]

"I support what is being said here because it in turn does not deny uncertainty. Does it not remain to know what you mean by 'from the beginning', and therefore to determine the time and place of departure? The verb 'to determine' becomes enigmatic too, since to determine, we need a determining property that remains outside of your observation as well as outside of mine. I think we meet on the final point in which you resign yourself to saying that the existence of these entanglements implies that the pair of photons constitutes a non-separable whole.

"From my understanding of quantum mechanics, I admit that even the observer, the instruments, and the laws hidden behind the correlation *per se* make a non-separable whole. It is on this criterion that I have based the paradox of my 'cat'. It is on purpose that I have chosen a living cat to highlight the natural bases of uncertainty and unpredictability. These bases escape my control and submit either to the instinct of the cat that could find a way to escape or to its sudden death before the release of the fatal device. I have replaced on purpose the mathematical state vectors with a living cat, to highlight the quantum unit of action which includes, in its dimensions, the very act of knowing. And to know, one must be alive! By correlation, the state of the cat inside the box must correspond to the state of the observer of the box: either positive–positive, or negative–negative (+, + or well –, –), which means life–life, death–death, uncertainty–uncertainty, etc. And, positively speaking, in spite of all the perfection that comes with Science, there is no guarantee that the very system of poisoning meets 100% of our expectations. Therefore, dear friends present on this side of the 'Wall' and, thanks to the unity of quantum action, to the principle of non-separability as well as to all the laws of quantum mechanics, in perfect correlation with our Friend Who is supposed to be on the other side of the 'Wall', I state the following: I personally cannot consider my "being" apart from the state of His "being". My 'cat' behind the walls of the box looks like Him with one difference that worries me a lot. It is that I introduced my 'cat' alive; He was, practically speaking, introduced dead. How, in the light of the correlation principle, can I be alive if He is dead? Even your entanglement paradox (EPR), dear Einstein, would refuse that. If He is not alive, I would doubt not only my intellectual capacity, my *cogito*, but also my being in exist ence, my *sum*.[663] If dying means no longer "to be", more precisely, no more to be part of the

---

662   *Cf.* https://www.pdx.edu/nanogroup/sites/www.pdx.edu.nanogroup/files/

663   As a reminder of the Cartesian proposal *"Cogito ergo sum"* which means "I think, therefore I am."

verb "to be" forged from the name YHWH, nothing can give me certainty in this vast box that is the universe, full of poisonous systems, that I and all of us here, supposedly before the 'Wall', 'are' and that we indeed 'are' alive.

"The time has come, and I do not know what to say. But is it not to face a similar paradox that each of us individually and all of us collectively have come? If I had put a definitively dead cat in the box, I would have risked all the meaning of my life. Would this Person be, in the quantum sense, put dead behind the 'Wall'? Would He be an isolated, unparalleled photon that has no correlation or entanglement with anyone or anything that is part of our quantum reality? I can neither affirm it nor accept an answer that would lead to a contradiction, and consequently to my annihilation.[664] All this is part of a 'mystery' whose beginnings are felt in the hidden laws to which it is subject in our experiments and which, from now on, has become to us all, somehow, unveiled. Perhaps Jean Guitton, a supporter of the Bergsonian theory of the Spirit of Matter and who stands behind the theory of the 'spark of God in Man', could indicate to us some more reassuring lights."[665]

4.6. Guitton's intervention

"Deterministic or not, you have all, since your tender years, vowed to explore the adventure of the Myth of Prometheus in search of the "Firebird" and to scrutinize the fire and the light that he stole from the gods to put at the service of humans. However, for me, my dream begins where yours falls into embarrassment as Schrödinger just said, that is to say, in the 'mystery', more precisely, in the paradox of what I call meta-reality.

"That this Person should be the spark of which I spoke, this "I am the One Who is" Who signifies only Himself and refers only to Himself, hidden behind the 'Wall' or in a theoretical blackbody, quite honestly depends on my own "I am". The question that I came to after a long life of deliberation – Why is there something rather than nothing? – logically implies first my own being, not His. Thereupon, and although I have spoken of the "spark of God in Man", I am willing to affirm that I drew it from your Copenhagen Interpretation, as well as from the laws and principles of your quantum mechanics.[666]

"It was from this 'spark' that the 'mystery' took hold of me, that the discreet light illuminated me in my search for satisfactory information, rather than letting me fall into absurdity, and I found myself very interested in your science, especially

---

664 *Dieu et la Science,* p. 15.

665 *Ibid.* pp. 31; 134.

666 *Ibid.*

the phenomena of wave-particle duality and of correlation.[667] All philosophers would have liked to see what I see right now and hear what I hear. I name two in particular: Heidegger and Bergson. I speak of Heidegger, who was the first to apply the adaptation of language to facts, as his contemporary Wittgenstein will later suggest. Bergson, my master, had the prophetic presentiment that the universe is only the product of a rupture in the symmetry of perfection that already existed. From that perfection, the components of the resulting asymmetry kept a specific 'spirit' that never breaks, *i.e.,* the Information contained in some intrinsic *Noûs.*[668] It is to Bergson and Teilhard that I pay tribute for having emphasized the existence of the 'spirit of matter' of which I am convinced. And if I am, it is because this relationship led me directly to the contemplation of God.[669]

"Was it a rupture at the level of the material particle or of the immaterial wave? Bergson, on the one hand, could not say! On the other, in Teilhard's case, what I was able to see was that the 'mystery' seized him twice, already at the age of seven.[670] Already at this young age, he felt the absurdity of nothingness, and the great questions about 'the Being' emanated from his luminous mind. For him, at the peak of his professional performance, he saw that, "In every particle, every atom, every molecule, every cell of matter, live hidden and work, unbeknownst to all, the omniscience of the Eternal and the omnipotence of the Infinite."[671] Is this applicable, one way or another, to this 'Being' hidden behind the 'Wall'? Do not blame me if my answer can only be: "Yes, I believe."

"I ask you not to blame me for transgressing, by my faith, the principles of Philosophy. But, you, fathers of quantum mechanics, do you not preach a specific religious belief? Do you not assert that quantum physics reveals to us that nature is a unified whole where everything stands and that the totality of the universe appears present everywhere and at all times? And, therefore, do you not preach that the notion of space that separates two objects by a relatively large distance no longer seems to have a palpable sense?[672] Why, then, should this 'Wall' that separates us from this Person Who is Teilhard's *Yeh-šuah* have a palpable sense? If a particular 'sense' is to exist, it is rather His very Name, *Yeh-šuah*, the 'I am the One Who saves', Who makes 'all new' and 'reassures' in a world that you, yourself, declare uncertain and unpredictable. Yes, my reason affirms it: with His 'Spirit' and His 'spark' in every being, He reassures me.

---

667   *Ibid.* pp. 30; 94; 156.

668   *Ibid.* p. 57; 61; 77; 104; especially 114-118; and 173.

669   *Ibid.* pp. 89-90.

670   *Ibid.* p. 89.

671   *Ibid.* p. 154.

672   *Ibid.*

"I see that I have to stop here, since I feel that I am so fascinated by the 'mystery' that I start to theologize and, as a result, begin to lose my professionalism as a philosopher. I apologize to Schrödinger and all of you because you expected philosophy to satisfy and reassure you without resorting to metaphysics, and specifically to theology. But as you must have sensed, I found myself unwittingly in a position to extrapolate to theology which is not at all my field.

"I believe that among us there is a person well placed to reassure you, since reassuring you is, from now on, possible only through 'mystics', and many of you have already called for their help.[673]

"If you, physicists, have control over the 'spark' whose presence and role I evoked *a priori,* he, a 'mystic', seems to possess an accentuated perceptive mind that he presents as a "luminous eye". Thanks to this 'eye', he seems to be able to see Schrödinger's cat and to know the state of its being without even opening the box. He can also meticulously observe at the same time the particle and its wave at the edge of the slit, without hindering the formation of the fringes on the walls of the universe, fringes that remain, for those who have only eyes like our own, the only evidence of a correlation between the "I AM" and our respective "I am". I give the floor to Ephrem of Nisibis."

4.7. Ephrem's intervention

> One day, my brethren, I took a pearl into my hand;
> In it, I beheld symbols which told of the Kingdom,
> Images and figures of God's majesty.
> It became a fountain from which
> I drank the mysteries (*râzê*) of the Son. [674]

"Then I went to Daniel in Babylon, where the first writing appeared but also where the languages were first confused, to borrow a term, *râzâ*, through the 'prism' of which the rays of the 'mystery' of the Being are diffracted. They diffract in a rainbow, a phenomenon that could not be expressed by any of the terms of my Syriac language. This indeed supports your theories on the continual linguistic insufficiency *vis-à-vis* scientific progress, and, therefore, I concur that it

---

673   Davies, Paul. *The Mind of God, The Scientific Basis for a Rational World.* New York, London, Toronto, Sydney, ed. Touchstone. Simon & Schuster, 1993, pp. 231-232: "We are barred from ultimate knowledge, from ultimate explanation, by the very rules of reasoning that prompt us to seek such an explanation in the first place. If we wish to progress beyond, we have to embrace a different concept of 'understanding' from that of rational explanation. Possibly the mystical path is a way to such an understanding…" From now on we will refer to this book by *The Mind of God.*

674   *H Fid* 81, 1; *Cf. TLE,* p. 106.

is up to languages to adapt to the facts, and not the contrary. In my lifetime, my Syriac language had to adapt to the facts of the progress in the understanding of the union between the Spirit of Christ, the Matter of man, and His salvific union with the entire creation. For the time being, we are all before this tomb to deal with the case of this same Christ Whose disciples' writings presented Him, from the beginning, as God's Word and Light. Let us exclude for the moment all the other qualifications. I, the 'mystic', put myself in your place, so I will be interested only in what interests you, in this case, the theoretical blackbody, particles, waves, fringes and the 'Mystery' you are discussing.

"I will be as brief as possible because I see that you are worried and expect to be reassured that, even if you die before Schrödinger's 'cat', the experiment will continue somewhere, in some way, and will lead finally to some optimistic result.

"Before making assertions that I have been able to accumulate thanks to the "luminous eye" that I have enjoyed, I wish to convey to you what I have learned from the "Mirror". It touches on method under its fundamental character of induction and deduction. It is an essential distinction between our method, as Christian 'mystics', and yours, as men of Science. We, as religious mystics, begin from the end, from the other side of the 'Mirror', to induce the beginnings, while you follow the opposite path.[675] By 'end', I mean the very *Ontos*, the 'I Am the One Who Is' that Guitton spoke of in the light of Moses' vision, and not any god. With this, I say that if you take into account our method with yours, that is to say, if you manage to find the explanations that are necessary for your satisfaction in both directions consecutively, you physicists and we mystics will meet 'in phase' on many themes and reassuring deductions. Allow me to explain myself:

"For example, what would the result of the double-slit experiment be if you were to inspect the route of the light from the fringes that are within the reach of your instruments to the light's source? Would you be able to distinguish, on the other side of the slits, in the opposite path of the light, the particles and the

---

675 "In the mirror of the commandments, I will behold my interior face..." *i.e.,* The Scriptures are placed like a mirror; the 'luminous eye' sees the Truth in them. (*Armenian Hymns* 6, lines 42 - 47). *Cf. TLE,* p. 129.
"A large number of synonyms express this function of sign, image, parable, or mirror. The term 'mystery' (*râzâ*), in which Ephrem sums up all the others, is used regularly to designate the symbols under which invisible reality reveals itself through a veiled manifestation, including the sacraments of the Church." (*Cf. La Pensée,* p. 71.) (Translated by the author) Moreover, "The mirror communicates knowledge, the grasping of things that are invisible and hidden. (*Virg* 7, 14) Thus, the knowledge and the words of Christ, which Ephrem describes as mirrors, are now considered as *râzê.* However, even more, oil, called a 'mirror', is a *râzâ* of Christ." (*La Pensée,* p. 63.)

waves at the same time, without preventing the formation of the source light that generated them?

"And you, Mr. Schrödinger, you have considered uncertainty with a living cat that you expose to an undetermined death. Would you not have been quieter while waiting, if you had considered it with a dead cat for which there is a device to revive it? Because, by the pessimistic course of events you envisioned, you exposed your own being to the same indeterminacy, and by that very fact, ours too. If, instead, you had chosen an optimistic uncertainty you would at least have been sure that the only solution that would come afterward would be happier. Then, even if you died, you would have had more of a chance that the cat, once revived, would put you in the same transitory box from which it would have exited. This box remains, after all, a little different from what is called today 'a cloning laboratory'.

"If I have come to these two propositions, it is not because I surpass you, dear friends, in intelligence and knowledge. Rather, it is simply thanks to this "luminous eye" that I wish each person would receive from its only donor. What follows gives an idea of the way in which, through it, I have been able to read the signs, types and symbols:

"Please know, Mr. Schrödinger, that your 'cat paradox' has already been solved by this Christ Who lives and watches us, right now, as He observed Lazarus behind his 'wall'. Therefore, I wonder who, for the moment, is the one who observes the other? Who deliberates on the case of the other, and who is the one who opts for the best outcome of the experiment in favor of the other? Do we, who are now in our 'true' tombs, or does He, Who is where the eternal 'I Am', the primordial creative Light, is?

"Moreover, with regard to the theoretical blackbody experiment of which you are so proud and which allowed Mr. Planck to discover his constant, Christ has proved by His resurrection its effectiveness and its applicability within His Economy of Salvation of which your science now forms a part.

"In His "theoretical blackbody", His tomb, before the opening of which we are now standing, He does nothing but emit a specific kind of energy, a specific sort of light, particles and waves at the same time, to the universe from which He exhaustively absorbed them. He returns them in a better condition than the original one, energized by a Paraclete Who makes them immune to disorder, entropy and absurdity. This is why, to believe my sensibility, I found myself so enamored of the 'pearl' that artisans pierce for a commercial purpose. Its hole evokes for me the one opened by the spear in the side of the 'Word' on the Cross. To use your words, it has reduced the body of Christ to a theoretical blackbody, "differAnt", which absorbs all the evil of humans to return it to them in blessings: [676]

> Your nature is like the silent Lamb,
>   in His (great) sweetness.
> Even if you are pierced and then put in a stud for the ear,
>   as on Mount Golgotha you throw more brightly all
>   your fire for those who look at you.
> Through your beauty is painted that of the Son when, dressed
>   in pain, the nails pierced Him.
> The needle pierced you, just as His hands were.
>   By His suffering, He reigns; by your suffering, your beauty
>   too has increased.[677]

"Can you do me a favor, Mr. Planck, and tell me, for example, what a theoretical blackbody that absorbs the entropic waves of hatred re-emits to humans?"

"As for the fringes involved in the experiments mentioned above, this 'Word' has proved that the opposite path is possible and has affirmed that the fringes of which we are a part may re-enter through the same slits to re-form the Light from which they emanated, or at least to find themselves in Him.

"Believe me. This Christ has allowed Himself to endure in His Own being all these experiments to teach us even what Science is. He is the only One to have seen and known what is on both sides of the 'Wall'. He told us all about it using our own words. If we have not understood Him and still cannot, it is for two reasons. First, it is because our eyes are unable to fix on the light until we recognize its primordial constitution and distinguish between the shadows and reality. Second, it is because our linguistic abilities, our words that He has used, have been and still are too limited to verbalize intuitions and discoveries. It has always been the case on both the religious side and the scientific side, in my time as well as in yours. This latter phenomenon stands behind all the problems of interpretation from which theology as well as Christology has suffered and from which Science can suffer. However, if humans understand Him better today, it is thanks to your scientific, philosophical and linguistic explorations, and all that resulted from your work, you fathers of quantum mechanics. I am sure and

---

[676] *Cf. La Pensée*, p.91, commentary on *H Fid* 82.

[677] *H Fid* 82, 11-12. *Cf.* Brock; *Cf. CSCO* vol. 154, pp. 253-254.(Translated by the author)

Kyânêk domê l'êmrâ šatqâ	*b'basimûteh dên nâš baz^ceh*
wašqal zaqfeh ^cal mašma^ctâ	*ak gogûltâ yatir šâdê*
kûl zaliqao ^cal ḥazâyao	*ṣoyar bšûfrêk šûfreh dabrâ*
dalbêš ḥašâ ṣêṣê ^cbar beh	*^cûqsâ ^cbar bêk dâf lêk baza^c*
ak dal'idao wadḥaš amlêk	*ak dabḥašêk saghê šûfrêk*

confident that the better you understand Him, the better you scientists and we theologians understand each other.

"It is by thinking a little like you and using means specific to my profession as 'mystic' that I was able, one day, to probe the heart of the sun. It is no different from the core of the atom or of the quantum of energy you have penetrated.

"Moreover, from the heart of the sun, from its triune constitution, I was able to see the "I am He Who is" in all His Trinitarian splendor and to worship Him. Believe me, once our 'eyes' become capable of seeing the particle, the wave and the fringes as a non-separable unit, they reveal to us the causing Cause of their *yât* and get closer to the real symbolism capable of reassuring us.

"A third point of great significance urges me to reveal it to you: it is, to be brief, my admiration for the discoveries of the two colleagues Planck and Einstein. Once the nature of light was proved quantic, it became applicable to Him, the Light of the world. Yes, I, Ephrem, affirm that it is appropriate and I assume, together with the Apostle John, the responsibility for my words.

"He, the Christ Who observes us from the other side of the 'Wall' with the purpose of bringing us back to the Source of His Light, is not, if we understand Him as He wills, of a past event that continues. He left our world without fearing to leave a void behind Him or to be unable to repeat the experiment of the Incarnation. If we interpret His event as an unrepeatable past one, He and we will have missed everything. Paul and then John of the Apocalypse have seized on this information and have affirmed that this Christ is in a 'continual' and 'discrete' state of 'coming', as it were, in successive quanta. This implies that the Father is in continuous process of generating the Son. Then, Father and Son together, as Source of *râzâtic energy*, hence of wave-particles of Their Love, make Their Love proceed by waves: the Holy Spirit, the Paraclete. This explains the Spirit's effusion on the Apostles and Mary in the form of tongues of fire, and therein, wave-particles of phlogistic *râzâtic* Love. The apostle Paul even implored Him not to stop coming, *"Maranatha"*,[678] and John did the same using the imperative form of imploration, "Amen, come Lord Jesus."[679] How similar is this to your theory of the quantum field or even the boson field. It respects space-time theory and, at the same time, coincides with general relativity included in Fuller's exhaustive dimensions to cover likewise the hereafter that is our present field of exploration, "On earth as it is in Heaven". Christ taught us this in the *Pater Noster* that He wants us to recite daily so that His Kingdom comes. All 'is' present. The divine Word/Verb comes conjugated only in the present tense of the divine Light that

---

[678] 1 Cor 16:22.
[679] Rev 22:20.

is always newly present. Past and future tenses belong solely to our contingent history, we who spend a few ephemeral years in bodies made according to a well-defined DNA map now considered one of the imprints of God.[680]

"Yes, as a 'Father' of the Church, I rejoice to see the 'Laws' discovered by Planck and Einstein, and by all their successors, applied to this incarnate Christ, this divine 'Word/Verb' and 'Light'. It is a must to tell humans how many times this Christ comes to meet them per second. So how much would the effect of His Love be, since His Light proceeds at a much higher speed than that of the light of your experiments?

"I will stop here while remaining at hand for any future intervention, with the hope of having somehow allayed your worries. From now on, we will meet, with greater ease, around the informative Love that we have received and will continue to receive from Him, either through our ears or through our eyes, but primarily through a 'luminous eye' and an 'illuminated heart'."

Conclusion

No one has the right to deny the role in our lives of any enlightening information that today's humanity, thanks to its astonishing technology, can recover either from the distant past or from the very depths of matter, and more specifically, information coming from the extreme dimensions of the human imagination, like Einstein's relativity and Schrödinger's paradox, that are enlightening as *râzâtic* photons. It is thanks to such information, that reassures us how extremely we are loved or reminds us of Love, that the power of the human understanding continues to work with common sense.

The negatively-charged and the positively-charged information, as well as anti-information, prove more and more that they are a necessity for knowing oneself as well as for having an excellent understanding of any reality. As we have seen, it turns out that there are natural Laws that, in our opinion, form the "gauge fields" of which Science speaks. Their existence, all together, is a *sine qua non* condition to ensure a kind of balance without which the "broken symmetry" at the origin of the universe would cease to be productive. It would become barren. All present information would be lost, and Love would vanish.

It is from this scene unfolded in front of the tomb, somewhere in the *râzâtic* space-time suitable for Newton's theories as well as for Einstein's, that we were able, thanks to Saint Ephrem, to recognize the importance of the information that the incarnate Word, Love, represents. This also applies to the other scenes of

---

680   Collins, Francis. *The Language of God, a Scientist Presents Evidence for Belief*, Free Press, New York.; London; Toronto; Sydney; 2006. p. 240.

the Christian faith: the Virginal Conception, the Birth, the Epiphany, the Eucharistic Supper, and the Crucifixion.

What we have tried to make clear is that, according to our theory, all these scenes and the possible and imaginable dialogues that accompany them took place and continue to take place in a time and a space placed between the physical world and the metaphysical one, which we have named *râzâtic*. They will continue to happen as long as there are persons in the physical world who act in correlation with Persons of the metaphysical realm and vice versa. They have proven and will prove, forever, to be valid for a 'reality' that is not fictitious. This pushes us, after having recognized the power of the 'information', to search for the data that informs us about the nature of the 'milieu', *i.e.,* the 'space-time' in which that 'reality' develops by expansion, and about its location in relation to the Being Who created it.

And if the 'space-time' which forms the aforementioned exhaustive *maškanzabnâ* (ܡܫܟܢܙܒܢܐ) where the dialogue takes place is taken itself for enlightening 'information', it would be our responsibility as Christian 'mystics' to grasp it and to discover by what right Einstein has molded into a single concept *'Raumzeit'* two concepts which are of divine right.

Besides, to take the Bible literally, the letter of the Bible does not express itself clearly on the creation of these two beings: space and time. Even the father of the Big Bang theory, Georges Lemaître (a Catholic priest), affirms, without the Catholic Church objecting, that they did not exist before the Big Bang. What are they, then? Are they really created? From what substance, from what *yât*, would they be? Christianity should pronounce its teaching on the 'time' and 'space' that form the 'milieu' where Paul's *Maranatha* should take place.[681] The spreading of creative and reassuring Love depends on it. This is what we will try to elucidate in the next chapter. It may sound crazy, but it will not intimidate us.

---

681   1 Cor 16:22.

PART IV

Creation out of Love:
Eternal Motherhood and Childhood's Tent

Chapter VI

Are space and time created?
The Creation out of Love

I know that I am mortal by nature and ephemeral; but when I trace, according to my pleasure, the meanders from and to the celestial bodies, I do not touch the ground with my feet: I am in the presence of Zeus himself, and I take my ration of ambrosia.

Claudius Ptolemaeus [682]

To seek and find the Tent of Meeting (*Maškanzabnâ*), in the exhaustive sense, is the primary purpose of our book. It does not lack adventures. We assume that in that Tent, Science and Religion can be in *šekinah* together. Etymological, linguistic and epistemological efforts are needed to explore the two concepts *maškan* (habitat-space) and *zabnâ* (time), and then to build out of them, as out of two constructive waves, the new Semitic-Biblical concept *maškanzabnâ* which can very well be considered the precursor of Einstein's space-time.

Maškanzabnâ is a concept supposed to represent the stability *šekinah* (שכ'נה) where time *zamen* (זמן) dwells, and also time that dwells in the stability (*šekinah*). Physically speaking, in both cases, it brings us back to the time zero (T∅), which is the state of energy in *šekinah*, waiting, just as when it was 'pregnant' with the Big Bang.

This exhaustive, unlimited *maškanzabnâ* which we highlight represents the fundamental state of creation long before the process of existence begins. It must also symbolize the state of symmetry and balance discussed by Guitton following his master Bergson. Indeed, Bergson assumes that this state was 'broken' so that

---

682 Ptolemy. This charming epigram comes from *Greek Anthology*, book IX, no. 577. *Cf.* http://rd.uqam.ca/Ptolemy/Rhosos.html [Accessed January 20, 2020]. Also *Cf. La Pensée,* p. 381; "Commentaire sur le pain Azyme" (*Az* 17, 17).

the creation could have started.[683] It is a symmetry that already existed, one way or another, in *šekinah*.[684] From that moment, the place, *atrâ* (ܐܬܪܐ), and the time, *zabnâ* (ܙܒܢܐ), began to move from the situation of 'potency' to one of 'act'.[685] The enigma that stands behind this *maškanzabnâ* is to see if there can be enough room for the one and the other of rational "beings", the self and the other, even the "Wholly Other" (*das Ganz Andere*). Is this 'Habitat', in fact, the subject of dispute between two egos, equal and jealous, Man and God? Or, better said, between two 'gods' since the two share the same thought, the first as the receiver and the second as the transmitter? Where does this space-time called *maškanzabnâ* exist? What are its confines and the conditions of entering it, so that we can inform scientists and religious?

Our stepping-stone in this adventure is the brilliant insight we find, with tens of centuries of difference, between the spirit that created the 'word' *maškanzabnâ* and the one that molded the 'word' space-time, *Raumzeit*. Can we, from the core of the Incarnation, the focal center of the Ephremian spirit, elucidate this enigma that even today shakes the relationship between Faith and Reason as if they were incompatible under the same sky, and then move this 'milieu' from the domain of the mysterious to the *râzâtic*? To do this, it is necessary to delimit, or at least, as Ephrem advises, to sketch this indeterminable and mystically geo-temporal place, in order to facilitate the communication between the Living One Who is in the tomb and the mortals, scientists and religious together, who consider themselves, until proven otherwise, outside the tomb.

To explain our point of view in a better way, we will develop it according to the following plan:

---

683 *Dieu et la Science,* p. 88.

684 *Dieu et la Science,* pp. 110-111.

685 The Kabbalah creation theory speaks about *En-Sof* (אין סוף), *En-Sof-Or* (אין סוף אור) and *Tehiru.* We will return to this in the next chapter.

1. The space-time problematic

    1.1. The space problematic

    1.2. The time problematic

    1.3. The problematic issue of space-time (*Raumzeit*)

2. Epistemological conceptualization and linguistic impact

    2.1. A better awareness

    2.2. The linguistic phenomenon

    2.3. Nominalism

3. The Kabbalah theory: *zimzum* (צמצום)[686]

    3.1. Moltmann's contribution

    3.2. Interpretation of *zimzum* theory

    3.3. Religious and theological dimensions of *zimzum*

    3.4. The *Maškanzabnâ* and the Creation out of Love

    3.5. The Divine Motherhood

    3.6. God and Mothers

Conclusion

---

686 Often 'Kabbalah' represents: 1- A body of mystical teachings of rabbinical origin, often based on an esoteric interpretation of the Hebrew Scriptures. 2. A secret doctrine resembling these teachings. [Medieval Latin *cabala*, from Hebrew *qabbālâ*, received doctrine, tradition, from *qibbēl*, to receive; see *qbl* in Semitic roots.] (*Cf.* http://www.thefreedictionary.com, *sv.* Cabalah) [Accessed January 20, 2020]. We opted for the spelling Kabbalah as the majority of scholars and websites use it. *Zimzum* is also found written *Zim Zum* or *Tsim Tsum*. We opted for the compound *zimzum* for its proximity to its Arabic homonym.

1. The space-time problematic

"When I trace!" said Ptolemy. "I"! The "I" here may well be represented by a particle such as those discussed in the previous chapter, a particle that can only be detected in a determined space and time, both subject to Einstein's general and special relativity.

Whether this "I" is an atom, a photon or an electron, it must be rational to be able to present itself. If it can "trace", it is only relative to a specific present space and time. There, in the case of Ptolemy, relativity goes beyond its two classic characteristics of "special" and "general". In this case, Ptolemy established the coordinates of his position by himself. So, we have to specify where Zeus is to know where Ptolemy is! Therefore, for such a shocking proposition to be digestible, a pioneer of the reconciliation movement between Religion and Science, Ian Barbour, comes to our aid.[687] He says:

> Science raises questions that cannot find answers in science itself. These are borderline questions. For example, the theories about the Big Bang and the origins of the universe raise questions about temporal, spatial, and conceptual boundaries to the point of questioning the very existence of the universe: "Why, in the absolute, is there a universe?"[688]

Space and time that represent, *par excellence*, the category of these enigmas in question do not only affect all those who take part in the dialogue, including Zeus or God, but also the very subject/object of their discussion which is the Truth itself. They all find themselves immersed in it, time and space included, in the same totality where the limits seem blurred.

Time, which is passing, is an integral part of the Truth. Another integral part of the same Truth is the idea of space, which we now measure by units of time (*e.g.*, light-years) as well as in meters. The divine Word (*Verbum Dei*), declaring Himself "the Alpha and the Omega, the One Who was, Who is, and Who comes", gives unmistakable proof of this fact.[689] Of which nature would the Alpha and the

---

687 Ian Graeme Barbour (October 5, 1923 – December 24, 2013) was an American scholar of the relationship between Science and Religion. According to the Public Broadcasting Service his mid-1960s publication in Science and Religion "… has been credited with literally creating the contemporary field of Science and Religion…" John B. Cobb wrote, "With respect to the breadth of topics and fields brought into this integration, Barbour has no equal." *Cf.* https://en.wikipedia.org/wiki/Ian_Barbour [Accessed Jan 20, 2020].

688 Barbour, Ian G., *When Science Meets Religion: Enemies, Strangers, or Partners*? Harper Collins, 2000. p. 198.

689 Rev. 8:1.

Omega be? Temporal? Spatial? Or both at the same time?

It is true that the Evangelist John clarified the enigma by adding "the beginning and the end", but he did not specify whether these two concepts indicate the beginning and the end of a specific time, of a path, or of the two at once since the Master should be Master of everything: almighty and omnipotent.[690]

We support the last assumption that it should be the two at once, the beginning and the end of a time frame as well as of a path. Our Maronite Liturgy emphasizes this at the opening of each of the Office prayers. Whenever the assembly stands before the Lord and the place shifts into "Liturgical Space", it stands before the Alpha and Omega [in Syriac 'Olaf' (ܐ) and 'Tao' (ܬ)]: the celebrant starts the prayer saying, "Glory be to the Father, to the Son, and to the Holy Spirit, at our beginning and our end..." (*Shubhâ l'Abo w l'Abro wal Ruhâ dukdshâ ʿal shuroyan w shumloyan ...*)[691] This opening Glorification does not specify whether what is meant is the beginning and end of the prayer or the beginning and end of life. Moreover, the opening continues with a supplication saying: "And abound on us, weak and sinners, mercy and compassion, in both worlds (*batrayhûn ʿâlmeh*) that You have created by Your goodness. O our Lord and God, to You be glory now and until the end of the (*ʿâlmin*) 'worlds' / ages / eternity." In Syriac, the term *ʿâlmin* (ܥܠܡܝܢ) supports the three interpretations.

So He must be the A and Ω of all times, of the two 'worlds', of eternity and all space, even the so-called 'heavens', being in fact the beginning and the end of the Phoenician alphabet itself which was the alphabet in use at that period of the fullness of time; so indeed He must be the beginning and end of Language itself.[692]

---

690   He encompasses Zeus, Cronos, and all the mythological gods possible and imaginable at the same time.

691   *Cf. Šḥimtâ Mârunâytâ*; the complete sentence in Syriac:

ܫܘܒܚܐ ܠܐܒܐ ܘܠܒܪܐ ܘܠܪܘܚܐ ܩܕܝܫܐ ܥܠ ܫܘܪܝܢ ܘܫܘܡܠܝܢ ܘܐܫܦܥ ܥܠܝܢ ܡܚܝܠܐ ܘܚܛܝܐ ܪܚܡܐ ܘܚܢܢܐ ܒܬܪܝܗܘܢ ܥܠܡܐ ܕܒܛܝܒܘܬܟ، ܡܪܢ ܘܐܠܗܢ ܠܟ ܫܘܒܚܐ ܠܥܠܡܝ

692   He is the 'Thau' of Ezekiel (Ez 9:4), the sign of salvation that some translations of the Bible call 'Cross'. Indeed, by comparing the various translations of the Bible, we find that only the Latin translation retains the origin Thau (Tav; Tao) which is the last letter of the Phoenician alphabet also used by the scribes of yesteryear:

‏...אָמֶר יְהוָה אֵלֹו אֵלָיו עֲבֹר בְּתֹוךְ הָעִיר בְּתֹוךְ יְרוּשָׁלָם וְהִתְוִיתָ תָּו עַל־מִצְחֹות הָאֲנָשִׁים הַנֶּאֱנָחִים‏
... et dixit Dominus ad eum transi per mediam civitatem in medio Hierusalem et signa thau super frontes virorum.
The Arabic texts vary between 'sign' and 'symbol' (سمة أو علامة) in non-Catholic translations and 'Cross' (صليب) in the Jerusalem Bible text. We find the same options in modern French and English translations. The fact remains that it is a privilege for religious orders that adopt this sign-symbol as their armor and present it to others as the symbol of the Savior and Salvation. A master key has ensured for more than

The relational movement between time and space and their coordination with the verb 'to be' is evident and forms the basis of the grammar of all languages, at least of those we possess. These languages cannot err regarding this truth since the very Name of the Being (*Ehieh*) depends on it. In this case, we can consider space and time as two entities 'created' and 'not created' at the same time, supposed to belong to two or more different 'worlds' and not only to one. Starting this vision, we find ourselves oriented towards interpretations that we will elaborate as we progress in our analysis. We will count, on the one hand, on the wisdom of Ptolemy who ventured into the field of the stars until he stood before Zeus and, on the other hand, on the wisdom of John Paul II (now canonized) in his encyclical *Fides et Ratio*. For both, the goal of the move out of "orbit" is to be able to eat ambrosia cooked on the divine Fire. The latter Fire is what compelled the god Prometheus, a representative of Greek mythology, to venture far, well before those two, for the purpose of stealing a 'flame' and passing it to humans for their benefit. We pray the Spirit of Creation to help us to be an instrument of better awareness of what space and time are.

1.1. The space problematic

Where can we find the Church's position on this epistemological problem, so universally discussed, if not in her kerygma? Starting from the encyclical mentioned above, let us return to the era of the Syriac fathers, precisely to Saint Ephrem, to supply ourselves with the necessary 'fire' and shed light on this point and on the ones that follow.

In his encyclical *Fides et Ratio*, which marked the beginning of the third millennium, St. John Paul II tried to bring the collaboration of Faith and Reason to a new orbit by writing:

> The preaching of Christ crucified and risen is the **reef** upon which the link between Faith and Philosophy can break up, but it is also the **reef** beyond which the two can set forth upon the boundless ocean of truth. Here we see not only the **border** between reason and Faith, but also the **space** where the two may meet.[693]

Although this quotation emphasizes the relationship between Faith and Philosophy, we consider it of great importance for the link between Faith and

---

six thousand years the mutual opening of the borders between civilizations as well as between religions. *Cf.* "Saint Jerome", Migne, *Patristica Latina*, vol. 25, col 88, Paris, 1845; Damien Vorreux, *A Franciscan Symbol, the Tau*, Paris, 1977, p. 35. In Arabic, for example, the Ω corresponds to the Ya', the last letter of the alphabet.

693   John-Paul II, *Fides et Ratio*, end of §23.

Science. Let us not forget that Science was part of Philosophy until almost the end of the 18th century. We will now use this quote as the main 'landmark' in our search for the substantial Tent of Meeting, the *'synesseral maškanzabnâ'* for both parties.[694]

The keywords of this short and precise quote (such as link, break up, reef, beyond, boundless, ocean, border and space) describe with high precision the scope of our action in search of understanding the space problematic. All these words belong to the geographical lexicon, and the comparison between the term 'Christ' and the term 'reef' suffices to initiate a rational shifting of paradigms from the level of physical geography to that of the *râzâtic* plane. Thus, the ocean becomes 'heaven', and space becomes 'the heavens' (*ha-šamayim* הַשָּׁמַיִם), and both encompass the human and the divine in an activity similar to that of the Epiphany, or of the Transfiguration, or of the final minutes on the Cross.[695]

Why do they become 'heaven' and 'heavens'? It is because, in Christ, the Incarnate God, Science and Religion can find common ground of conflict and separation, as well as mutuality of deliberation and understanding.[696] If the latter case is what prevails, then something "good", well founded on Christ, must be able to simultaneously embrace what separates Science and Religion from Him and what unites them to Him.

Here we find ourselves jumping ahead urgently to declare for the first time that this something "good" can only be 'Love' (written with â, Lâve, submitting it to the same pronunciation principles as in *râzâ*). It should be a love with capital A taken from the Derridean "differAnce". In this case, it is a *râzâtic* love, neither divine nor human, for divine Love does not belong to our categories nor our love to those of God. We will come back to this later, especially in the next chapter in which we will discuss spin and gravity. For the moment, let us continue with the problematics of space.

Therefore, because of the common ground and the understanding that Religion and Science can find in Christ, we enter into a union with the 'One' while preserving our own identity, thanks to the Lâve that, in quantum terms, plays the role of the gluon. In this case, the difference with the divine remains unshakeable, but the "space" that divides the inner unity loses its meaning in 'act',

---

694 *Synesseral, Cf.* the "Linguistic Preliminary" on page 9.

695 We will use 'Heavens' instead of Heaven as expressed in the French, Syriac, and Arabic *Pater Noster* to avoid any confusion between the habitat of the divine and the habitat of waters, stars, comets, and galaxies known biblically as the 'Firmament' or 'Sky'.

696 It was, moreover, on the basis of the first Christological conflicts that the Church was torn apart and thereby Christ Whose "mystical" body it is.

keeps it in 'potency', and becomes unlimited, that is, "heavens".[697] The situation of Moses, as described in the following biblical paragraph, is more than sufficient to support this conception of intimate space that separates human geography from that of God:

> And the Lord continued, "See, there is a place by me where you shall stand on the rock, and while my glory passes by I will put you in a cleft of the rock, and I will cover you with my hand until I have passed by; then I will take away my hand, and you shall see my back, but my face shall not be seen."[698]

What does this space represent? Is it the cavity in the rock or the hollow of God's Hand?[699] The answer to this question is intimately linked to the problem of time which this same reference highlights and in which the Glory of God needs to pass by. Let us now consider the problematic of time.

1.2. The time problematic

Saint John Paul II shows his interest in the time problematic by writing:

> God's Revelation is therefore immersed in time and history. Jesus Christ took flesh in the "fullness of time" (Gal 4:4); and two thousand years later, I feel bound to restate forcefully "in Christianity time has a fundamental importance". It is within time that the whole work of creation and salvation comes to light; and it emerges clearly above all that, with the Incarnation of the Son of God, our life is even now a foretaste of the fulfillment of time which is to come.[700]

Further, in the same encyclical, he states, "Truth can never be confined to **time** and culture; in history, it is known, but it also reaches beyond history."[701]

This is said based on a specific vision of time, and of times, that God and

---

697 *Cf.* Deut 4:7; Lk 10:9,11; Mk 1:15.

698 Ex 33:21-23.

699 We have to note that as long as we cannot fix with our eyes the Face of God, and since there is an existential separation between God's Way and ours, the Love of God existing on both sides provides the bridge between human and divine geography. As we see through the loving Hand of God protecting Moses, this Love overcomes distance and time, and the Hand of God transcends separation. The same can be said about the divine Hand when God closed the door of Noah's Ark. (Gen 7:16)

700 John-Paul II, *Fides et Ratio*, §11. *Cf.* Heb 1:2.

701 *Ibid.* §95.

humans inhabit. We would say that the distinction between the fullness of time concerning the concept of "God", and its fullness concerning us, humans, must resemble precisely the distinction between divinity and humanity in the concept "Jesus Christ".

Since the Incarnation of the divine Word represents the fullness of time in its dual human and divine dimension, He, the Incarnate Word, could also represent either a "reef of time" on which Faith and Reason always flow and sink or a "space of time", as in music, without which 'harmony' remains elusive, without a concert hall (musical *maškanzabnâ*) under whose roof Science and Religion can intertwine and interconnect, in an eternal symphony.[702]

In these two passages, the author uses keywords (such as immersed, time, history, fullness of time, two thousand years, fundamental, foretaste, even now, fulfillment of time, time to come) that are quite specific and distinct from each other. However, all these keywords are taken empirically from the collective psychological and spiritual experience of humans.[703]

Other keywords belonging either to space or to time can be deduced from the previous ones, this time to express what goes beyond human experience, "what reaches beyond history" as said by John Paul II: eternity and heavens. This transcendence could be subject to a logic of extension that in turn goes beyond the logical domain towards the meta-logical one, to take up the Guittonian expression. The poetic domain, to respect the strategy of a so-called Ephremian context, could also be considered. It is more didactic and better suited to our subject. "Eternity" should not be understood as a *nihil* of time, rather a "differAnt" time, and "heavens" are not something beyond all space, but just a "differAnt" space. In "differAnce", Derrida's linguistic lever, language does not change essence; it merely opens up to a dimension corresponding to that of general space-time relativity. It is indeed, as Ludwig Wittgenstein says, language that has to adapt to facts and not the other way around. The facts require that one can, at least, locate the position of Ptolemy indicated in the epigraph and then express mathematically and verbally its relative coordinates. It is true that his feet no longer reach the earth because he is present before Zeus, but there "present" and "front" include at Zeus' level all kinds of time and space. That is why, there, they eat 'ambrosia'.

To write such a thing, Ptolemy, like any man, would have to transmute to the

---

702   *Cf.* https://www.journals.uchicago.edu/doi/abs/10.1086/448063?journalCode=ci [Accessed Oct 2019]

703   Everything that belongs to the Parousia falls into the psycho-spiritual category. Consequently, the spiritual there belongs to the lowest level of recognition, which leads to considering a solve-it-all God.

state of a non-man, which is a "differAnt" man; otherwise, everything falls into the absurd, which is not the case here.[704] What Ptolemy wrote continues to make sense to him as well as to us, as if the verb 'to continue' applies 'here', concomitantly, to the present-continuous tense as to all tenses, in all space, eternity and heavens included.

Another way of expressing the present-continuous tense would be by applying it on the essential dimension of these beings, their *forma*, that is to say in our case, the *yât* (אֶת ‎ ܐ) of Ptolemy, of eternity, and of time.[705] Indeed, Ephrem emphasizes it since the Bible says *b-riš-ït*, where *riš-ït* (ית - רֵאשִׁ) indicates the beginning of all *yât*, of all essence, as explained in previous chapters. So, eternity and heavens, in their essence, can well be considered the essence of the 'mass of energy' that will ensure in the first place the potential space and time for the Big Bang to start and to allow them be in act.

Ian Barbour, under the subtitle "Chance and Law" in his book *Religion and Science*, quotes the famous American poet T.S. Eliot who, by a poetic expression, indicated the importance of a movement of the encounter between past and future in an eternally progressive present, eager for redemption. He wrote:

> Time present and time past
> Are both perhaps present in our future
> And time future contained in time past.
> If all time is eternally present
> All time is unredeemable.[706]

By these lines, rather this sapiential discourse, the poet seems willing to express a theory of the redemption of time. Studies in physics concerning optics, light, and waves, as explained in Chapter V, in our opinion support this theory. Out of these lines, we deduce that, if the first one means the flow of time as if it were a ray whose past (the source) cannot be distinguished from its present (light and heat), the second line highlights the unique probability of finding the 'whole' in a personal future. The latter plays the role of the screen that reveals the fringes that in turn reveal the nature of time and its dichotomous constitution. Therefore, the third line states, as if by correlation, that the past contains the future. Tragically, these lines draw an ideal circle closed in a theoretical present such that the author concludes with the following words: "If all time is eternally present, all time is

---

704   *Cf.* 2 Cor 12:1-6.

705   Which is to say, after the Latin expression *'hic et nunc'*, that the *yât* of space is here *'hic'*, and that the infinity of *'hics'* relative to all existing beings forms space. Similarly, for time, its *yât* is now *'nunc'*, and the infinity of *'nuncs'* forms time.

706   *Religion and Science*, p. 193.

unredeemable."

The question that arises here, for Science as well as for Religion, is how to redeem a fallen time, an eternally barren time. The answer is unique. Chained by the expression "eternally present", humans must bombard the word "eternally" to open it to a "differAnt" eternal which presents itself as a 'window' and not as a obstructive reef. Through this 'window', Einstein's general relativity could reach fuller dimensions and the Big Bang could reveal what preceded it. How can our friends, the scientists mentioned above, say anything about the end of the universe, whether it be by Big Explosion or Big Crunch, if they are unable to know what preceded the Big Bang?

So far, nothing new. Philosophy and wisdom endorse the general truth that we have just described by considering time from the angle of the propagation of light. However, the inspiration of the poet T.S. Eliot refuses the sterile repetitiveness, and we give him reason because, to take the behavior of time as he describes it, we see only a historical reality, a Platonic one of closed circular movement similar to that represented by the alchemists of yore.

Figure 33 : Closed eternity. The snake of the Alchemists

The theory of the poet T.S. Eliot, which assumes in the second line a probable presence of the present and the past in the future, makes the future indicate a place, a dam, which halts the flow of time or, in the best case, halts the possibility of measuring it since this future is unreachable. The adverb 'perhaps' indicates uncertainty. The imprisonment of the future in the past joins this uncertainty and seals the possibility of exploring the space/time relational enigma as well as its origin. This impediment implies for the poet the need for an event outside this closed circle to redeem time and give it 'sense', *i.e.,* directionality, by opening it to an eschatological 'there'.

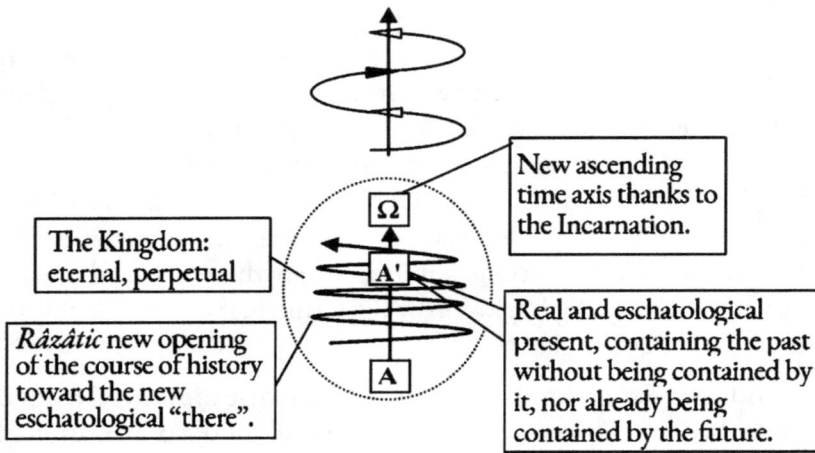

Figure 34: Open eternity: form and application

By this opening, nothing will change for time as the eschatological 'here' makes part of the present 'here', even if this present is in a way progressive and transcendent, as mentioned above. What will change, instead, is simply the dimension of the concept 'creation' that rediscovers its ontological meaning, 'kingdom'. It could never have been recognized as such before the redemption of time, as well as of space, from humans' existential materialistic mentality. It is the "differAnt" kingdom of the Book of Genesis, that Eden, that state of symmetry, equilibrium, serenity and joy, where God and Man (male and female made in His image and likeness), Creator and creature, were meeting around a Word, a Cup, and an indescribable happiness.[707]

A future contained in the past is a future without kingdom because being held in the past is also being restrained by the past, no matter whether it is in the preterite or perfect tense. As a result, everything remains in a state of imprisonment that sends us back to the Platonic Cave. This case of slavery to which human fate is subjected by several concepts that imprison the fluidity of the temporal significance in the rigidity of the Newtonian spatial connotation has always shaken the psyche of the 'rational animal'. This collective state of mind is what required the ardent desire for redemption, for the filling of the gap that had prevented the daily meetings between Heaven and Earth since the rupture, since the breakup of the symmetry and the harmony between creature and Creator.

---

707   1 Cor 2:9.

Redemption, in this case, is just the rediscovery of the happy time, the time of joy of which we have always kept, God and humans, good memories and for which we have always long. The circle of time mythologically expressed by the snake that locks on itself by biting its tail, and also by the *Mécanique Cleste* of Laplace, will be obliged to yield its tail and have its head taken towards its Creator by the natural movement of the four-dimensional creation, under the effect of the 'gluon' of the affinity that directs it.[708]

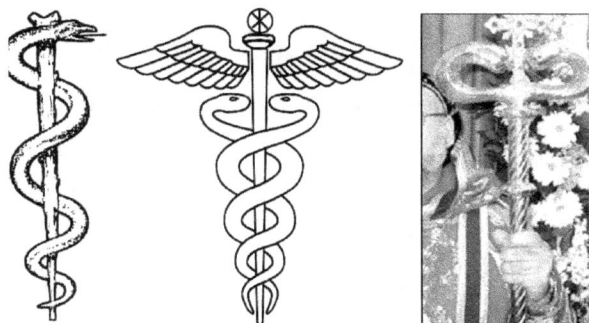

Figure 35: Moses' Snake, a symbol of Wisdom
Syriac Orthodox bishops use it as an ornament of their pastoral staff.

Under this four-dimensional spiral motion, where time plays a primordial role, we understand the Pauline concept of *Maranatha*.[709] It means that we totally deny the idea that the Lord comes, as in an elevator, straight and softly over our heads. This is what we think Paul meant too, he who received Him the first time in a violent way on the road to Damascus. We classify the *Maranatha* under *râzâtic* reality since what is realized in this action between Faith and Reason joins time and eternity in the same undulating present, as it merges at the same time the fringes on the same screen. By this, it allows considering the past and the future in a quantum *Hic et Nunc*, split between Verb, Action, and "differAnce".[710]

---

708 We limit ourselves to the four dimensions of general relativity so as not to overwhelm the imagination.

709 1 Cor 16:22.

710 There, the quantum expresses a successively continuous event. It takes place in quanta according to the unit of time of Planck.

Figure 36: Waves caused by each coming of the Lord

The imperative of the verb 'come' falls like a mass of energy somewhere in the temporal "river of life" and causes waves. Those waves that interfere with ours, those that we provoke by our agitation, become personal to each of us and make a "differAnce". It is that what is proper to redemption is to redeem fallen time.[711] "Fallen" here means desecrated by the spirit of the world, by the materialistic happiness which weighs on our existence as a Freudian libido which is a natural, primary datum for fertility, generation and, consequently, for temporal continuity.[712] This *Maranatha,* as a *râzâtic* expression, helps us become aware of the interweaving of the waves of temporal life and those of eternal life and to realize, in the *Hic et Nunc,* the distinction between them. It also participates in capturing the fringes and watching over the harmony of their phases without which there would be no happiness either physically or metaphysically.[713] Moreover, it seems superfluous to underline, with Guitton, that we cannot find waves in the absurd.

Standard terms and concepts that are related to both time and space in their two physical and metaphysical states may inspire a game of hide-and-seek coming from God's desire to have fun.[714] They may be better represented by a partition, similar to the Tent of the Old Testament or the Veil of Moses, since no human would be able to see the face of God and continue to live.[715] This hide-and-seek

---

[711] Eph 5:16.

[712] *Cf.* Col 3. With these verses, Saint Paul joins the danger caused by material-sensual pleasure to that of the intellectual (*libido sciendi*) which was behind the first fall of humanity.

[713] Consider the dialogue between Jesus and Philip (Jn 14:8-10). As, for example, in the case of a surgeon who, before starting the operation, pronounces 'Maranatha', with full conviction that the *Maran* will come to his help somewhere during his work... He will show him His Face which will reassure him and guide him towards certainties inaccessible even to the best of surgeons and technology.

[714] Gen 3:8*ff*; Ex 33:20-23.

[715] *Ibid.*

suggested by the possibility of being able to see the face of the Truth in the space and time proper to humans, without risking death, can be exceptionally reduced by the reconfiguration of space and time in the light of the Light brought by the Incarnation. On this basis there was the necessity to forge new concepts like transcendence, eschatology, and similar others.

This reconfiguration supports, as John Paul II mentioned, the transmutation of the obstructive reef into a meeting space and, in biblical terms, the conversion of the reductionist Yahwistic 'I' and 'you' into Jesus Christ's relational and connectional 'We'.[716]

Quoting Paul's oath to the Areopagites according to Acts, the Pope mentions this kind of hide-and-seek in search of the 'other' as described in Genesis, in the scene of God's search for Adam and Eve:

> If from a unique principle, He made all nations to inhabit the whole earth and He allotted the times of their existence and the boundaries of the places where they would live, so that they would search for Him [God, Lord] and perhaps grope for Him and find Him—though indeed He is not far from each one of us. For 'In him we live and move and have our being'; as even some of your own poets have said: "For we too are his offspring."[717]

Indeed, Jesus tried to relieve this enigmatic plague with the words that He first addressed to the disciples: "A little while, and you will no longer see me, and again a little while, and you will see me";[718] and then to Philip: "Have I been with you all this time, Philip, and you still do not know me? Whoever has seen me has seen the Father..."[719] The disciples remained amazed, and we together with them ask ourselves about this "little while", which only accentuated the enigma instead of solving it. How long would this waiting be, this "little while" or "lapse of time", between the act of seeing the face of the Son in a physical place and that of His Father in a metaphysical place? Once again, the 'gear' connecting space and time is asserted and, at this instant, in a much more sophisticated way to make it possible to decide whether they are created or not. This leads us to start a more in-depth inspection of this problematic in the light of Einstein's *Raumzeit*.

---

716   Jn 6:37-57; 8:18-19; 8:54-55; 10:30-38; 14:23; 16:15-32.

717   Acts 17:26-28.

718   Jn 16:16. Note the resemblance between this phenomenon and that of the wave dance with the aforementioned fringes.

719   Jn 14:9.

1.3. The problematic issue of space-time (Raumzeit)

Today, theology can no longer conceal the profound anthropomorphic aspect of Semitic descriptions that are based on the natural spontaneity peculiar to childhood. After all, childhood spontaneity is not so bad. Childhood, this state-of-being on which Jesus relied heavily, seems to offer a potential for the probable global solutions, or more precisely, to use the Fullerian terminology, exhaustive ones. In our case, childhood has the potential to specify whether or not space and time have been created.

In its fetal state as well as in its first moments of encounter with reality, childhood represents the most appropriate expression of existence "outside of time" and also "outside of space". It is to this state that the Son of Man invites humans to return, to realize the transcendence: "Truly I tell you, unless you change and become like children, you will never enter the kingdom of heaven."[720]

Time and space, in the absolute, as well as in their essence (their *yât*), are signs and symbols of the One in Whom every beginning (Alpha) and every end (Omega) meet in a non-beginning and a non-end.[721] He is the same *Unum* of the philosophers Who is Alpha and Omega at the same time and in the same space. Einstein was most probably inspired by this same *Unum* based on Zeno of Elea's paradoxes when he developed his relativistic *Raumzeit*. In this sense, like any closed rhythm, the Platonic millennial cycle, whether of history or eternity, opens to a positive eternity, subject to our concepts. In this aspect the part and the whole meet to make us understand the words of Jesus' saying: "Whoever welcomes one such child in my name welcomes me, and whoever welcomes me welcomes not me but the One who sent me."[722] The whole Gospel is a reflection of this desire to make time meet eternity and the limited meet the unlimited. These two dichotomies that appear to be peculiar to childhood make it possible for young scientists who maintain a specific purity of mind to be creative and to realize, to invent, what is now almost impossible for their adult teachers to do.[723] Jesus expressed this by saying, "Blessed are the pure in heart, for they will see God."

---

720 Mt 18:3; *Cf.* Mk 9:36; 10:15; Lk 9:47-48.

721 Indeed, even when we say that the Big Bang is the beginning, we can never say when this process started and when it will end.

722 Mk 9:37.

723 It is common among scientists that the best age for great discoveries would be between 20 and 26 years of age. Why is it so? Better not to ask rhetorical questions. On the other hand, in the field of scholars, an answer must be provided to the reader. In a text to the general public, it may be answered with another question to be pondered by the reader, but scientists do not like this kind of method.

Ian Barbour shed greater light on the problematic issue of space and time by revealing Kant's theory on the subject as follows:

> Among the forms of understanding imposed by the mind are space and time. We organize experience, Kant says, by spatial and temporal relations, and we cannot imagine a world without them, even though space and time cannot be directly perceived.[724]

Barbour adds that according to Kant, "Space and time are forms of thought that determine the way we both perceive and conceive things."[725] Anyway, since Einstein opposed Kant at this very point, he sharpened in a much better way this problematic by submitting time and space to his laboratory and bringing out of them the *Raumzeit*, an unknown phenomenon inconceivable to Kant.

We would say that this idealistic positioning of the problem of knowledge of reality, as well as of all that supports this knowledge, must be reviewed. From the didactic point of view, the Hegelian current, for example, assigns a determinative role to the influence of the human mind in all knowledge. A contingent of American scientists, represented by James Jeans and Arthur Eddington, do the same. These last two note that, based on Einstein's relativity, "In relativity, all the basic properties of objects, such as length, time, and mass are relative to the observer."[726]

Obviously, idealism opposes all materialist and pragmatist ways of thinking. However, Ian Barbour's school admits no extremism in any knowledge. It is fatal to any progress, he asserts. His school appeals to moderation and dialogue from the point of view that affirms room for all currents, and that it is not necessary to compromise or to exclude oneself nor to exclude the other, and that "time" is flexible and enjoys patience. For this school, the conflicts experienced in the 19th century can be summed up in the fact that the men of Science, Philosophy and Theology did not support each other and that the problem was, and always is, that of a challenge for supremacy between Reason and Faith, Science and Religion: which comes first, and the question "To which does the final truth belong?"

As for us, we stand with Barbour's school. We indeed support the idea that, on the same plane of existence, the rational psyche, this ability of *cogitatio*, can never precede the physical brain, its container, with its gray matter. However, we affirm, thanks to the fabulous discoveries of today's sciences, that there can be

---

724 *Religion and Science*, p. 45.

725 *Ibid.*

726 *Ibid.* p. 185.

no limit either to what precedes or to what follows any particular time. Even the famous Big Bang is now said to be preceded by something. Besides, if one day a Big Crunch will take place, it will certainly not be the end. To be seen exhaustively, something must follow.

On this, and for our approach to this issue to be insightful, we hold to the Ephremian counsels to sketch the information poetically, without ever limiting the illimitable or defining the indefinable. This method finds effective support in Heisenberg's uncertainty principle and in the principle of non-separability at the same time.

Time and space, taken under the aegis of the Cartesian condition that defines the fact of being according to cogitative ability, need nine months to be born, in the case of the fetal brain. It is only with the birth of the brain that barely reaches the fullness of its physical form that time and space are born, and they are just scarcely born. To begin to conceive the space and time that are born with him/her, the newborn needs a long experience under the lights and shadows of the human firmament. What precedes this step does not represent a cognitive problem. It represents, first, an issue of connectivity between the container and its content; second, it becomes a concern of self-consciousness relative to the two phenomena of time and space in which connectedness takes place; and, third, it becomes a challenge of comprehension of all that is gradually realized at the level of the psyche. These are the three essential operations by which '*numena*' are transmuted into '*phenomena*', and it is from the latter that flows incrementally everything that could be called *cogitatio*.[727]

It is further proof that childhood cannot think as adulthood does. It cogitates, however, in the manner of the *Little Prince* of Saint-Exupéry, in other words, in a way that surprises adults and that adults, generally, find hard to adopt. At this level, the Derridean concept "differAnce" comes into effect. With childhood, knowledge seems to lean more toward the Kantian theory which implies that the information that comes from the *a priori* – from the Platonic world of Ideas – joins, by connection, its root located in the subconscious of the child, to be transmuted subsequently by observation into a conscious foundation.

Moreover, since Platonic-Kantian ideas are included by concomitance in the specific space and time of their nature, children, rebellious to every 'space' as well as to every 'time', would seem to belong to a dimension beyond Kant's. If Guitton

---

727 An excellent expression in used within the Lebanese community that comes addressed to persons who show behavioral disorientation and inconsistency: "He/she doesn't know where God has placed him/her."

classified them in what he called "metalogic", we prefer, *râzâtically* speaking, to refer them to the realm to which the angels belong, as Christ has suggested.[728]

Connectivity remains, *par excellence*, the key to the relationship with time and space. However, what surprises us profoundly is to see that, in the text of Genesis, Adam does not name these two elements as he did all other creatures: "...and whatever the man called every living creature, that was its name."[729] No one mentioned them, not even God. They were there, as were the Word of God and His Wisdom. The whole, even revelation, was immersed in time and space as mentioned by the reference borrowed from Pope John Paul II. Here, the verb "to immerse" is very significant. It evokes, in our mind, the space in which the human intellect – rationality – was always immersed, as well as the time during which the immersion continued, awaiting the appearance of the consciousness with which they will be born.

According to creationist theory, the question "Which comes first, and to which does the final truth belong?" is as old as the beginnings of human self-awareness. Pioneers, such as Ian Barbour and Jürgen Moltmann and other theologians, philosophers and scientists, especially the supporters of quantum reality, have not hesitated to make this dilemma a welcoming Tent in which they can meet rather than the obstructive reef of their continual shipwreck. Hawking gives a fascinating example of this way of thinking. He asks: "Where does the universe come from and where does it go? Was there a beginning and if so, what existed before it? What is the nature of time? Is there going to be an end?"[730]

Such questions that we keep asking, or at least rephrasing differently, push people to feel caught in a continuously worrying reality, pregnant with entropy.[731] It is with similar terms that the American Kenneth Ford, an expert in quantum physics, describes the feeling that permeates developed societies, especially America.[732] This kind of pessimism (eeriness) toward the current reality, as depicted by extremists, atheistic scientists, is fed by the anxiety of parents towards the future of their offspring. At this level of uncertainty and fear, even Christological concepts such as the Incarnation, the Kenosis and the 'Suffering of God' make the situation even worse, especially with the failure of the ideal God, omnipotent,

---

728 Mt 18:10.

729 Gen 2:19.

730 Hawking, p. 17.

731 Natural disasters and terrorizing political conflicts from September 11, 2001, to the present, mark extreme and disturbing situations fueled by the media around the world. Recently, on the one hand, the repercussions of climate change, and on the other, the appearance of global terrorist movements like Daesh, called ISIS, which humanity will not soon forget, continue to pour oil on the fire. How can humans be reassured?

732 *The Quantum World*, p. 247.

invincible, capable of saving from all things sinister and of safeguarding lives in this world.[733]

The idea of a God immersed in time and space, as we have painted it through the typology of 'ember', is in itself a sufficient cause of this problematic. A God subject to our conditions including the tomb cannot inspire security in this world. And, in case He cannot go beyond our terms, as the thief challenged Jesus to, "Save yourself and us," He will not be able to accomplish anything until His death and resurrection take place in a different space and time, nor can He inspire as much security and predictability for mortal humans as mentioned in the psalter.[734]

Until now, the existential problem of space, time, and space-time has not known a reassuring horizon. Instead, it gains in complexity as it increases with the progress of Science, in enigmatic dimensions. According to our *râzâtic* approach, it is indispensable for humans to situate their selves relative to the Being's Self, as we were going to do for Ptolemy. The most important is that humans know where this "Father-Mother" Creator idea has placed them, to become able, as observers, to identify the coordinates of these three concepts that weigh on the meaning of their life and of their longevity, and to tame them.

We will then try to deconstruct this anxiety, with the support of the brilliant Semitic *zimzum* theory of the Kabbalah School put into relief by Jürgen Moltmann. We will do it by starting to address the two phenomena of epistemological conceptualization and linguistic impact, which will facilitate the acceptance of a concept like *zimzum*, foreign to European languages, but familiar to the Arabic and Syriac languages that share the same Aramaic roots with Hebrew.

2. Epistemological conceptualization and linguistic impact

"God loved us first."[735] He then declared clearly, and in a manner indispensable to human happiness, that we must return this Love to Him, not only with all our heart but also with all our mind and soul.[736]

---

733 It is, moreover, what lately each 'superpower', specifically the United States, is pushing to relieve. It tries to do this by pushing to the maximum, through the media, so that the "pessimists" become convinced that the solution is in its system and its values. Thus, it suffices for people to enroll under its colors to feel safe and reach the desired optimism. It is indeed what we consider a way to substitute the "ideal god".

734 Ps 91:1-6 "He who dwells in the shelter of the Most High, abides in the shadow of the Almighty... under his wings you will find refuge... You will not fear the terror of the night, nor the arrow that flies by day..."

735 *Cf.* 1 Jn 4:19. Benedict XVI, *Deus Caritas Est*, Introduction.

736 First commandment. *Cf.* Lk 10:27.

Taken in the scientific sense of the term "love", this exchange should, to be adequate, encompass in its expression the quantity as well as the quality. That is, He has created us free; we must, in turn, make Him free from all servitude to our categories and our definitions. Love demands this expression, especially when 'It' is conditioned only by 'Itself'. Could this exchange be part of the quantum reality as described above? At the same time, could we base it on the propagation of wave-particles as well as on general relativity? Could we finally establish this exchange on all those already mentioned principles, laws and constants of quantum mechanics, where all of them collaborate in harmony and synchronization to make possible an epistemology of this Love?[737]

Jürgen Moltmann introduces a solution to this epistemological problem. He writes:

> The Creator suffers the contradiction of his creatures. He endures their closure on themselves. He lays the sins, the bad deeds, the sufferings and the diseases of his creatures upon the new servant of God: "By his sufferings we are healed." (Is 53, 5) The creation of justice proceeds from the suffering of injustice. This is why historical divine action involves both **action** and **passion**.[738]

These words show the depth of God's love for His creatures. He expressed Himself by His Son and through Him, offering Him to share exhaustively with humans their condition, except by becoming a sinner, specifically to share in what broke their relationship with His Father. The Father has purposely sent Him to transmute our mode of conceptualizing from the deductive logic (positive), whether it is syllogistic or Boolean, to the inductive one, called *metaphysical* by some, *metalogical* by others, and by us, from now on, *râzâtical, i.e.*, a logical mode extracted from the very Event of the Incarnation.

Due to the Incarnation Event, the deductive conceptualization of the created reality has become inductive without ever ceasing to be deductive, and that is because of the ontological primacy: on the one hand, because of the divine Breath planted in the human, called the Spark, the receptacle of the Incarnation; on the other hand, due to the promise of salvation made to Adam and Eve that we humans should always recognize and respect. Consequently, thanks to the information that the Incarnate One gave us, the Church has been able to give birth

---

737 In the final chapter, we will introduce Bell's Theorem and also the string theory which both give a considerable verve to the flow and reflow of quantum reality.

738 Moltmann, Jürgen. *Le rire de l'univers*, CERF, Paris, 2004, p. 46. From now on, we will refer to this book by Moltmann. (Translation made by the author)

to the "Christology from above",[739] without ever neglecting the "Christology from below" embedded in the lineage of Jesus Christ from Joseph, the spouse of Mary. The movement inspired by Jesus' words (such as, "All things have been committed to me by my Father. No one knows the Son except the Father, and no one knows the Father except the Son and those to whom the Son chooses to reveal Him.")[740] proves what we are asserting. However, the event of the Incarnation itself and the meditations that Ephrem writes around the presence of the 'God-Son-Word' in the womb of Mary are now soliciting us to approach a "Christology from within", from the Core.[741] The Core in the Augustinian sense means a specific environment that joins the 'above' and the 'below' in a "familial-sacred state" that knows neither above nor below, but rather equality, explicable only under the *râzâtic* category that 'glues' the human to the divine in an indescribable Love. This Christological conceptualization, from the Core, which introduces a new *râzâtic* partner to the divine perichoresis; a woman pregnant with the Son by the Holy Spirit will play the central role in our new reading of the scene of creation. It becomes evident that this kind of knowing can no longer be accomplished in one of the two directions without the other direction or at one time without the other time, including Eternity.

As soon as we base knowledge on Love, it requires relationality and connectedness, and therefore action, passion, opposition, contraction, suffering, birthing, procession, emanation, breathing, sending, etc. The creation thus becomes made of each quantum of information picked up from one specific dictionary, that of maternal conception rather than simply feminine. Derrida has well cherished this dictionary for its linguistic philosophy and has used it, even in the Latin language, while imitating Saint Augustine's relationship with his mother, Saint Monica, to make a "differAnce".[742]

Where and how did the exercise of this logico-epistemological Love begin? It is none other than the challenge known as the 'mystery' of the Holy Trinity, declared

---

[739]   Jn 1:18; 1 Cor 2:11.

[740]   Mt 11:27.

[741]   A Christology that springs from the bosom of humanity, specifically from the maternal tabernacle, as Christ's Light shone from the tomb, Him in Whom the Law and the Prophets found their fulfillment. It is that He is indeed the Prophecy *in se*, which was contained in the words of the prophets. He is also the Justice *in se* on which the Mosaic Law was erected.

[742]   *Cf. Circumfession.* This philosophical-linguistic project of Derrida turns in a spiral around 'G', *i.e.*, his mother Georgette, and the projection that he makes of it on Monica, Saint Augustine's mother, demonstrates the destructive role of the *machismo* of the Old Testament Judeo-Israelite mentality. In support of his position Jacques Derrida did not circumcise his two male children and ended his life by expressing his love to all those who were surrounding his bed saying "I love you!", words he had never used before.

so precious by Moltmann who evokes it in the third chapter of his book entitled *Trinitarian Creation*, where he wrote:

> If the Father creates the world by virtue of His love for the Son, the world becomes, by virtue of the love given back by the Son, the happiness of God the Father and the Son ... The Father thus creates the world through his Son. He creates by the power of the Holy Spirit.[743]

This conceptualization of the relationship between God and the human intellect, that is to say, to know the Father through the Love that the Son has for Him and vice versa, – something that requires knowing Him pathetically, by the 'heart' and not by the intellect – may cause us to feel somehow frustrated. This frustration is due, in large part, to the quantity of Love placed at the disposition of humans and which remains unexplored. It is so either because they ignore it or, at least, are unaware that God ardently desires to fill them with happiness. The reason might also be a lack of human self-consciousness. Benedict XVI, introducing his first encyclical *Deus Caritas Est*, clarified this point saying:

> In the same verse [1 Jn 4: 16], Saint John also offers a kind of summary of the Christian life: "We have come to know and to believe in the love God has for us." We have come to believe in God's love: in these words, the Christian can express the fundamental decision of his life.[744]

Such words, so often repeated in homilies and in biblical studies without ever being well absorbed, once deeply meditated and assimilated by the mind and by the heart, add new elements of frustration: "How can we love Him if we do not know Him?" we ask with Saint Augustine. It is with this question that we face, once again, the frustration that takes root within the linguistic limits of humanity. How can we express a relationship based on feelings, especially when these feelings do not belong to the same reality, as in the case where we are invited to love the Lord our God with all our heart, soul and mind? How do we get out of this dilemma?

On the one hand, Saint Paul understood that each reality must be treated according to its own kind. He wrote well to the Corinthians:

> How foolish! What you sow does not come to life unless it dies. When you sow, you do not plant the body that will be, but just a seed, perhaps of wheat or of something else. But God gives it a body as he has determined, and to each kind

---

743   Moltmann, p. 61. John Polkinghorne, throughout his book *Science and the Trinity,* allies with Moltmann.

744   *Deus Caritas Est,* Introduction. *Cf.* 1 Jn 4:19.

of seed he gives its own body. Not all flesh is the same: People have one kind of flesh, animals have another, birds another and fish another. There are also heavenly bodies and there are earthly bodies; but the splendor of the heavenly bodies is one kind, and the splendor of the earthly bodies is another. The sun has one kind of splendor, the moon another and the stars another; and star differs from star in splendor.[745]

But he also understood that language itself, like all other means of expression, follows the same criterion of value *vis-à-vis* its comprehensibility. He added to the Corinthians in the same epistle:

Undoubtedly, there are all sorts of languages in the world, yet none of them is without meaning. If then I do not grasp the meaning of what someone is saying, I am a foreigner to the speaker, and the speaker is a foreigner to me. So it is with you. Since you are eager for gifts of the Spirit, try to excel in those that build up the church. For this reason, the one who speaks in a tongue should pray that they may interpret what they say. For if I pray in a tongue, my spirit prays, but my mind is unfruitful.[746]

This leads us to suppose why the Areopagites could not grasp what Paul, once in Athens, was advancing to them. It was not due to a phonetic problem in the Greek language, since it was a second mother tongue for Paul. They had understood what Paul was saying but could not grasp the meaning since Paul was not at the same orbit of signifiers and signified as they were. Could he have shared better with them at the level of referents and paradigms? Nevertheless, this meeting resulted in an exception that the Book of Acts accurately described: "Some people yet believed."[747]

---

745   1 Cor 15:36-50.

746   1 Cor 14:11-14.

747   Acts 17:34. This phenomenon of 'veil' is just a continuation of what happened with all the prophets before Jesus Christ. It will continue with Jesus as we see in Lk 9:45, and so on with the apostles and the Church. That is why Jesus promised to send the Holy Spirit to help everyone to understand. (Jn 14:26) The fact remains that understanding requires intellectual, psychological and spiritual maturation. This was already underway: on the one hand, thanks to the Wisdom books of the Bible, and on the other hand, thanks to the philosophies which had preceded the Advent and which continued to influence the kerygma of the Church as we see, at first sight, with Saint Paul. The gift of the Holy Spirit becomes equivalent to the Grace of Faith which was considered, by the unanimity of the Fathers, as a *sine qua non* condition to understand or, better said, for the veil to be removed. "*Credo ut intelligam*," Saint Augustine will say, and Ephrem will emphasize the need to acquire the ʿaynâ šafyâ, the 'Luminous Eye'.

On the other hand, Saint Augustine did well to solve this linguistic enigma, in his *De Trinitate*, by approaching it with the old categories of signs and symbols. However, we ask ourselves: is it enough to call God "Father" and write His name with a capital letter to reach the maximum expression of our love and respect for Him? Can this scriptural style help us to devote more attention to a Father Whom we do not see, when we do not give enough attention to our fathers whom we do see? Moreover, we humans use the same vocabulary to express what belongs to our senses as well as what is beyond them, and we call this "analogy". To what extent would this method be correct and effective, especially in this scientific era where the human eye has been able to see, for almost a century, what has been considered metaphysical? How true would it be in terms of the problematic of the identification of time and space?

Augustine's "*Credo ut intelligam*" means, unquestionably, that neither clarity nor syllogistic or Cartesian precision is required to understand. Dionysius and Damaris, who according to the Areopagus account believed Paul, are good models of common believers. However, could such believers ever support the idea of having missed most of the Faith? Would they admit to having thus missed the Love which they were supposed to enjoy just because they had locked themselves into a framework of controversy that suited the Greeks?

In this regard, we ask ourselves whether the best expression of Love contained in the word 'Father', which Jesus emphasized so much, remains, despite everything, obstructed by an imperfect faith distorted by any analogy, anthropomorphism and credulity. So too, we wonder if it is not the same for the two terms 'time' and 'space' that both stand behind all order, all symmetry and all synchronization. The fact is that it is their function to separate between creatures in general, between creature and Creator and, precisely, between humans and God; between the past, the present and the future; between the here, the there and the beyond, etc. However, would they not both be obstructed by an imperfect science, distorted by anthropomorphism as well as by consumption?[748] What does John Paul II try to clarify by affirming, "There, we do not only see the reef that separates reason from Faith, but also the 'space' where the two meet"? Does the holy Pope try to point to the relationship between physics and metaphysics? Indeed, we believe that he does because, in another place of his encyclical, he wrote in defense of metaphysics, accusing:

---

[748] We understand by consumption the human tendency to reduce everything into market products, based on the materialistic philosophy and the globalization of trade. All technological progress is used to control it at the "speed of light". Even time and space are submitted to this human commercial inclination and, nowadays, consumers are paying for them dearly in the fields of communication and e-commerce.

... the deep-seated distrust of reason which has surfaced in the most recent developments of much of philosophical research, to the point where there is talk at times of "the end of metaphysics".[749]

And, in another place, he adds:

... But the use of a hermeneutic open to the appeal of metaphysics can show how it is possible to move from the historical and contingent circumstances in which the texts developed to the truth which they express, a truth transcending those circumstances... Human language may be conditioned by history and constricted in other ways, but the human being can still express truths which surpass the phenomenon of language.[750]

We conclude this point by admitting that once a language is invited to break the chains under the 'groan' of the Holy Spirit, the linguistic impact, especially the conditioned and constricted one, weighs heavily on the conceptualization. "Time" and "space" are common words, common names of this "conditioned and constricted language" that John Paul II emphasizes, and yet, despite their union in *Raumzeit*, they are still unsettled. Would Jesus have wanted to make us understand that He, as the Word *par excellence*, is the Alpha and the Omega between which any conceptualization, any word, any name, as well as any linguistic impact should be settled?

2.1. A better awareness

It remains for us to see that the barriers of language itself can be accused of making it difficult for us to know and understand the Truth. These barriers are detrimental to the evolution of a conditioned and constricted language, but dynamically comprehensive. If Saint John Paul II emphasized this problematic, he who possessed fluently a good number of languages, it is only to praise, on the one hand, this phenomenal capacity and, on the other hand, to delineate its own function in the extent of one of the two aforementioned realms, that of Faith and that of Science, or in both at once. He does so in such a way as to make it clear that the 'edges', 'boundaries' and 'reefs' between these languages are nothing but words which in turn form a concomitant 'language', a different way of understanding and perceiving the Truth. It is through the different cultures of the nations that have dispersed from the "curse of Babel" that the so-desired Truth can be detected,

---

749 John Paul II, *Fides et Ratio,* §55.

750 *Ibid.* §95.

no matter what 'Babel' could have been.[751]

What took place in that 'Babel' resembles, in a way, an experimental nuclear bombardment. The Word of God has bombarded the so-called 'unified language' of humans and the 'fission' of this 'language', just as the fission of a heavy atom nucleus, began. Consequently, if there was also, during that event, a Grand Unified Theory (G.U.T.) for which scientists are striving today, it should have suffered the same fate. We can say the same of the Truth itself as of anything that was united in its parts, once these parts had planned, pretentiously, to join God's realm.[752]

God "broke" language itself, thus breaking the language of humans in an irrevocable way.[753] This paragraph of Genesis has always been a core of reflection for language philosophers and recently, as mentioned above, for Jacques Derrida, a Jew to whom the Torah is addressed by default.

What surprised us the most was to discover that from St. Augustine to Derrida, the work carried out to develop a unified linguistic rule that binds the signs, signifieds, and their referents to each other, as the fathers of the Copenhagen Interpretation did for modern Science, is negligible. However, lately, and under the requirements of a better awareness of what is happening at the level of our intellection of reality itself, and of the various realities that keep coming up as we go along and that impose themselves on our minds, an ascendant linguistic activity has taken place. It is now imposed multi-dimensionally on all other areas of Science and knowledge. We can read about one of the most important goals of this linguistic awakening in the works of the Oxford philosopher J.R. Lucas. Paul Davies pointed it out in his book *God and the New Physics* in the chapter "The Self":

> To say that a conscious being knows something, we are saying not only that he knows it, but that he knows he knows it, and that he knows he knows he knows it, and so on... The paradox of consciousness arises because a conscious being can be aware of itself, as well as of other things, and yet cannot really be construed as being divisible into parts.[754]

In addition, Paul Davies quotes John Locke who asserted: "It is impossible to

---

751 'Babel' is a composed name formed of 'Bab' and 'El'. One of its possible meanings is the 'House of El', the first God of the Hebrews and the God of Abraham who came from the Chaldean city Ur. *Cf.* Paul Feghali, *Al Muhit al Jame*ᶜ. http://www.albishara.net/dictionary/m/read/1003?nav_show=; [Accessed Jan 22, 2020].

752 Gen 11:4.

753 Taking into account the "shibboletic" difference between groups that share the same language. *Cf.* Judg 12:6.

754 Davies, Paul. *God and the New Physics*. New York, London, Toronto, Sydney, ed. Touchstone. Simon & Schuster, 1984, p. 92. From now on, we will refer to this book by *God and the New Physics*.

perceive without perceiving that we perceive."[755]

Here, as a result of this better awareness, an added-value question comes to mind: is it possible to perceive beyond language, and how many languages are necessary for a human to perceive that he is really perceiving as much as the broadest spectrum of collective perception demands? This question takes us back to the very heart of the linguistic phenomenon. Let us consider immediately this phenomenon which shakes knowledge and perception at the same time.

2.2. The linguistic phenomenon

It becomes challenging for us to know how two words like "space" and "time", once used to determine the distances between spaces and times, put us in a continuous half-way and half-time, according to Zeno's famous paradox of Achilles and the turtle, as if they put us in a no-where and outside of any time.[756]

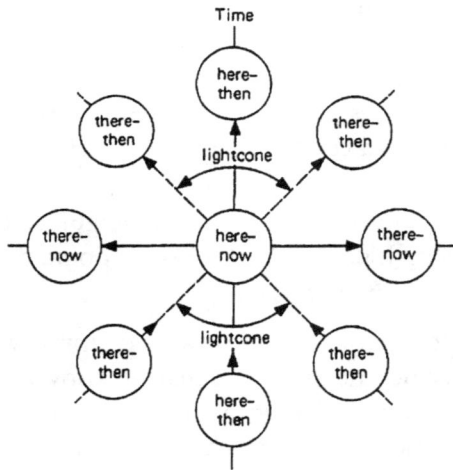

Figure 37: Light cones, the ultimate unit of cosmological measurement

This figure above now forms part of the quantum Bible. It shows how space and time are conceived and illustrated. It turns out that at the level of space-time (*Raumzeit*), the distance according to the light cones between all 'here-then', all 'there-then', and all 'there-now' is zero, as all of them are supposed to pass by the 'here-now'.[757]

---

755   *Ibid.*

756   *Ibid.* pp. 14-15.

757   Harrison, Edward R. *Cosmology*. Cambridge University Press, London, 1981, p. 135.

As for this phenomenon, Kant clarifies it the most. He considers that: "Time, space and causality are categories of human thought that we impose on nature. We can never know things as they are in themselves."[758] Quantum mechanics is no less influential in this sense, especially in its relation with mathematics, a language in its own right. This is why a good number of physicists suggest that space and time are not fundamental concepts, but simply approximations. And, as it turned out, matter despite seeming continuous is made up of quanta, so *Raumzeit* may also be built of some more primitive, more abstract entity.[759]

On this, we comment: if Kant insists on the idealistic *a priori* of these two terms, physicists, especially Einstein, support neither mythization nor sentimentalization (*sic*) as suggested by the novelist Patrick Cauvin in his work *E=MC² Mon Amour*.[760] It seems that physicists, evoking a more primitive, more abstract entity, display total disarray in the heart of the languages they possess and seek, at any price, a language at the level of the *Raumzeit* concept. Would it be mathematics? If this is not the case, one would wonder how Einstein could have explained the quantic nature of these two 'beings', which in Newtonian terms are continuous, divisible, and dissectible matters? Under Einstein's watchful eye, they proved to be so immaterial, so etheric, that he was able to reshape them and come out with a new entity, a new being, just as God made mud from water and earth to model Adam.

The *Raumzeit* is as alive as Adam is and contributes with him to the discovery of the phenomenal language of certainty and predictability that should inspire the Goodness, Justice, and Satisfaction (*šekinah* שכנה) of the One who created their first constituents, just as He did Adam's elements, starting with the Alphabet. It is a language whose disclosure, as well as that of its relative reality, is in continual becoming. What would that language be?

The answer sought by any linguist, philosopher and theologian interested in this phenomenon should henceforth also cover the following question: How could entities such as 'space' and 'time' have been transmuted from two *numena* to one *phenomenon*? How could they have been shifted from the ideal realm to the empirical one and become subjected to our measurement systems, to mathematics, to psychology, to the Boolean logic, as well as to a last, specific thing?[761] What could that "last, specific thing" be in this case, other than the

---

758 *Cf. Religion and Science*, p. 183.

759 *Cf. God and the New Physics*, p. 40.

760 Cauvin, Patrick. *E=MC², Mon Amour.* Lattès, 1977. A film adaptation was created by director George Roy Hill under the title *I love you, je t'aime* in 1979.

761 As Eugene Wigner put it forward. *Cf. Religion and Science*, p. 185

nominalist discipline, because according to nominalists, by the very fact that these two entities have names, they exist, and by the very fact that we name their measuring instruments, they become measurable?[762] After all, what did Zeno of Elea measure by his paradoxes and where did he find his words? Let us then consult the nominalists.

2.3. Nominalism

Even beginners in modern physics note that mathematics, 3D (three dimensionality), virtual reality, psychology, awareness, etc., all played a role in Einstein's laboratory to give birth to this famous discovery of special relativity as well as general relativity. It left the two beings, 'space' and 'time', suspended in a 'milieu' that is neither physical nor metaphysical, neither ideal nor empirical, neither of Newtonian logic nor of a metalogical one. That is why, we would say, neither the adaptation nor the development of this relation between 'space' and 'time' would have been possible, nor their positioning in the universe of humans, if Einstein had not individualized them with a proper name and baptized them *Raumzeit*.[763] Thus, he established them as a dimension never conceived before him: the fourth dimension.

Here we cite three sources that support this kind of nominalism. The first is John Wheeler who introduces us to a new theory called *An Observer Created Universe* where things are nothing until we give a name to each of them respectively.[764] The second is Derrida who has been able to read, in his own way, the power of writing so that God's eternity becomes dependent on it.[765] And the third is the Bible, specifically the Book of Genesis, where we find Adam, the Man, naming all creatures as a sign of dominion and lordship over them, giving them with their names existence, meaning, and finality.[766]

On the basis of these last three references, categories such as 'space' and 'time' seem to be transmuted from an idealistic reality to an empirical one. They become subject to our scientific calculations by the mere fact of naming and writing them because this means observing them, being conscious, and conscious of being conscious of their existence and their finality. But to the discontent

---

762 The Derridian vocabulary, Latin more than any other, would prefer the writing: "*Verba volant, scripta manent.*" We write them; they are here. We write 'tomorrow', and here it is. But, who of humans was practically able to reach "tomorrow"?

763 As it would please Ockham to say.

764 *Cf. Religion and Science*, p. 185.

765 Derrida, Jacques. *Circumfession.* The University of Chicago Press Ltd, London 1993, p. 263. "*Je t'écris Dieu, je t'éternise,*" which means, "I write you God. I make You Eternal."

766 Gen 2:19-25. Was the final purpose to keep Adam company?

of this nominalist theory, the collective consciousness of humanity was forged, throughout history, in several languages at once. Humans do not all share the same roots of names. The event of 'Babel' had to take place effectively, so to speak, in human culture. Could the consequences of 'Babel' be behind any other existential relativities in the understanding of the space-time relationship?

Humans, under pressure from their needs and abilities, sought a solution to the "curse of Babel". The human mind has developed the two phenomena of translation and interpretation that might be the foundation of an answer to the last question. The following is an example of how these two phenomena work.

Always in the shadow of nominalism, we will consider the verse of the Bible that deals with space and time *par excellence,* Genesis 1:1.[767] This verse is presented here, first in Hebrew, then in Syriac:

ב/ראש/ יח אלהים ברא אח שמים אח ארץ

B/rêš / it Êlohîm bara êt h'šamaïm êt h'êrêtz

(ܐܪܥܐ) ܝܬ ܐܦ (ܘ) ܫܡܝܐ ܝܬ ܒܪܐ ܐܠܗܐ ܝܬ / ܪܝܫ / ܒ

B/riš / it Alâhâ brâ yât šmayâ w yât arᶜâ

Then, with the same meaning, in Arabic:

في البدء خلق الله السماوات والأرض

Fi-l bad'i ḳalaka Llahou - ssamaouat wa-l ard

And, on purpose, we add some other non-Semitic translations:

> In the beginning, God created the heavens and the earth. (3G, Darby; JCB; etc.)
> *Au commencement, Dieu créa le ciel et la terre.* (BJ and BG, BDS)
> *Am Anfang schuf Gott Himmel und Erde* (BG, HOF),
> *In principio creavit Deus cœlum et terram* (BG, Vulgate)
> *Nel principio Dio creò i cieli e la terra.*(BG, NR2006 and LND)

Without considering our contribution of the *yât* in Chapter IV, we notice that the majority of English versions say that God created 'the heavens', and not 'heaven', to emphasize the plurality. French and German do the opposite by stressing singularity: *le ciel* and *Himmel*, while the Semitic languages, *i.e.*, Hebrew, Syriac and Arabic, ensure that God suddenly created 'the heavens'. Italian and Latin are exceptional for, despite their common roots, we see a contrast between the Italian plural *cieli* and the Latin singular *cœlum*.

In what concerns "the earth", all languages respect its unicity.

---

767   For this exercise, we counted on the translation closest to our Syriac and Arabic ones, which share the same roots as Hebrew. Furthermore, as with most of our biblical references in this book, we picked the most fitting translations from www.biblegateway. com. For French verses, we used also *La Bible de Jérusalem*.

Coming to what Jesus repeated so often, and finally taught us in the 'Lord's Prayer', we notice a language unanimity that God dwells in 'the heavens', as given in the Greek original of the Gospels: Πάτερ ἡμῶν ὁ ἐν τοῖς οὐρανοῖς. (*Pater imon o in tis ouranis*).

This is just one example of the added difficulty in translating a 'name', whether definite or indefinite. The tiniest variation directly affects the localization of God's Space that we have taken as an example, *i.e.*, the Space of His Kingdom, the Environment in which we strive to situate ourselves. What then would be the situation if we consider, in the same orbit, the words of Christ Who assures us, "Behold, the kingdom of God is in your midst" (other versions say "between you")?[768] Would this dilemma "heaven - heavens" be a compound allegory, an expression of a kind of particularity that we will always be invited to discover? What would this particularity look like? What would be its *Gestalt*, and under which *Zeit*, under which *Raum* would we find it? What would be its conditions, circumstances and coordinates, etc.? It is, in fact, only one example of the additional deviation that translation might cause in the search for an answer to the question concerning the understanding of the space-time relationship and the relativity that manages them. As we can see from this small example, translation has not profoundly reduced the "curse of Babel". Would the role of interpretation be more encouraging? Philosopher Jacques Derrida is not so optimistic in this sense, and we discussed this point in previous chapters.[769] However, it turns out that, just as in quantum reality, crystal-clear and precise answers concerning the *Raumzeit* phenomenon and all that is of its realm, are difficult to specify distinctly.

In addition to nominalism, translation and interpretation problems which are supposed to represent the opposite path to the "curse of Babel", it is also evident from biblical tradition that God Who created light and darkness, day and night, morning and evening, is the One Who created the dichotomies that will be for humans the basis of a great number of noetic problems. Do we not have in the text of Genesis all the necessary components for a kind of dualism, even at the fundamental level of the creation of 'space' and 'time'? Would these two concepts be the product of dichotomous effects or paradoxical ones? How could 'time' and 'space', 'sky' included, be plural? Where was God's habitat when He commanded the heavens to be? Moltmann, based on the *De Trinitate* of Saint Augustine, finally asks himself: Where did God place creation? *ad extra*, outside of Himself? or *ad intra*, inside Himself, since nothing could be outside

---

768 Lk 17:21.

769 Bennington, Geoffrey. *DerridaBase*. The University of Chicago Press Ltd., London, 1993, p. 175.

of Him? Or *ad extra* and *ad intra* at the same time, since according to biblical terminology, it is deduced that there are different spaces for the habitat of God and, concurrently, one unique space?[770]

The frustration holds tight because nothing discussed to this point has helped to clarify whether 'time', 'space', and their corollary 'space-time', are created or not. During our research, we came across a theory called *zimzum* from Isaac Luria's Kabbalah School. It seemed valid since a large number of specialists support it, especially Jürgen Moltmann. It appears, as far as possible, to contain the most reassuring and satisfactory solution to our problematic and to all the questions that have arisen from it. What is this theory, and does it indeed possess what will satisfy us? This is what we will examine in the following section.

3. The Kabbalah theory: *zimzum* (צמצום).

The first time we heard about this compound term '*zim zum*', which has gradually become a single word *zimzum*, we found it very strange as an English term. It was during a conversation with a biblical scholar friend that this Kabbalist creation theory came up. What is meant by this word, this concept from Semitic roots? (It is frequently used in the Middle East, especially by Muslims under its derivative '*zamzam*' in reference to the water of the sacred well '*Zamzam*'.)[771] Moreover, how could this creation theory serve our purpose with regard to 'space' and 'time'?

Zimzum is a creation theory developed by Kabbalah Jews to solve the problem of the location of creation, proving that God realized it outside of His Entity, *ad extra*. This theory seems to require that God create something outside of Himself; in other words, God had to prepare a space outside of Himself where creation could take place. Since the localization of this site by Kabbalah scholars, or at least of the space where the Big Bang had to take place, imposes itself as a solution to our problematic of the creation of space and time, we went in pursuit of information that relates to this theory.

We noted that something similar to this *zimzum* has been repeated in the Syriac liturgical texts that insistently praise the kenosis of God, for centuries. It is found in the Syriac term *zʿûrûtâ* (ܙܥܘܪܘܬܐ) of the God-Son Who, as Ephrem says, has shrunk, has reduced Himself infinitely to enter the womb of Mary and

---

770   Moltmann, p. 57. *Cf.* Augustine, *De Trinitate,* XV 4, 7.

771   The most sacred water in Islam. It arises from beneath the *Kaaba.* All pilgrims are expected to drink of it. When a friend offered us some one day, I noted that it was yellowish in color and had a sulfuric, bitter taste.

has taken our humanity.[772] Then follows some result of what we have discovered. Could it serve our fundamental purpose of solving the enigma of 'space' and 'time' and, consequently, finding the greatly desired *maškanzabnâ* where Science and Religion dwell with the divine Word?

3.1. Moltmann's contribution

Under the subtitle "The Self-Limitation of God", Moltmann accuses Christian theology of not having devoted enough attention to the idea of creation as an action towards the outside or towards the inside, in contrast to the Jewish Kabbalah tradition. He describes the Christian theological current as being the one that privileged a single divine action of creation towards the outside. He draws, for this point of view, on the Augustinian formulation which was decisive for the theology of the Western Church: "*Opera Trinitatis ad extra indivisa esse, tribus personis communia, salvo tamen earum ordine et discrimine.*" [773] Moltmann nevertheless objected to this limitation by asking himself: "Can the almighty and omnipresent God have an exterior? Is there, after all, an *extra Deum* for these *opera ad extra*?"[774]

We do not hide the fact that this surprising problem is new to us. We believe that Holy Providence put Moltmann and the Kabbalah School on our way in search of answers to the above questions.

Moltmann, surprised like us, has wondered about the same issue and, while being inspired by the same School, generated suppositions and other questions. He wrote:

> Only a return of God in Himself frees the space for that *nihil*
> in which God can then exercise His creative activity … Should
> we not say that this "creation outside of God" exists at the

---

772   *Cf. La Pensée,* pp. 242-3, and precisely p. 137, where Bou Mansour writes, inspired by Ephrem: "A God who does not love, a God who refuses all kenosis, is a God jealous of his identity to the point of refusing any representation of Himself. This would be in opposition to the Christian conception of God which was mainly and absolutely represented in his Son, and secondarily in man and the cosmos. Thus, the movement of return to God is made possible, thanks to a divine initiative in favor of creation."

773   Moltmann, p. 56. For our best understanding, it should mean that the works of the Trinity on the "outside", *i.e.,* the universe, are indivisible, saving however the order between the Three Persons and what is critically proper to each of them. Some talk about "terminative issue", which means that "special works are attributed to each Person *terminatively*: *e.g.,* the atonement is the work of the whole Trinity, but it 'terminates', or finds its completion, in the second Person; and the special divine presence in the Church is the work of the whole Trinity, but it terminates in the third Person." *Cf.* http://assets.newscriptorium.com/anglican/39-articles/litton39.htm [Accessed Jan 22, 2020].

774   *Ibid.*

same time in God; namely, in the space that God has prepared for it in His omnipresence? Therefore, did not God create the worldinHim?[775]

To this dilemma that shakes every traditionalist theologian, he finds a way out by formulating a trinitarian synthesis as defiant as the previous one. He says, and we try to translate his words as faithfully as possible for fear that the impact of their vibrations on the mind of the alert reader will be diminished (we quote him *verbatim* in the footnote):

> The Trinitarian relationship of the Father, the Son and the Holy Spirit is so vast that all of creation can find, within it, space, time and freedom. Creation as the action of God in nothingness and as an order of God in chaos is a male conception. Creation as an action of God in God and from God would rather be a feminine conception: God creates the world, by letting become, and be, a world 'in Him': let it be![776]

Amen. What an excellent synthesis for our research based on the two dichotomies, *ad extra–ad intra*; feminine–masculine! By this assertion, Moltmann highlights the Motherhood of God. It is as if he is adding to the fourth dimension of Einstein's relativity the 'feminine–masculine' bipolarity, which like the electron–positron one, the primary constituent of created matter, becomes unavoidable in any analysis of creation. Will this transmutation in the paternal paradigm of 'God' help us have a better understanding of the nature and behavior of time, space, and the space-time corollary? Will it assist us in answering the unlimited number of questions that emerge and in better situating ourselves in front of Zeus?

The answer seems to be positive. These words of Moltmann are an echo of Hermaphroditus of Greek mythology as well as that of the Phoenician Astarte and her son Adonis, and any similar deity found in ancient civilizations. This

---

775   *Ibid*. p. 57. (Translated by the author.) Original French text: "Seul un retour de Dieu en lui-même libère l'espace pour ce nihil dans lequel Dieu peut ensuite exercer son activité créatrice... Ne doit-on pas dire que cette 'création en dehors de Dieu' existe en même temps en Dieu ; à savoir, dans l'espace que Dieu lui a préparé dans son omniprésence? Par conséquent, Dieu n'a-t-il pas créé le monde en lui?"

776   *Ibid*. (Translated by the author.) Original French text: "La relation trinitaire du Père, du fils et de l'Esprit-Saint est si vaste que toute la création peut y trouver espace, temps et liberté. La création comme action de Dieu dans le néant et comme ordre ce Dieu dans le chaos est une conception masculine. La création comme action de Dieu en Dieu et à partir de Dieu serait plutôt une conception féminine: Dieu crée le monde, en laissant devenir et être un monde 'en lui': let it be!"

mythological figure should no longer be overlooked. It is, after all, the product of the human imagination meditating on the beginnings of humanity. Moreover, if we rethink the Platonic realm of ideas, we can say that nothing can be created from nothing, from *nihil*, especially not ideas. Therefore, this idea of feminine begetting the masculine from itself, in itself, as a different divinity, to later offer it to the world as a source of light and fertility, thus suffering its loss for six months to be resuscitated for another six months (as in the case of Astarte and Adonis), is not a negligible idea. The following interpretation of *zimzum* theory will prove it.

3.2. Interpretation of *zimzum* theory

Our traditional formation in the Magisterium of the Church triggered our first reaction to the *zimzum* theory. It is about the principle of the 'Immutability of God', well-rooted in our mind by philosophical and theological education, and conservatism does not allow tolerance, especially in such thorny cases. Moltmann, moreover, as mentioned above, stressed his concern about this difficulty: how can God be subject to changes or movements, even if they are considered free self-limitation? However, once convinced to tolerate this kind of self-motivated movement, based on the principles of omnipotence and freedom in God, there comes our second reaction: what measure of anthropomorphism is included in this theory? Do Western theologians who find this word *zimzum* so biblical, so archaic and/or so inspiring know that its namesake is in daily use in the Middle East? We felt that it is absolutely necessary to shed light on the epistemological roots of this word and on its present uses, its meanings in general, and its sacredness in Islam.

First of all, the verb *zamma, yazimmu* (زَمّ - يـزِمّ), coming from the concept *zimzum,* means "to reduce, shrink, contract, flatten, minimize, flow, drip," etc. As this range of meanings suggests, it can be transitive as well as reflexive.

When the transitive form is applied to humans, it often includes a derogatory meaning. This verb is used when a person is lowered, marginalized by another, by social, religious or military obligation, especially when this lowering is imposed by an authority. Affectively, it applies in the case of passionate love where it often takes on a subjective value, with the exception of the case of maternal love, said to be unconditional.

In its reflexive form, it also applies in psychological cases, in the sense of backing down, reluctance or repugnance; therefore, it is related to words in the lexical field of fear, modesty, and precaution, etc. Morally speaking, in this same form, it means to humble oneself in front of persons of superior dignity, to give them vital space, sometimes to the point of self-erasure. Subjectivity is not negligible in this case. If a person succeeds in applying the principle of *zimzum* freely and reasonably, he/she is considered wise and well aware of his/her limits. Such a person knows how to bow like the reed before the storm, no matter how slight it is; he/she remains immune to any possible contamination by its violence. Evangelically speaking, this person is said to be in accordance with Christ's invitation to be "cunning as serpents yet as harmless as doves", advice that Christ (the Servant of God)[777] applied to the Cross.[778] Such a person demonstrates being aware, and aware of being aware of moral and ethical, social and/or religious values. He/she adopts the "golden rule" and above all recognizes limits which requires respect for the vital space shared with others.

In general, in its reflexive form, this verb *zamma* means to apply to oneself any contraction, even a kenosis, for the good of others, especially when it requires the reduction of one's own ego compared to another ego which represents more vital or more sacred values. John the Baptist applied it wonderfully and openly expressed it before the Son of Man: "He must increase, and I must decrease."[779] We are thus faced with the phenomenon 'kenosis–plerosis' which signifies, symbolically speaking, by measuring things in decibels, that the volume of the sound waves of the voice of the Baptist, in the Economy of Salvation, must be reduced in order that those of the Voice of the divine Word increase. The divine waves, rather than the Baptist's, have the power to reach by expansion all creation. This is how Moltmann preferred to say it in order to support the theology of Nature.

As a result of this concise analysis, it becomes clear that the act of Redemption could not have been accomplished without the *zimzum* principle in its sense of kenosis and plerosis ($\pi\lambda\acute{\eta}\rho\omega\sigma\iota\varsigma$).[780]

Accordingly, the Baptist silenced himself. Jesus also diminished Himself, giving primacy to His Father. The role of continuing the redemptive *zimzum* was entrusted by the Holy Spirit to the Church itself, and to all Christians, in order

---

777 *Cf. La Pensée.* pp. 41, 91, 121-149, 248, etc. The work of Bou Mansour highlights all the figures that Ephrem used to emphasize the importance of kenosis. He dedicates to this Ephremian innovation all of Chapter II, entitled "Structure symbolique du Cosmos et de l'Écriture". He subdivides it into three subtitles: "Le cosmos et le Père", "Le cosmos et le Fils", "Le cosmos, le Paradis et les enfers".

778 Mt 10:16; In the Maronite Liturgy, this attitude is highlighted in the preparation of the Oblations. The priest says, while preparing the hosts for the consecration: "Like a lamb, He was guided to slaughter and like a sheep before the butcher He did not say a word." سيق كالحمل إلى الذبح، وكالشاة أمام الجزار لم يفتح فاه.

779 Jn 3:30. In the Arabic translation of this verse we use "minimize" (صَغُرَ). In our case we prefer to use "zamma" (زَمَّ). The proverb "*Ubi major, minor cessat*" supports this choice.

780 Lk 1:52; 14:11 and 18:14.

to continue to assimilate this process and to fill all the vital space inaugurated on Golgotha until reaching its fullness, that of the stature of Christ.[781]

Ephrem sings this, saying:

> Blessed are you, O Church, whose congregation sings with
> three harps.
> Your finger plucks the harp of Moses and [the harp] of our
> Savior and [the harp] of Nature.
> Your faith plays the three [harps], for three Names baptized
> you.
> You were not able to be baptized in one Name nor to play
> on one harp.[782]

Then the Syriac liturgy praises the kenosis of Christ with wonder as Saint Ephrem described it:

> Your mother is a cause of wonder!
> He, the Lord, entered into her and became a Servant;
> He Who is the Word entered – and became Silent within her;
> Thunder entered her – and made no sound;
> There entered the Shepherd of all, and, in her, He became
> the Lamb, bleating as He comes forth.[783]

If, in Mary and Jesus, we return to the mythological scene of Astarte and Adonis, we find that this phenomenon has always been a *pre-evangelion* in various mythologies. It originated from the interpretation of the four seasons: autumn and winter for the death period of nature, spring and summer for its life period expressed by fertility and fruits. The power of a *zimzum* creational understanding sits in the latter. Since humans have become aware of their rationality, they have understood that the voluntary self-limitation of egocentrism, especially in the maternal case, to make space for harmony between action and passion, and therefore for peace, is the unique sign of the humanization of all creation, both rational beings and nature.

If Jesus highlighted the data of this dilemma through washing the feet of his disciples, specifically through the dialogue with Peter, it is just because He was aware that all the possible space of a life and of a world would not be enough for one 'egocentricity' (*e.g.,* Cain's murder of his brother) especially when it is

---

781 Eph 4:13.

782 *H Virg.* 27, 4. Cf. *La Pensée,* p. 123; *CSCO* vol. 223, p. 100; Ephrem *Syrian Hymns, the classics of Western Spirituality,* p. 383.

783 *H Nat* 11, 6. Cf. *CSCO* vol. 186, p. 70; Cf. *The Harp of the Spirit,* p. 49.

measured against the divine. A common expression used by Middle Easterners, "I only shrink before God", is sufficient proof. We wonder, therefore, about the role that this verb plays at the level of the phenomenon of Creation?

We already know that its transitive version *zamma* and *zamzama* applies in a more practical and/or pragmatic sense, and in this case, this verb will be synonymous with 'to accumulate' or even 'to shrink' something in an orderly way to make room for another.[784] A fabric curtain, for example, is hung in a way that makes it easier to 'shrink' it in an orderly way, to let light in. Moreover, a garment once ample or relatively wide for one who wears it must be narrowed or shrunk, *zammahu/ha* or *zamzamtahu/ha* (*ha* replaces *hu* for feminine), to coincide with the desired size.

Of all this, nothing is logically refutable. We are rather faced with practical examples which would help us to understand the theory of *zimzum* in its religious dimension. What is this dimension, especially in Islam, and to what extent does it serve our problematic and our fundamental purpose mentioned above?

3.3. Religious and theological dimensions of zimzum

The term *zimzum* written as one word to correspond with the Arabic *zamzam* (which is in turn the proper name of a sacred well in Mecca, *Bîr Zamzam*, close to the *Kaaba*) has its roots in the Hebrew Scriptures. This term is deduced from the story of Ishmael and his mother Hagar, the servant who once fled her mistress Sarah after having been mistreated by her and later was dismissed by Abraham with her son thirteen or fourteen years later, just after the birth of Isaac.[785] According to the text, God-Ël came twice to the aid of Hagar for the sake of the child, the first time, when she fled, just pregnant, and the second, when Abraham dismissed her by order of Sarah and with the consent of God.

The first time, "The Angel of Yahweh met her near a certain source of water in the desert, the source which is on the way to Shur." She called it *Bîr Lahaï-Roï*. The angel entered into a covenant with her, ordering her to call her son, once born, Ishmael (*Išma-Ël* ישמעאל), which means "God-Ël will listen". The second time, he showed her water where no one expected to see it, in the desert of the Well of Satiation (*Bîr Šebaa* : שבע באר), and in accordance with the Ishmaelite

---

784 Even about the lips, it is said that under the effect of surprise or anxiety, a person who does not know what to say contracts the lips (*zamzama shafatayh*) and consequently leaves the word to others.

785 Gen 16:7-15.

covenant, the life of the child was saved.[786] This time no name was given to the place where the water appeared. Everything took place in a desert, just as the place where the Islamic religion began and where water equals life.

However, it is easy to say or write "a well was there" or "water was shown", but for the inhabitants of this region who were in strict and continual contact with the context of the Old Testament, as well as with its text, the idea of water appearing in the middle of the desert was of great importance. It forms a crucial common denominator between the different civilizations of the region. Hence, they typically named things and places, either according to the description of the phenomenon that arose out of them or according to the way in which those places affected their popular belief. This is what we call the power or the authority to give names, as we pointed out above.

From the Arabic accounts mentioned in references drawn from Islamic culture, it is evident that the two biblical scenes concerning Hagar and her son were intertwined. When Hagar was dismissed the second time, Ishmael was almost thirteen years old. Hagar had weaned him a long time ago, and he could no longer be carried on his mother's arms or shoulders as the Islamic texts imply. These last accounts emphasize the innocence of the child, as if he were still a newborn, so as to highlight the divine mercy. It is thus, we would say, that the site called *Bîr Šebaa* sums up in itself the relationship between the *'zamzam'* concept and the well of *Bîr Lahaï-Roï* (Living One Who sees me). That is why, when we ask Muslims about the Well *Zamzam*, they answer that it was at the plea of Hagar, the mother of Ishmael, that God sent the angel Gabriel who struck the sand of the desert – some say with his foot, others with his wing – and opened a cavity from which water gushed out. And, in order to retain water for her child, Hagar began to surround it with sand, repeating, *"Zam, zam,"* which gradually caused the shrinking of the sand to complete the cavity and to conserve more water.[787] Thus, the sand struck by Gabriel and added by Hagar formed a cavity by contracting

---

786   Gen 21:14-20. *Cf.* http://islamport.com/k/ser/4405/221.htm [Accessed Jan. 2020].

إنها طعام طعم وشفاء سقم » : قال أبو ذر، رضي الله تعالى عنه : قال رسول الله صلّ الله عليه وسلّم.

Cf. https://ar.wikipedia.org/wiki/

قصة بئر زمزم : قدِمَ إبراهيم - عليه السلام - إلى مكة هو وأم إسماعيل، وكان إسماعيل طفلاً رضيعًا، وترك أم إسماعيل وابنها... فظنّت أم إسماعيل أنه يموت...فلما وصلت المروة في المرة الأخيرة سمعت صوتا فقالت أغث إن كان عندك خير، فقام صاحب الصوت وهو جبريل بضرب موضع البئر بعقب قدمه فانفجرت المياه من باطن الأرض وظلّت هاجر تحيط الرمال وتكومها لتحفظ الماء وكانت تقول وهي تحثو الرمال زم زم، زم زم، أي تجمع باللغة السريانية، وهو ما يعود إليه السبب في التسمية.

787   This is the concise paraphrase of the Arabic text. We noted a basic error in it. It refers to the Syriac language what is Hebrew. Neither the word *zimzum, tsim tsum* nor the verb *zamma* exists in Syriac. These words may, however, have existed in ancient Aramaic.

(*zamzama*), retaining more vital water.

Symbolically speaking, this means that God ordered life to come out of death, as when Isaac came out of his mother Sarah's 'dead' womb. The same can be said of Joseph who came out of the empty well where his brothers had almost buried him, and above all, of Jesus Christ, the Water of Life, Who after leaving the virgin and fertile uterus of Mary will spring one last time from the tomb.[788] All these places are cavities, and obviously the Water, the Life, must somehow accompany the presence of God Who is the "Living One Who sees me". The biblical literary parallels are based on a fertile imagination and analytical simplicity. This is where the sacred name ***Bîr Zamzam*** comes from. Etymologically speaking, it undoubtedly inspired Luria in his ***Zimzum*** or ***Tsimtsum*** (צמצום) theory.[789]

All this is said to prove that ***zimzum*** does not only mean to contract, to shrink to the left or the right, down or up, in a two-dimensional plane, even if this plane is circular. It means also to shrink in three dimensions to cause a partial or more or less complete spherical cavity, ***e.g.***, a theoretical black body. This is the main idea that we expected when looking for the religious and theological dimensions of ***zimzum***. However, this creation idea was not unexpected, ***i.e.***, it did not surprise us. Moreover, by taking the experience of motherhood previously mentioned by Moltmann under his feminist conception of creation, we can well deduce that this theory was already present in the human core since the first centuries of homo sapiens. Rather, we were genuinely amazed by the insight received from this way of conceiving creation, which is "***ad extra***" and "***ad intra***" at the same time, without ever compromising the principle of the ***Šekinah*** of God, even His immutability.

It is to this amazement that we hold and will hold since in it we foresee the beginnings of our theory of ***Creatio Ex Amore***. Before moving on to the next point and introducing ***Creatio Ex Amore***, we question if there is a scientific

---

788 *Šḥimtâ Morûnoytâ*, The Annunciation Sunday, Evening prayer:

ܘܫܠܚܗ ܕܐܣܒܪ ܒܛܢܗ ܘܗܐ ܫܪܐ ܒܒܝܪܬ ܗܕܘܡܝܟܝ
ܫܠܡ ܠܟܝ ܒܝܪܬܐ ܡܫܟܠܠܬܐ ܕܐܡܠܟ ܡܠܟܐ ܫܪܐ ܒܓܘ ܥܘܒܟܝ
ܠܥܠ ܒܬܪܘܢܘܣ ܐܪܦܝܬܗ ܘܗܪܟܐ ܨܝܕܝܟܝ ܐܫܟܚܬܗ
ܡܪܗܒ ܠܟܪܘܒܐ ܥܘܒܐ ܒܬܘܠܬܐ ܐܚܕܬܗ ܗܘܐ

w šalḥan d'sabar baṭnêh w hâ šrê b-birath hadomayk,
šlâm lêḳ birtâ mšaklaltâ damleḳ malkê šârê bgao ᶜûbêḳ.
Lᶜêl batrûnâs arfiteh w horḳâ ṣêdayk êškaḥteh
marhêb lakrûbê ᶜûbâ btûlâ aḥdeh wo.

789 *Cf.* John Bowjer. "Zimzum". *The Concise Oxford Dictionary of World Religions*. 1997. http://www.encyclopedia.com/religion/dictionaries-thesauruses-pictures-and-press-releases/zimzum [Accessed Jan. 2020].

reference that supports this key idea.

Indeed, the American theologian and physicist Paul Davies places great emphasis on *Šekinah* in his writings. In one of the chapters of his book *God and the New Physics* entitled "Did God Create the Universe?" Davies mentions some new concepts that consider more flexibility in the human way of reading physics.[790] We highlight how Davies tried to resolve the deadlock of the 'gears' from the Big Bang, with respect to the appearance of space, time and matter, with the intention of paying tribute to Einstein for his discovery of 'space warps' as well as 'time warps'.[791] Davies's input consists of concepts like elasticity, gum sheet (membrane), bump, and ball. He quotes Einstein, saying: "Gravity expands or distorts space and time ... the elasticity of time can also be demonstrated."[792] After using graphic art to make what he says clearer, Davies adds:

> If we envisage our universe – all of the space to which we can possibly have physical access – as the 'new balloon' then it is certainly the case that this universe has not always existed: it was created. However, its creator can still be found within the scope of natural physical process, namely a creation mechanism with its origin in the 'mother sheet'.

Figure 38: Cosmic membrane[793]

'Mother' membrane! 'Daughter' universe! After all that has been said above, we feel that there is not much left to say to reveal the basis of our theory, which is the location of the creation made from Love, the place (*maškanzabnâ*) where God placed it. What remains is to say things in a different way, in the direction opposite to what is common and traditional in Western culture, from right to left for example, according to the mores and habits of the Lebanese.

---

790   *God and the New Physics*, p. 25.

791   *Ibid.*, p. 13.

792   *Ibid.*

793   *Ibid.*, p. 41.

3.4. The *Maškanzabnâ* and the Creation out of Love

Let us start by drawing a figure, as it pleased Davies to do and, well before him, Antoine de Saint-Exupéry in his *Le Petit Prince*. This will make what we are going to advance more admissible to the mind as well as to the imagination.

Figure 39: Perfect globe

Perfect sphere, perfect globe – this is how God is classically represented: omnipresent, omnipotent, eternal, beyond time, beyond space, unmovable and moving everything. According to the comprehensive design of Figures 2 and 3 of the first chapter, what could be the edges, the limits of this God, other than temporal infinity (eternity), spatial infinity (beyond borders), as well as the 'non-submission' to any gravity (reduction of absolute freedom)? God, indeed, as traditionally known, is Trinity and enjoys attractive and gravitational intra-Trinitarian self-satisfaction, indescribable even if approachable by analogy, thanks to signs and symbols that He has sown everywhere in creation. [794]

How and where could He then make room outside of Himself, outside of His Fullness, outside of His Time (Eternity), so that the concept "creature", deriving from the concept "creator", can be materialized in a creation? In which place, in which time, under which gravity can it be, respecting at the same time the principle of non-contradiction? According to the *zimzum* theory we ask ourselves: could it be in the following way?

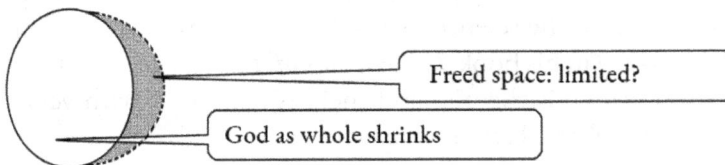

Freed space: limited?

God as whole shrinks

Figure 40: God shrinks *ad intra* (centripetal)

---

[794] *Les Origines*, pp. 77, 88; *ODL* p. 62, mainly p. 66.

This figure represents a *zimzum* in oval by contraction of the whole of the sphere and the displacement of its center. It could even be circular – nothing prevents it! However, in both cases, God 'suffers' a change in shape and/or mass, and this to evacuate the much-desired place. This is theologically unacceptable. Where could this liberated place be said to exist, and what distinguishes it from God? Would it not continue to be God since it stays in the same sphere?

Let us suppose, still according to Kabbalah theory, that God does not change shape or mass but decides to "suffer" a *zimzum* in movement, in plane or in space. This sounds absurd since God would cease to be God. According to the principle of immutability without which the concept of God loses its meaning, He cannot go through such stages; otherwise, the worry of Moltmann about the whole kerygma would be licit. In short, for this kind of *zimzum*, once we deal with concepts like God, omnipotence, omnipresence, etc., it is absurd to admit movement in any direction.

So, could God, still according to the Kabbalah conceptualization, be able to undergo an action such as allowing the formation of a cavity in Himself, at His circumference, as the story of *Bîr Zamzam* suggests, and as the theory of the mother membrane also does, without going through acrobatics like those described above? This seems more in line with the principle of *Šekinah* and the Mover which moves everything and is not moved by anything.

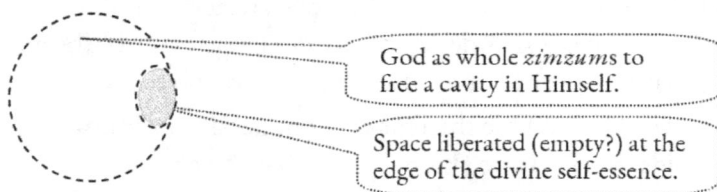

God as whole *zimzum*s to free a cavity in Himself.

Space liberated (empty?) at the edge of the divine self-essence.

Figure 41: Would it be this way?

Apparently, this way works better than the previous one. It is in God but just on the extreme edge, towards a supposed exterior of God. The idea of the sphere is also more common in the sciences, since the time of the Greeks.[795] Indeed, we find Stephen Hawking in his book *The History of Time From the Big Bang to the Black Holes* supporting his theories and their explanations with very significant figures based on the sphere. Here is an example:

---

795   Hawking, p. 18.

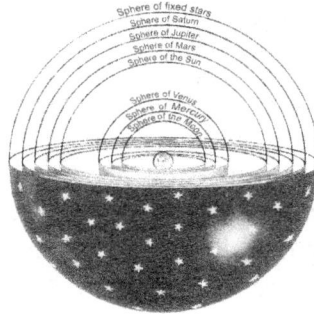

Figure 42: The cosmos [796]

Such a figure makes it essential to understand the universe in a spherical form. Can we find within this sphere the answers to Hawking's questions: "Where does the universe come from? Was there a beginning, and if so, what was before it? What is the nature of time? Would it have an end?"[797]

The most common Semitic explanation of *zimzum* theory as set out above, based on the ancient maternal conception of creation that Davies addressed and that Moltmann explicitly evoked in his famous book *The Laughter of the Universe*, highlights a sure and certain difference in understanding the phenomenon of creation. Yet scientists continue to calculate and meditate on the nature of the universe in a state of expansion, beyond any probable border or limit, beyond any likely end or finality, *ad infinitum* as it pleased Davies to represent it in the following figure:[798]

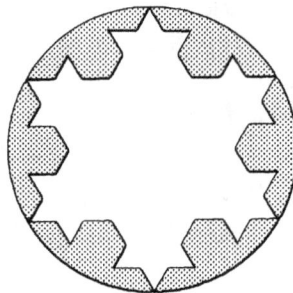

The irregular perimeter in this figure is constructed by raising equilateral triangles on the sides of larger triangles in a sequence of steps. The third step is shown in the figure. As the number of steps increases, so the perimeter becomes longer and more 'spikey'. The length of the perimeter grows without limit as the number of steps is increased indefinitely, but the perimeter never protrudes outside the enclosing circle. The area enclosed by the irregular perimeter is therefore finite, even though the length of the perimeter approaches infinity in the limit of an infinite number of steps.

Figure 43: Centrifugal expansion (first kenosis)

---

796 *Ibid.*, p. 129.

797 *Ibid.*, Introduction.

798 *God and the New Physics*, p. 15.

A theory like that of *zimzum*, considered to be subject to a centrifugal movement, from the core of the divine essence towards the limitlessness of this essence or, to use our poor expressions, towards the 'external limits' which supposedly preserve the *Šekinah*, inspires a more satisfactory solution. It is worth meditating on the maternal process of pregnancy, from time zero until delivery, to see if the 'conception' and 'creation' could be accomplished outside the 'maternal womb' of the Creator, outside of His 'Uterus', as it were. Analogically speaking, it seems also convenient to distinguish between the 'milieu' (central point) of God and His 'circumferential extremes'.

While respecting the stability of the latter, God liberated a space in Himself, in His divine substance, at the center of His 'space', with great respect for symmetry and harmony that are both fundamental to divine perfection. This is why we preferred to use the verb 'to liberate' instead of 'to empty' because what would have been called 'emptied' remains filled with the 'act' itself of emptying, of the potential of emptying, which is specific to 'passion', even to love. We opted for 'to liberate', thus 'to free', because it makes it easier to build the link between the spontaneous filling and the creation theory of emanation, without it being involuntary.

As we have just demonstrated, the idea of 'God' cannot bear emptiness or *nihil*. It is the strength of kenosis, but not of any kenosis or any "self-limitation". Here again, with this theological concept so often used in Christology, we underline a significant "differAnce" in its understanding between the Arabic language and the Greco-Latin ones.

Traditional Western Christology understands kenosis as previously described: humility, self-limitation, minimization, shrinking, etc. It is what fits the description that the prophet Isaiah made of the Servant of Yahweh. However, the way the Arabic language understands it embraces all the Greco-Latin dimensions, with a slight "differAnce". Kenosis (*akla* أخلى) indicates, according to the Arabic and other Semitic languages, a conscious act of freeing space in the substance of oneself to allow another self to be within it, to exist practically in it, to find within it space, life, and capacity for growth. It is a description of the Trinitarian relationship.

In the case of God where the substance and the essence are both Love, this results in liberating in one's Love a space of Love to create by Love.

It is according to this conception that we dared to mention above *zimzum* in the sense of "liberating". The root *rhêm* (رَحَم), around which orbits everything that relates to uterus, entrails, mercy (*caritas*), unconditional Love, supports the specificity of this understanding.

Khalil Gibran, a Lebanese Christian (Maronite), understood this well when he wrote in his *Prophet*, speaking of the virtue of Giving: "You give but little when you give of your possessions. It is when you give of yourself that you truly give."[799] That means opening the entrails where the other finds hospitality, refuge, life, and "life in abundance".

These are, in general, the axes along which we read the *zimzum* theory concerning creation. Consequently, the question Nicodemus addressed to Christ concerning rebirth from the maternal womb to gain access to the Kingdom no longer seems absurd. If it were, the Lord would not have bothered to respond. After all, He did affirm that, by the Spirit, baptismal water will be that 'uterus' from which we are born anew; furthermore, that we come out as new creatures for a more abundant life and for the Truth that liberates.

In this baptismal scene, kenosis would also mean to make room, in the sense of creating a space between the core of the self's substance and its circumferential extremes, and not outside of them, in imitation of the same divine act, according to our understanding. Love cannot be limited or blocked. It cannot even stiffen. Kenosis is a loving act which empowers the loving person to an invisible and tacit metamorphosis, to open a "womb" in oneself, symbolized by the "Well" in the abovementioned biblical text, so that life becomes continuously regenerated. This is the *Bîr Zamzam, Bîr Lahaï-Roï* (באר לחי ראי). It is what Jesus meant when He spoke to the Samaritan woman, saying, "But whoever drinks of the water that I shall give him will never thirst; the water that I shall give him will become in him a spring of water welling up to eternal life."[800] This is what could also be understood, in this unique reference from the Old Testament, where the male God uses feminine language to compare Himself to a 'mother' who bears a womb.[801] On this, the representation of the *zimzum* theory of Figure 42 finds its fulfillment in the following one:

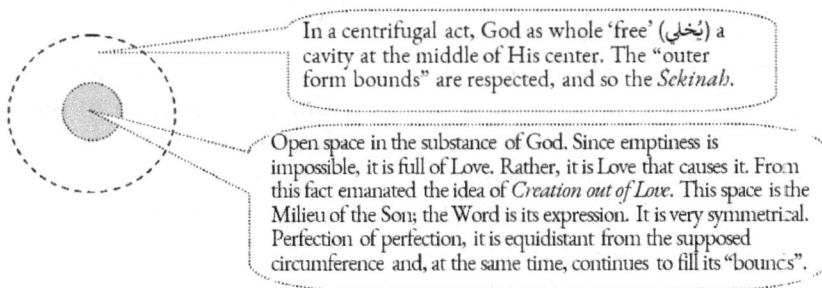

In a centrifugal act, God as whole 'free' (يُخلي) a cavity at the middle of His center. The "outer form bounds" are respected, and so the *Sekinah*.

Open space in the substance of God. Since emptiness is impossible, it is full of Love. Rather, it is Love that causes it. From this fact emanated the idea of *Creation out of Love*. This space is the Milieu of the Son; the Word is its expression. It is very symmetrical. Perfection of perfection, it is equidistant from the supposed circumference and, at the same time, continues to fill its "bounds".

Figure 44: Perfection of Perfection

---

799 *Cf.* http://www.katsandogz.com/ongiving.html [Accessed Jan 17, 2020].

800 Jn 4:14.

801 Is 49:15. *Cf.* Hos 11:8.

This figure represents a cross section of the divine 'Globe'. The cavity could be considered as a theoretical black hole whose opening at the reception of energy is not towards an outside, since there is none. The energy that heats it is, by default, intrinsic. Did the creation take place in a divine Black Hole, something that allowed God to command the Big Bang to happen, the first photon to be born and the Light to be? We give no answer at the moment; we will dwell on that later.

So far, we can say that there have been three ways to represent the *zimzum* theory of the Kabbalah School:

- The first, to shrink the globe totally in one way or another, which contradicts the *Šekinah*;

- The second, to contract to make a cavity from the wall open to a certain exterior, which sins against the fundamental symmetry and harmony of divine perfection, without neglecting the absurdity of all that is said to be exterior;

- The third, which wants the whole to create a uterus-cavity in its core, in full symmetry and harmony with its whole, which is filled, by default, with the same whole, this time called Love, and which is nothing but the very essence of God as humanity has come to know Him through the Incarnation.

And it is on this third way that we will focus.[802]

3.5. The Divine Motherhood

Moltmann discusses the subject of maternal divinity, natural femininity, and male divinity among the Hebrews. He explains why it was necessary to pass from the feminine process to the masculine to understand the divinity. It was to avoid pantheism and immersion because it would later be difficult to distinguish between mother and child. Well, this view of the priestly source of the Bible was behind all conflicts between the Canaanite deities, specifically the maternal ones and the Hebrew deities. However, despite all male deity chauvinism, the paradox of a single 'god', male and female at the same time, as expressed in the various mythologies of the biblical region, was unavoidable. Despite all the concentration on the centrality of masculine creation in the Old Testament and its repercussions on the religious, moral and educational system of the Israelite community (which can be summed up as *Šamaᶜ Israel* שְׁמַע יִשְׂרָאֵל), all traces of motherhood in the God Yahweh could not be eliminated. Not even the categorical rupture with the

---

802　Because only Love requires kenosis and suffering. This is how we understand it, we humans, and so we express it and write it, especially when we talk about these two issues at an ontological level. Khalil Gibran agrees with this by distinguishing between ontological love and pathetic love. In the first, we give of ourselves; in the second, of what the self has or possesses. In the first case, the suffering is ontological; in the second case, it is accidental.

Canaanite deities, especially those of fertility, that the priestly source imposed on the Jews by order of God was able to achieve this elimination. This is why important references, such as Isaiah 42:14, 49:15, 66:13 and Hosea 11:3-4, are considered keys in exegesis.

Based on this, if *zimzum* means letting a space come to be, why should this place necessarily be the result of shrinking or moving away or contracting from somewhere to create a cavity? All mammals coming from the act of creation, especially humans made in the image and likeness of God, ensure in themselves, in their interior, room for their offspring, their perpetually redesigned entity, to become their other self, and thereby, the 'other'. Humans "create" the other in their own feminine part, the womb of their maternal core, their entrails prepared by nature to achieve it.[803] The 'I', 'you', 'him', 'her', the 'all of us' came from this maternal cavity, where hollowing was the result of a certain liberating centrifugal *zimzum* (first kenosis), voluntary and involuntary at the same time. Voluntary, thanks to the act of Love that generates, and involuntary, thanks to the natural disposition of the cavity to adapt proportionally to the expansion of life in it, energy, wave, and particle at the same time.[804] This life incorporates itself into this "liberated" space, with a tempered growth until its deliverance, its birth, without anything being able to exist outside the main Generator from which everything proceeds, *i.e.*, the Creator.[805] Everything proceeds from Love and through Love, – without Love, nothing could have been, – but always with suffering, the Cross, which is the inevitable consequence of all mutation, liberation, transition and transmutation, actions that are always intrinsic to the Creator's sphere, intrinsic and extrinsic to the created one. A new graph is emerging in our mind which does not hide some amount of anthropomorphism, but which is nevertheless supported by Religion as well as by Science:

---

803 Abraham's faith and vocation story, considered in a realistic way, could well affirm the ontological unity in complementarity between man and woman, as expressed in the first account of the creation of the human being. (Gen 1:27-28) While the male is made to create god, the female is made to create man. Here, to create means to make space in oneself for the other, even for the "Wholly Other".

804 We consider this phase a second kenosis since the first should be experienced when the "female" opens the door of her "body" to receive the "sperm" of the "male". There, the act is also voluntary and involuntary at the same time, with an execution of a centrifugal *zimzum*.

805 The application of this point of view to the "Mother Nature" phenomenon explains what mythologies have tried to emphasize in what concerns the four seasons.

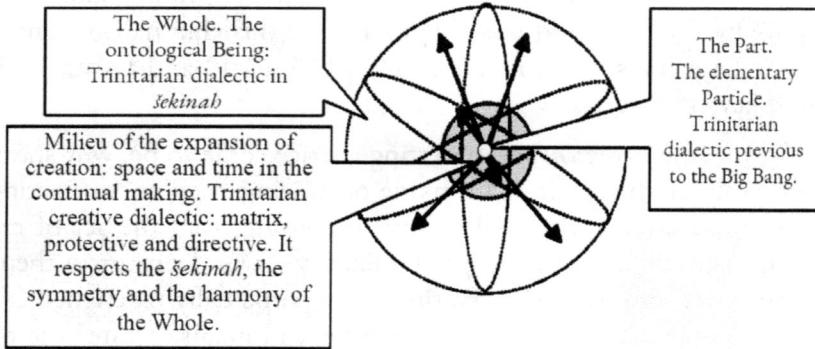

Figure 45: God "kenoses Himself" from the core of His Love towards His
infinity and His eternity and frees a place for creation.

Here, it is no longer the cosmos that grows and pushes a meta-logical
previously-prepared membrane to expand. Rather, it is God Himself Who
continues, beforehand, the intimately maternal divine process, and at the same
time providentially paternal kenosis, as much as the expansion of the creation,
which is in Him, necessitates.

Once convinced of this, can we still allow ourselves to ask if there is enough room
in God for an infinite expansion? We have no answer or comment for now. Rather,
we wonder simply, with all those who wonder, if some component of time and space,
objects of this chapter, would be perceptible on the horizon of this figure and help us
to note if they are both created or not? Alternatively, more precisely, if they are created
and 'not created' at the same time? We wonder if our proposal would similarly satisfy
the two problematics of transcendence and immanence, as well as that of emanation.
If yes (and we assert it is yes), we can indeed consider accomplished a conceptual shift
concerning the dogma of creation. Hence, we can answer the questions left pending
and affirm that all the answers come from a single interpretation of the phenomenon
of creation, that of creation out of Love: *Creatio Ex Amore*.

This is how the *zimzum* theory led us to the conception of an exhaustive
creation inside God, immersed in Him, and at the same time 'outside', on the other
side of a certain membrane, intrinsic, osmotic, which makes God transcendent
and immanent at the same time. This conception is not pantheism, and at the
same time it is not separation since the very idea of separation has become absurd.
The "whole" and the "part", being equal and at the same time "differAnt", are
respected. Even the paradoxes of Zeno of Elea find their satisfaction here, finding
in the same 'milieu' the extremes and the 'milieu' (central point) of all. We add, as
additional proof of our sound vision, that this is what any fetus experiences during

its residence in the maternal uterus where 'time' and 'space' become both its own and those of its mother. Let us add that even the gravitational energy of the fetus that pushes its body out of the folds of DNA in perfect harmony is, at the same time, that of its mother.

All the theorems, principles, axioms and constants of quantum mechanics, especially entanglement, EPR, the 'butterfly effect', etc., find their effective applications between the effects of the energy of the heart of any mother and the reactions of that of any fetus.[806] We leave to young scientists the task of exploring these bio-quantum phenomena. We content ourselves to meditate *râzâtically* on the similarity between God and mothers *vis-à-vis* these effects and reactions between 'heart' and 'fetus'.

3.6. God and Mothers

It is necessary to complete this panorama to be able to provide a better inspiration to Theology as well as to Science. No one should see it as dividing between two realities as if what is said of God belongs to a specific framework and what is said of a human mother belongs to another. It is true that nothing can be outside of God, but it also belongs to our human limitations not to be able to see reality as He sees it and/or precisely as it is in Him. Still, if humans could remember the days they spent in their mother's womb, they would form an idea of what the divine Word meditated on during His residence in the womb of Mary. Moreover, if we join Saint Ephrem in his Marian hymns, we can imagine what Mary was meditating on as she became more and more aware that she was carrying within her the One Who carries all of creation. While taking into consideration the perception of Davies' maternal membrane, we conclude, after adopting the spherical form of eternity in place of the flat one, that those elastic membranes exist but in a concentric position that inspires perfection.

Figure 46: Maternal creation according to Davies

However, in place of the conception of Davies' cosmic membrane which we reproduce here to facilitate the comparison, we prefer the following:

---

806 What to say then about the father's heart?

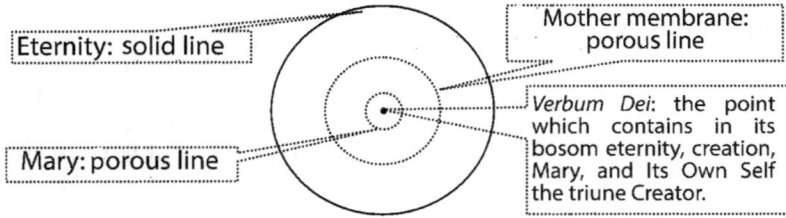

Figure 47: Intermediary membranes between the center and the circumference

On this, we wish to refer again to the Delphic saying mentioned by St. John Paul II in his encyclical[807] that we summarize as: "To know yourself well, you have to become aware of being aware, and aware of being aware of what you are aware of." Whereas to say seriously that we understand creation in one way or another, the human mind must feel, on the one hand, how this creation carries Love, and on the other hand, how it is carried by Love.

Figure 44, taken from Davies, and which we reproduce here as a word in a sentence, ⊙, invites us to interpret it in reverse mode this time, not as being an expanding universe in a previously prepared space, but rather as symbolizing the divine Maternal Kenosis freeing up as much space and time necessary for the 'Fruit' of Her Womb, the 'Fruit' of Her Love. The creation is, in this case, similarly to a human fetus, in the image and likeness of its *Matrix*.

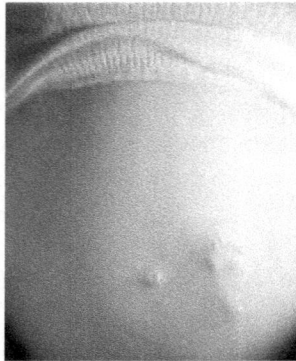

Figure 48: Courtesy of Providence[808]

---

807   As well as by Eugene Wigner quoted by Ian Barbour. *Cf. Religion and Science*, p. 185.

808   Chance and probability made it possible for all of us. Unexpectedly, when I was composing the first version of this book in 2011, I received this photo in an email message. I was very surprised to see the footprint of a being inside the womb. I felt like it was providentially sent to me.

Even the tomb that accompanies us from the last chapter joins, in this context, all its dimensions.[809]

Conclusion

At the end of this chapter, we deduce that if on the physical plane space and time are said to be created, separated when Newtonian and united when Einsteinian, they do not cease at the philosophical level to be perceived as two pure, logical and/or metalogical ideas. At the psychological level, they remain perceived as pathetic and syndromic. At the religious level, they are truly 'mysterious' or 'mystical', while being taken for spiritual, without ever ceasing to be part of the created realm. However, once both step beyond physics to reach metaphysics in the service of dimensions like those of mathematics, transcendence and even theology, they adorn themselves with the quality of being 'not-created'. What is interesting is that, in all cases, they keep the same linguistic concepts as names, but they change paradigm. They metamorphose into prose, verse, notes and measures, numbers, and into numerical letters, to suit any kind of event and circumstance, especially those related to the process of the Economy of Salvation. There, according to Saint Ephrem, they become considered *râzâtic*. The two seem to be "alive" and endowed with a purpose that serves to make sense where the absurd risks putting everything in doubt, including the Creator, the only vital Source. Their qualities, strength, and finality imply that they are, at the same time, created and not created. It means that they are made, produced, but not created. The elementary, personal and collective instrument to explore, tame and make good use of them is called 'Patience'. It is an instrument well recognized by Science, and also by Philosophy and Religion, essential for any successful outcome. It belongs to the nature of these disciplines and is measured by quanta, by that 'little' that Jesus spoke of and which is just enough so that what is in potency comes to be in act (reality), and vice versa.[810] Patience is the *sine qua non* condition to become better aware of those two "beings" who defy humanity so much. Would they not also be a *maškanzabnâ* where Science, Religion, and everything that humanizes can meet?

The Kabbalah theory of *zimzum* has been of great use to us in revealing the nature of time and space and giving them their own value in 'created' as well as in

---

809   *Cf.* Saint Cyril of Jerusalem, *Homily on the Resurrection.*

زمن الدنح المجيد، قراءة من القديس كيرلس الأورشليمي، ص.٤٠٣؛ جامعة الروح القدس - الكسليك، لبنان ١٩٧٨

810   Mk 13:13; Mt 10:22 and 24:13.

'not-created' realms. The 'mystery' has gradually passed from the Greek level to the *râzâtic* one, without which, in our opinion, no possible insight could have been made to solve this enigma of space, time and space-time.

The fact remains that the explanation of the phenomenon of the transmutation of these two terms, between 'created' and 'not-created', was wonderfully made through the theory of 'act' and 'potency' according to St. Thomas Aquinas. Indeed, these two terms, obviously specific to the divinity, helped us to conceive this process. This is why our understanding is fixed on time and space as beings in potency, since they are in God, and are realized in act, are concretized in the created universe, all the more as creation comes to be in act.

The accumulation of successive acts that require clean, consecutive and discrete spaces forms the distances and measurements between incidents, actions, and all their derivatives, and takes shape in physics, first of all in quantum physics based on bosons, before taking shape in Newtonian physics. These "space-times" are also ensured by the successive kenoses that accompany the acts. This leads to 'predicting' and 'affirming' that if, according to pessimistic physicists, the universe one day might undergo the Big Crunch in a movement opposite and/or complementary to that of the Big Bang and space and time disappear, they will not be annihilated.

They will resume their place, in potency, in the Almighty who manages them in His maternal *šekinah* (שכינה), by His paternal Providence, until they resume their role in the service of creative Love, of order and harmony, directed by their own final cause that is Gratitude, Eucharist (*Efkharistia*), the exchange of Love by Love. This will not be the first time because the Flood of Noah with its covenantal Bow was followed by a renewed earth. This will only require the same 'short while' mentioned above or the same blink (*rfâf'aynâ*) mentioned in the first chapter.

From this given viewpoint, even Laplace could regain his 'redemption', since he would only have to admit that he could not be aware in his lifetime of being in the very 'womb' of the same Hypothesis of which he denied the necessity.

Space and time are of quantum nature where quantum mechanics requires it and of Newtonian nature where the correct course of things, at the service of human well-being, requires it. The whole, finally, should be at the service of the fair relationship between creation and Creator, between the 'Daughter Universe' and the 'Mother Eternity'. The whole space-time relationality should be serving the glory of the rational, just, and good human beings until they reach "the measure of the stature of the fullness of Christ" in *râzâtic* Love.

We close by saying that God is Love and all of creation, including humans, arises from Him, immersed in Him without ever being confused with Him, just as Christological hypostasis reflects. It was not by absurdity that Karl Rahner once wrote: "When God wants to be what is not God, Man comes to be." [811]

For the next and final chapter, we have to analyze the nature and the process of this divine movement that leads to kenosis. How did it come about, and what made it happen and repeat itself? Why is everything in the universe round, spherical, turning, spinning? Why is all not cubic as we pointed out above? This topic we will consider under the title "Spin and the Trinity: the relation between cosmic gravity and Love".

---

811 Rahner, *Foundations*, p. 225.

CHAPTER VII

Spin and the Trinity :
The relation between cosmic gravity and Love

Little by little, we begin to understand that reality is veiled, inaccessible, that we barely perceive its drop shadow, under the temporarily convincing shape of a mirage. But then, what is there under the veil?

Jean Guitton.

Preamble

In tribute to Antoine de Saint-Exupéry's *The Little Prince*:

At twelve years of age, thanks to the curriculum of my school, I have been introduced to The Little Prince of Antoine de Saint-Exupéry. Then, I could not grasp the symbolism it contained, yet, I still remember what pleased me the most: the verb 'to tame'. That was to allow me to have friends among the animals. As for the images that accompanied the text, I found them magical because their colors were so appealing, and their characters, the Prince, the Boa, the Hat, the Elephant, the Fox, the Flower, the Planet, etc.,very friendly to me. Since then, I never stopped blessing the author saying: « May God bless Antoine de Saint-Exupéry and the civilization that brought him to humanity. »

At that time, I understood everything a child had to understand and everything Saint-Exupéry expected a child of my age to follow. Sadly, I was not yet able to compare what I was getting as a child and what I would one day get as a 'fully-grown' person.812 Was I going to become like the 'fully-grown' person of the story itself, or was I going to preserve, despite maturity, a child's understanding, so lauded by the author? Today, I hope through this work to succeed in discussing this comparison but with a change at the level of the 'characters': Universe will replace the Planet; creation, the Hat; cosmos will replace the Elephant; time, the Boa; and the human being, the Beast, etc. What hasn't changed at all is childhood that still enjoys the same strength, the adults who never understand anything by themselves, and the fatigue that it costs children to give them explanations all the time.[813]

Is it not that similar to what I discussed while tackling interpretation, angles of view, fields of view, and the invitation made to every person by the Master of Childhood to become children anew?

M.C.R.

---

812 By this concept, Saint-Exupéry refers to adults who detached themselves totally from the spirit of childhood.

813 Saint-Exupéry, Antoine De, *The Little Prince;* Chapter 1. http://www.cmls.polytechnique.fr/ perso/tringali/documents/st_exupery_le_petit_prince.pdf [Accessed the 5th Dec 2017]. We will refer to this book as *The Little Prince*.

Figure 49: Figures of the beginning followed by those of the taming:

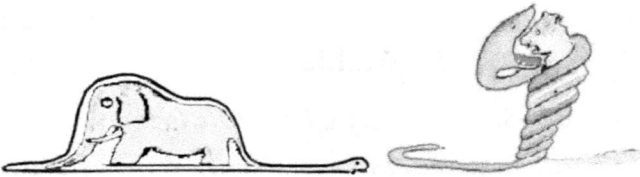

Figure 50: The consuming time: can it be tamed? .

Figure 51: The fox and the taming

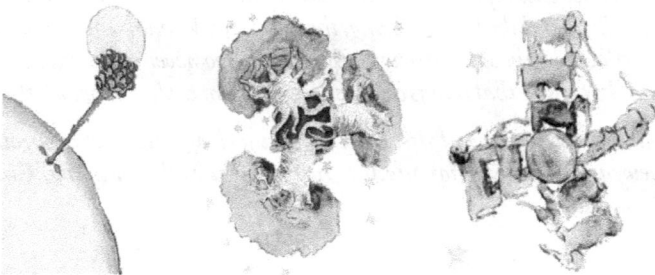

Figure 52: Planet of the Little Prince, populated beyond
Einstein's relativity

Introduction

In this final chapter, we are going to tackle four notions: two of them are from the field of Science, spin and gravity, and the other two, from the realm of Theology, Trinity and Love. They represent for us, four new bases of a comprehensive, multidimensional *râzâtic* reality. In this reality, Creator and creature seem to share the same habitat that contains them, according to a given gravity and a given spin, in a way already described above. It is a 'milieu' which, because of the importance of freedom, dignity, and love that its thinking residents enjoy, knows neither beginning nor end. It also requires, simultaneously or unilaterally, that one agrees to undergo *zimzum* so that the other can grow till reaching the most beneficial fullness for the good of all partners. Here is the core of the Messianic principle, a "life for a life".

We will deal with these four bases because it is in them, and through them, that kenosis took place and still takes place. It is just at their level, to believe the theories of Bergson and Guitton, that the perfect symmetry 'has broken', which still has to be proven. Yet, what remains enigmatic – although, until now, we have taken care to reduce the riddles – is determining the method by which *zimzum* could have happened in God without undermining His immutability.[814] It is this divinely existential question that we will answer by shedding light on the two scientific 'mysteries', spin and cosmic gravity, studied in the same 'milieu' that we suggested for the location of creation, namely, the divine milieu, the womb of God.

To dwell on the secret of this kenosis is to investigate all the secrets of nature. We are convinced that the answer to our question lies in the Christian root of the concept 'Creator', Who created all that is veiled and unveiled, the Trinity.[815]

We will do our best to describe the possible solution based on the combination of the two theories of *šekinah* and *zimzum*. We do so to elucidate, at the same time, the dogma of the kenosis of God and the two phenomena of cosmic gravity and spin, both of which are still not understood in the present day. In what follows, we will try to find answers to multiple questions humanity has asked since Aristotle, such as: Why does everything revolve (spin)? Why is everything round, spherical, or almost spherical? To do this, we will stick to the following plan:

---

814 The problem does not arise at the level of humans because, as all the proverbs of human wisdom teach, if humans fill the universe with their vanity and each moment build a new 'Babel', the very dialectic of their death will defeat them and will bring them back to what is the proper volume of this creation: a handful of dust.

815 This draws us to know how the Trinity 'kenosed' Itself and freed an enclosure to prepare a space for creation whose synthesis is the human being.

1. The positioning of the problem of cosmic gravity

2. The positioning of the problem of spin

    2.1. The role of imagination and feeling

    2.2. God and humans

    2.3. Moving from the speculative realm to the *râzâtic* one

3. The Kabbalist theory of *zimzum* (צמצום) and the creation of the Cosmic Milieu: The hidden God *"Deus Absconditus"*, *En-Sof* (אין סוף), Unlimited, *En-Sof'Or* (אין סוף אור) Unlimited Light[816]

4. The source of cosmic gravity and spin

    4.1. The source of cosmic gravity and gravitational force

    4.2. The source of spin and all rotation

        4.2.1. What is spin?

        4.2.2. The source of spin and every rotation: the Trinity

5. Spin and the cosmic Trinity

6. The nature of cosmic gravity and its relation with Love

7. The *yât* (ܝ) of Childhood and the Eternal Motherhood

    7.1. Creation of the alphabet and the first syllables

    7.2. Maturation of the alphabet: the beginnings of communication and kenosis

    7.3. Implications of this Name for Marian dogmas

        7.3.1. Regarding the two accounts of the creation of Adam and Eve and the promised redemption

        7.3.2. Regarding the Immaculate Conception

        7.3.3. Regarding the Assumption of Mary

Conclusion

N.B. We let the reader assess our understanding and our childhood because it is in this taming between 'childhood' and 'maturity', in the depths of things, that significant events happen in the world of the 'fully grown'.

---

816 *Absconditus* = Unintelligible; imperceptible, inaudible, incomprehensible, indistinct. We write Cosmic Milieu with capital letters in line with Chardin's 'Divine Milieu'.

1. The positioning of the problem of cosmic gravity

Regarding this point, Ian Barbour blames determinism which was the product of Newtonian physics:

> It reduces the world into a complicated machine that follows immutable laws, with the slightest predictable details... conceived by an intelligent Creator Who expresses His purposes.[817]

For his part, Paul Davies says in his book *God and the New Physics* that:

> The force of gravity activates the whole phenomenon of measurements on a cosmic scale. For astronomically measured objects in the universe, gravity trumps all other forces like magnetic and electric. Gravity shapes galaxies and controls intergalactic circulation. When it comes to explaining the expansion of the universe, Gravity is the key.[818]

Brian Greene, in his book *The Fabric of the Cosmos,* joins this problematic by informing us of the cause that stimulated Einstein's curiosity regarding relativity. Green says it evoked in the mind of Einstein a key question Newton had already raised two centuries before him: "How does gravity exert its influence over large expanses of space...how does gravity do its job?"[819]

Figure 53: The Universe as seen by the eyes of 'fully-grown' children

In our case, to talk about something, first of all, we have to see how it was epistemologically defined or conceived. In the book *Symmetry and the Beautiful Universe* by Leon M. Lederman and Christopher T. Hill, the authors seem to be

---

817 *Religion and Science*, p. 18.

818 *God and the New Physics,* p. 13.

819 Greene, Brian. *The Fabric of the Cosmos, Space, Time, and the Texture of Reality.* New York: Knopf, Random House, 2004, p. 64.

amazed by the beauty and power of the symmetry associated with gravity.[820] As a result of their analyses of the phenomenon of the force exerted by the sun on the planets to keep them in its orbit, they ask themselves the following questions: "What is this omnipresent force? Would it be gravity?"[821]

On this basis, we allow ourselves to say that gravity already emerges from the 'symmetry' present between the letters and the words that form such questions and that seems to come from a well-balanced and harmonious mind. However, since no one dares to push one's research forward once it faces a question without an answer, we see it necessary to take part in Lederman's and Hill's efforts and get ahead of their issues by asking: if gravity is an omnipresent force, what makes it so? Or still better: where does gravity get its ubiquity?

Further in their work, Lederman and Hill sharpen our curiosity even more and push us to ask the following questions with them: "What is the nature of the gravitational force between the planets and the sun, that shifts the movement of the planets from straight lines to elliptical orbits? Why elliptical? What is the precise mathematical form of the gravitational force?"[822]

These questions are providential for our project. Lederman,[823] undertook the answer to these questions using the famous Newtonian formula proper to the measurement of forces. Lederman warns the readers of the disturbance, the 'eye glaze', that the equation could produce in them. The formula is: $FA_B = G_N m_A m_B / R^2$

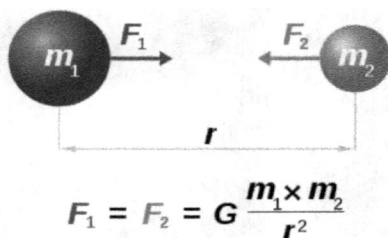

$$F_1 = F_2 = G \frac{m_1 \times m_2}{r^2}$$

Figure 54: The force of gravity

820 Lederman, Leon & Christopher T. Hill. *Symmetry and the Beautiful Universe*. Prometheus Books, NY, 2004. We will refer to this book as Lederman & Hill.

821 *Ibid.*, p. 136.

822 *Ibid.*

823 Lederman, Leon & Dick Teresi. *The God Particle – If Universe is the Answer, What is the Question?* Del Publishing, New York, 1993; reprinted by First Mariner Books Edition, 2006.
    N.B.: Leon Lederman is the co-winner of the Nobel prize, Director Emeritus of the Fermi National Accelerator Laboratory, and the author of *The God Particle book*.

Lederman did not introduce this formula to let it be followed by an explanation that helps make gravity understandable through its calculation, but rather to underline another conundrum that makes gravity still more enigmatic. It is the Newtonian constant G_N described as 'The Magical Number' of gravity,[824] and characterized by Lederman as follows:

> Note that G_N is not just a mathematical number like 3.1416,
> but rather a physical number, because it is a reference
> to a system of units, and its value is different in another
> system.[825]

The authors of *Symmetry and The Beautiful Universe* add: "Despite its ubiquitous nature, gravity is, by nature, a very weak force." [826]

For the moment, weak or strong are not in themselves what interest Religion nor the Science-Religion relation, we may say. What interests this relation, as far as we are concerned and according to our conception, is to discover and to know the origin of this force and the reason behind its omnipresence.[827]

2. The positioning of the problem of spin[828]

It is an old discovery of humanity that all planets revolve around themselves and that some rotate around others. According to Aristotle and the experience of humanity, everything turns. The American author Kenneth Ford says in this regard:

> Our Earth revolves around its axis once a day and around the

---

824 It is equivalent to 6.673×10^{-11} m³/(kg s²) or 0.00000000006673 m³ kg⁻¹ s⁻²
   In 1684 Newton started writing *De motu corporum in gyrum*, the first draft preparing the *Principia* that were completed in 1686. In his first approach, he had already corrected the Cartesian conservation of motion by taking direction into consideration, had redone the formulation of the principle of inertia, had conceived in terms of strength the composition of movements – that of inertia and those which alter it - and had formulated the law of centrifugal force independently of Huygens and in terms of centripetal force (*i.e.,* of the cause of the movement, and not only of the effect).
   Cf. E.U., "Newton, Isaac", article written by Michel Paty.

825 Lederman & Hill, p. 137.

826 *Ibid.*

827 "But what is gravity? Newton understood how the force of gravity affects the movement of objects but not why gravity works in such a way. Admitting the limits of his knowledge, he adopted an instrumentalist point of view: the task of the scientist is to capture the observations in precise mathematical equations; to explain the how and not the why. Only the experienced things of the world are eligible in science. Even if the 'why' is fascinating and many scientists would willingly spend years of their lives to find an answer, the scientific majority approve Newton's instrumentalist point of view."
   Cf. http://www.astronomynotes.com/gravappl/s7.htm#A6.2 [Accessed Dec. 5th 2017].

828 Spin is the internal property of elementary (subatomic) particles analogous but not identical to the concept of rotation.

sun once a year. Even the sun revolves around its axis once every twenty-six days and circulates the galaxies every 230 million years. At a longer time scale, galaxies rotate around each other by sector. The question [is]: Does the whole universeturn?[829]

Rotation of an object around its own axis or around another axis is not only the quality of what is visible. Ford affirmed this to his readers, as well as to us who are just novices in this area. The reality of spin and rotation is also applicable at the invisible level of the microscopic. Ford goes on, saying:

> While going down the scale of sizes, the molecules rotate (at a velocity that depends on the temperature), and the electrons inside the atoms revolve around the nucleus at a speed that varies between 1 and 10% of the speed of light. The nuclei spin – most of them do- and the protons, neutrons, quarks and *gluons,* all rotate inside the nucleus each about its relative axis. In fact, most of the particles, fundamental and compound alike, have the property of rotation.[830]

Thus this author "... rallied the mind of his students distributing to them a header emblazoned with the motto πάντα κυκλεῖ (*panta kuklei*), which means: everything revolves." [831]

Thereupon we say: to discover that everything revolves is like discovering that God exists, and as a result, the case becomes part of Mr. Ford's students' common sense. It comes to their intuition to generate questions such as: What kind of rotation is it? Around which axis does the rotation take place? In what direction does it happen, clockwise or counterclockwise? Questions will follow that also concern the shapes, the orbits, the kinetic momentum, the dynamics, the velocity, the symmetry, the 'ellipsoidality', etc.

All of these questions with those mentioned above about gravity, and all those that can result from them, are no longer the priority! The essential question becomes, henceforth, to know why "everything turns" and where this movement comes from? For that, we need a multi-dimensional imagination and extreme openness, as vast as the cosmos, if we dare say so.

---

[829]   *The Quantum World*, p. 24.

[830]   *Ibid.*

[831]   *Ibid.,* Footnote.

2.1. The role of imagination and feeling

> *"Imagination is more important than knowledge..."*
>
> Einstein

Antoine de Saint-Exupéry stresses the importance of imagination to pierce the veil, the skin of the boa, the *Hijab* of Moses, etc., in order to have access to the essence of truth. In this sense, Saint-Exupéry is consistent with a biblical worldview. He develops an occidental biblical version of the manner to see reality, a version quite analogous to the Middle Eastern one. According to this style, feelings and imagination are the two daily fertilizers of an analogical, metaphorical, symbolic, and almost mythological epistemology. For the peoples of the Middle East,[832] the act of guessing goes in the opposite direction to that of Europeans, just as their way of writing and reading requires. In the world of western 'fully-grown' persons, the mind stops seeing what is behind the shape of the 'hat' or behind 'the skin of the boa'; in the Middle East, however, whatever age the mind is, it will take great effort to see the 'hat' or the 'boa' as such.[833]

For example, the Semitic way of seeing things will strive to see, behind the appearance of the 'hat', all the different realities based on the probabilities that the omnipotence of God Who knows better than humans –*'Allah aʿlam'*– allows someone to imagine, all kind of realities with their causative cause, final cause, and intermediary cause, but never the true reality, the truth, as it presents itself. It is often an erroneous whole since it cannot escape the influence of the anthropological primordial cause which is the subjective satisfaction of the questioning person, especially if one is 'religious'. It is appropriate, then, that such a person gives a name to this reality.[834]

---

832 Khalil Gibran could be considered as the prototype of the Oriental person. André Siegfried wrote in 1935: "The mental structure of the oriental man gives the impression: 1- of his interest in speculation much more than in organization; 2- of his attraction to skill much more than to the rule; to see his benefits in life come from a succession of particular causes including flexibility, fortune and intrigue. *Cf. Al Assala*, MCDR, #22, October, Lebanon, 2016.

833 Just as what is happening now [*sc.* 2010-2015], in the region of the Middle East. Nobody wants to see the war that is threatening to devastate everything. On the contrary, all the groups of the population of this region, including the Hebrews, see only the will of God and some impending paradise. Later, they will only notice that it was a war in due form. ["Would they be able or would they want to have the good will to see behind or in the hat of the Little Prince?" Prof. Dennis O'Hara's commentary, University of Toronto, Canada]

834 A good example is the Middle Eastern custom to use coffee grounds to foretell the future and to repeat: "*Inshalla*" ("God willing").

This, in a way, is what happened to Jacob at *Penuel* and to Moses at *Horeb*, the Mountain of God.

This is also why the first accents of theology were founded on the final causes by naming visible signifiers after what is invisibly signified.[835] We would say that it is only thanks to Greek philosophy that the Semitic mind has more or less learned to deal objectively with visible phenomena and to apply the Aristotelian syllogism to deduce the essence and the purpose of what is invisible.

It is through this encounter between the Semitic and the Greek minds that the rational and symmetric conception of a Creator found a place in the Nicene Creed. The importance of this conception resides in the crucial fact of bringing back the whole creation to the same God. Its magnitude sits in the fact of emphasizing that there is nothing in this created universe – of which humans can be conscious or become conscious of – that does not refer to its causative cause (*a priori*) and final cause (transcendent), God, the Most High, *El Elohim* (איל אלוהים).[836]

After all, it is a question of 'feeling comfortable' in a context that instills imagination, common sense, and satisfaction. Is it not true, in the end, that anxiety and security provide conditions for the imagination to grow?

Psychology teaches that feelings and imagination are the instruments of a reality different from the objective one. Consequently, by reconsidering the formula of the Creed, "Creator of all that is visible and invisible", that is meant to be exhaustive, we can ask ourselves: where will God Himself, the Creator, be "placed"? Is He part of what is visible or of what is invisible, or of both at the same time, knowing that He presented Himself to Moses on Mount Sinai as an invisible being even if audible?[837]

Our Catechism has taught and still teaches that angels and demons, like any other spiritual creature, are invisible but are part of Creation. Now, is God, Who is also spiritual, part of the same category of the invisible Creation? In other words, would He have created Himself, or should there be a metaphysical category, specific to God, called the invisible-inaudible and non-created at the same time?

---

835 Gen 2:22 and 3:20. By naming his mate first, Woman and then Eve, Adam signs an epistemological masterpiece whose signature forms two signs in line with the Platonic theory of signs.

836 The bases of the Nicene Creed are said to have been developed by Origen in accordance with the theological School of Alexandria. The Holy Fathers, especially Paul and Ephrem, will complain about any mental rigidity impervious to divine grace. Putting aside all rigor, rigidity, and reductionism is a basic key to soften the difficulties between Science and Religion.

837 Ex 33:21-23.

With what kind of epistemological instruments are we equipped to evolve in this specific world other than imagination and feelings, faith and love included?

Ephrem talked of a 'luminous eye', but who among us can pretend to be Ephrem? Shouldn't we, theologians and scientists, instead build a more precise vocabulary able to help us progress at the level of a tri-reality dialectics, to move toward a better understanding of the 'power' that this world could exert on the other two worlds, as John Bell mentions in his demonstration of non-separability?[838]

This theorem generated a whole body of literature known as the *Butterfly Wing Effect*, which means that even a wing stroke of a butterfly is not lost in this universe because it makes a whirlwind somewhere. If, we suppose, the world of quantum reality replaces the invisible world of the Creed, what name is left that humanity could give to the created spiritual world? If metaphysics studies the natural environment of everything that is created and invisible, what attribute would remain to be provided to the 'milieu' that includes everything that is not created and thereby is invisible? Teilhard de Chardin was aware of this complexity when he introduced the concept of 'Divine Milieu' to the human consciousness. If this 'milieu' is considered metaphysical, would it then belong to what the Creed calls the visible or the invisible?

To solve this enigma of the visible and the invisible, the story of creation, fortunately, leaves some clues that support the fully-matured childhood.

2. 2. God and humans

"To cultivate the ground",[839] working the visible, was the purpose behind God's creation of the visible Man. Consequently, the interrelation and connectedness

---

838 "Non-separability: that in some very special situations, two photons that have interacted in the past, have properties that their mutual distance, however great, is not enough to separate. They form an inseparable whole even when very far from each other: what happens to one of the two, wherever it is in the universe, is irrevocably entangled with what happens to the other in another place in the universe, as if a quantum bond, immaterial and instantaneous, held them together. In this case, we speak of a "strong correlation". *Cf.* Klein, pp. 115-117. The term 'non-separability' is not ideal, even if consecrated by use, for two reasons judiciously noted by Jean-Marc Lévy-Leblond. First of all, this term is built on the negative mode when it would be advisable to give a positive account of this property of quantum reality. Second, it is deceptively concrete, since in everyday language, separation concerns ordinary space, whereas we are dealing here, in the quantum field, with the abstract space of states... Jean-Marc LévyLeblond has recently suggested substituting the word 'implexity' for the word 'non-separability'..." *Cf.* Jean-Marc Lévy-Leblond, "Mots et maux de la physique quantique", in *Revue internationale de philosophie, La Mécanique quantique*, n° 2, June 2000, pp. 243-265. (Translated by the author)

839 Gen 2:5 "... and there was no man to cultivate the ground."

between Adam and all that is visible is clear and strong. Thereupon, the question that imposes itself here would be: who or what did God create to work the invisible? We cannot postpone the answer to this question. **It is the same human being or, more specifically, the few abilities of the human mind, particularly the ability to write.** After all it is Man who forged from primitive sounds the first audible words and wrote them down. He gave meaning to each word, and since then, he has been refining their signification or creating new words and giving them new meanings so that hearing encounters sight in knowledge and recognition. Thus, the first words pronounced by God ("Let there be light," for example) are Man's product, which explains, on a linguistic basis, the way Man wrote 'Light' before God created the Sun.

In short, and while paying tribute to Saint-Exupéry for his critique of western reductionism, it is indeed Man – this 'Unknown' of Alexi Carrel – who invented his name to become the Man, known as Adam. It is to his own model that Man drew the word 'Adam' in the stomach of 'Time' which spirals like the boa of the above-placed drawing, to crush and digest everything that it swallows, even the soil-dust (*Adïm* ܐܕܝܡ) from which the word 'Adam' derives.[840]

Then, to answer a question similar to the one raised by the Little Prince and to make sure of what the fully-grown-up people see in his drawing, Man wrote the word 'God'. This is how those two concepts 'Man' and 'God', like all others, came into existence. There has never been any evidence, not even today, that language has been taught to humans. To believe Jung in his 'collective mind' theory, and Freud in his idea of the subconscious, humans must have been self-sufficient in this sense. Besides, they have not yet finished learning to speak and write.

This point also represents one of the challenges raised by Derrida against the reductionism of religions, especially his Jewish one, that attributes the creation of the name YHWH to the biblical God.

It is the right of human curiosity to seek why concepts, belonging to the Being Who is from the uncreated, invisible and inaudible 'milieu', must emerge in the 'milieu' of visible and audible beings, under names like Word, Light, Way, Truth, Life, Love, etc. There is no doubt that humans discovered that these names-concepts are signifiers for signifieds belonging to an elusive essential reality, evident only in the *a priori*. Therefore, they have been pushed to invent them.

New names and verbs are forged daily through imagination – which could also be the fruit of divine inspiration – to pave the bridge between the

---

840 In Aramaic as in Hebrew *Adam* (man ܐܕܡ) and *Adïm* (soil, clay ܐܕܝܡ) share the same paradigm. Similarly, it should be noted that the type of serpent, or giant dragon, that wants to swallow life, very primitive in the Bible, is also found in almost all civilizations.

contingent and the essential, between the created and the uncreated, between the temporal and the perpetual, as between the two shores of the created eternal and the uncreated Eternal. We already used the verb *synesserate* to determine the state of being substantially together (with the other), at the risk of 'being' no longer or at all once 'withing' is lost. We introduced it above for the first time, and it represents the first example of a verb with an exhaustive dimension that covers the different realities as well as the various environments with no hindrances.

This type of 'being', simple and sophisticated at the same time, always refers to a Triune Being, as a duality can never stand without a third intermediary element. The elements of this Triune Being can belong, at the same time, to realities of different kinds and can produce by specific chiasmus between them a new one, which remains "differAnt" from each of them. As an example, we can point to every child born in this world, particularly the case of Jesus of the Gospel, the '*Issa*' of the Quran.

Whenever we conceive of a being, any being, as a multiplicity in *synesseration*, we comprehend as well how it can be, at the same time, stable and mobile, far and near, dead and alive. Furthermore, we comprehend how it can be capable of ubiquity, not just at the level of the same reality or of the same 'milieu', but also at the level of "differAnt" realities and "differAnt" 'milieus'.

Taking into consideration the two principles of correlation and non-separability, we understand better how this being can be in the stomach of the boa – or of the tomb– and out of it at the same time, and furthermore, how it can enter and get out of that 'space' as if from death itself.

In fact, without the triunity, this multiplicity in unity, *i.e.*, in the case of definitive separation between the components of the being, all is likely to 'disappear'. It is from this perspective of conceiving of the 'Being' that the strength of the *râz* reaches its apex. By it, as a genuine concept, while defending the Trinity, the Incarnation, and especially the depth of the kenosis that accompanied Creation and Redemption, Ephrem communicated to humanity the "angle of view" under which his luminous eye perceived Creation.

From this perspective, Ephrem also understands the Emmanuel, the *Maranatha*. When reading him closely, we clearly grasp that, as soon as God commanded the Light to be, He, God, has come to be 'with us'. He has come to be with all Creation too. Not only with it, but as said before, it is in Him, from Him, through Him that everything comes to life, proceeds, becomes, emanates, is created and is made. Furthermore, by Him, everything is done and continues to be done, and lets itself be done, as a sign of the process of which the Big Bang theory speaks.[841]

He is always present in Space (*Atrâ* ܐܬܪܐ), in Time *(Zabnâ* ܙܒܢܐ), in the Always as in the "All the Time" (*ḳûl ʿedon* ܟܠ ܥܕܢ). The Incarnation was only the concretization or the objectification of this *synesseral* reality. It expressed the culmination of kenosis by the total embracing of the human state, without ever disturbing the freedom of humans, so that all justice can be done. Indeed, all justice has been done through that, because after that, the Just One unveiled one of the hidden divine Laws, whose one previous proof of existence was the experience of Cain with Abel. **This Law establishes that a human being kills him/herself by killing the 'other', any other, by ignoring the ontological equality between the other's dignity and his/her own. All justice has indeed been done thanks to the kenosis of the Word-God Who made Himself 'other', to suffer from the 'other' what any human being suffers and would suffer, to make this 'other' understand that the Kingdom He preaches is the Kingdom of Life.**

He went through kenosis to testify that in this *synesseral* unity between God and Humanity the last word belongs not to Humanity (the contingent), but rather to God (the permanent).[842] The culmination of this justice would be the synthesis in the dialogue between Jesus and the rich young man, *i.e.,* the two greatest commandments of the Law that share the same imperative verb "Love".

Since psychology emphasizes the inevitability of intergenerational conflict, similar to the struggle between the ego and the superego, the conflict in our case turns out to be between the different realities that the same Triune Being shares. Which one prevails? Is it the human reality, the divine reality, or rather the *râzâtic* one, in which the divine and the human intertwine as described above? Should both, the human and the divine, submit themselves to the Law of the creation of the universe that the first kenosis imposed? And scientifically speaking, which

---

841   The late Father Michel Hayek considers the *râzâ* as the 'mystery' of 'being with' that we translate, after our Canadian Prof. Jim Olthuis, by the verb "to with" *(Mitsein)*. Prof. Olthuis forged the English term 'withing', without knowing anything about Father Hayek. *Cf.* Olthuis, Jim. *The Beautiful Risk*, Wipf & Stock Publishers, Mar. 1, 2006. Yet, being a Maronite priest used to celebrate Mass on a daily basis, Father Hayek could not deny that this 'with' coming from the Syriac ʿ*aman* meant precisely "to be united to; to be able to remain 'with' for ever". This reflects, in the best way, the 'mystery' of the Church as 'Bride' of Christ (ʿ*arouss al Massih*), not His fiancée, nor His wife...
     We came across good support for this verb "to with", in Pope Benedict's book entitled *Jesus of Nazareth* which appeared in 2011, where he writes on page 247: "At the same time, Jesus' suffering is, however, a messianic passion – a suffering in communion with us and for us; a being-with *(Mitsein)* that stems from love and thus already bears in itself redemption, the victory of love."

842   "Fear Him Who, after your body has been killed, has authority to throw you outside, into the darkness." *Cf.* Lk 12:5; Mt 8:12..

one prevails? Is it the physical reality, the metaphysical reality, or the quantic one where reason and a specific aspect of faith meet with mutual respect? Let us note that without this respect, everything risks returning to the state of chaos, for the 'laws' governing the *synesseration* of matter and anti-matter in the same realm disappear.

We recall that opinions about the problematic of *synesseration* between "differAnt" realities, "differAnt" 'milieus', and "differAnt" natures that were almost unpredictable before this quantum era are now predictable. The same applies to the *râzâtic reality* that Ephremian mysticism was able to generate. It could never be as well understood as it is today in light of Einstein's photons and the principles and theorems of quantum mechanics. Even what the Maronite Eucharistic prayer describes by the words of the consecration ("You have united, O Lord, Your divinity with our humanity...") becomes more comprehensible.

This dialectic concerning the superiority of the ego and its consequences leads to different syntheses, including the following:

Although humans are indeed at the center of the universe, visible and invisible, created and uncreated, and given that they are, grammatically speaking, behind every personal pronoun, particularly the subject pronoun 'I', a distinction imposes itself. We have to distinguish between the 'Edenic' phase and the 'post-Edenic' one, because in both, the same humans are those who stand behind God and write about Him. Therefore, if these humans are honest, they must not contradict themselves. They have to distinguish between their history in Eden and that out of it, basing the comparison on the transmutation required for the passage of a human being from infancy to maturity, to the 'childhood' of whose loss Jesus, the *Râz*, warned us. The fact is that Humanity, fully-grown by its temporal experience, came out of the Eden that exists in the categories of non-time and non-space, to work the universe that exists under the laws of time and space, and recently, under the laws of space-time. This last law, *râzified* by the Incarnation, may cause the Einsteinian concept *Raumzeit* to be transmuted into *râz-Raumzeit*.

Thus, we deduce from what preceded that to work what is visible as well as what is invisible, in all their categories, the Creator of the universe did create humans: "Male and Female He created them". He, somehow, molded them. Humanity is unique, but, unfortunately, it must abide by the laws of the milieu that humans enter through birth and leaves through death. Within the limits of this temporal framework, humanity passes from the point which is the beginning of the closest phase to the source of its infancy to the most distant point, the stage of the 'fully-grown', that of the fullness of egocentrism. Age has nothing to do with this predestined process.

Between the two phases, humanity foresaw, while writing the books of the Old

Testament, that once childhood is lost, there must be a 'Flaming Sword' to prevent anyone from returning to it unless a similar 'Fire' purifies him or her.[843] To be able to realize this way back, an intermediary category of persons, now immune to the 'Fire of the Sword', capable henceforth of visiting and revisiting the two extreme phases, is needed. The primary mission of this intermediate category is to keep itself well informed and then to inform whoever desires it, very accurately, about what is visible, invisible, or meta-invisible on both sides. It is to this category of persons that experts in Science and Religion such as Barbour, Herbert and Lederman point.[844] It is the category of those who have been seized by the *Râz* and now lead a life of purification of the heart and the eye until they become 'signposts' for the 'luminous bridge' between realities that supposedly cannot be joined.

Ephrem and many other perceptive minds like him belong to this category. They have all affirmed, starting with Saint Paul himself, that they ought to acquire the 'luminous eye' thanks to a purification *ad intra* that meets the requirements of the symbolic 'Sword'.[845] This is the eye of spiritual childhood and purity, the eye proper to pure hearts of which Christ speaks in His Sermon on the Mount. With this luminous eye acquired, they can penetrate all places, even that of the uncreated milieu, as well as all the phases of the creation.

If scientists like those mentioned above and many others seek these perceptive minds, it is because they know that these people were one day able to dot the 'i's for the other inhabitants of the Platonic Cave. They have been able to affirm, from the invisible they have seen, that all that is created and is visible contains, from its 'name', a seed ('a spark') of what is due to it from the created and invisible end, as from the invisible uncreated Creator.[846] Consequently, this 'spark' directs the human spirit by default toward its created sources until it joins its ultimate invisible source, the uncreated Creator. Neither mathematics nor microscopes are needed to reach these certainties. What is required is just an intuition capable of simple and "differAnt" analysis, a dreaming childlike logic which cannot, however, be free of errors. This 'science of the mystics', to believe Ephrem, was promoted by the Incarnation from the purely speculative level to the *râzâtic* one. Let us examine, in what follows, how this promotion happened.

---

843   Gen 3:24.

844   *Religion and Science*, pp. 111-115 & 121. *Cf.* Herbert, Nick. *Quantum Reality: Beyond the New Physics*. Anchor Books, New York-London Toronto Sydney Auckland. Doubleday, 1987. (Quoting Einstein, p. 250). Lederman & Hill, p. 289.

845   Mt 5:8 "Blessed are the pure in heart, for they shall see God." Also see Lk 11:34.

846   1 Cor 15:37-44.

2.3. Moving from the speculative realm to the *râzâtic* one

Jesus Christ is not just the most perfect 'mystic' who has talked to us about the hereafter that is inaccessible to humans. Let us remember that one day, He said that whoever sees Him sees His Father and that He, from then on, considered His disciples as friends because He communicated to them all He had learned from the Father.[847] He is, according to Ephrem, the very 'Mystery', the *Râz* on Whom all mystics of humanity meditated until His advent, and Whom they will contemplate under His *râzâtic* aspect forever. A part of the promotion lies in this. However, for more precision, *i.e.*, for less confusion between 'secular mystery' and the Christological one, Ephrem used, as we have already seen, the concept *Râzâ*.

In this concept, the mystical current recognizes a landmark (the Incarnation) that connects not only what came before Christ to what came after Him, in the horizontal direction of time, but also what is before and after Him in the vertical direction, as the Cross symbolizes. Furthermore, that landmark includes the inter-human relationship that covers, at the same time, *Agape*'s *Raumzeit* as well as that of *Eros*. This current expresses the best knowledge of God's nature and the relationship with Him (1 Cor 2:2), the living *Râz,* Who came to encounter humans, to make Himself available to any mind eager to meet Him, and to meet in Him the Source of the divine seed ('the spark') that the human (the mind) recognizes or feels in Him.

And, if speaking of those who are "born of the Spirit" Whose movement He has compared to the wind saying, "… you do not know where it comes from or where it goes,"[848] Jesus said, "They are born from above, or born anew," it is of Himself that He also says it, Himself Who is born in this world by the Holy Spirit. Consequently, to consider His Cross only at the flat geographical level would be pure reductionism. The Cross of the *Râz* covers exhaustively, in the Fullerian manner, every corner and every angle of the visible, the invisible, and the meta-invisible universe. In this lies another part of the promotion, the movement from the speculative realm to the *râzâtic* one.

So although humanity has always aspired to see any veil, any doubt, in the knowledge of God dissipated, as the apostle Philip stated in Jn 14:9, it was not so. In fact, the concept 'mystery' was transmuted, at the epistemological level of the term, into *râzâ*, but still this new concept, born of the synergy between the incarnate Son and the Father in their continual trinitarian *šekinah*, keeps a specific 'filter', horizontally, vertically, and at all angles. It is the 'filter' of "differAnce" that Jesus imposed on Philip who insisted on seeing the Father to be satisfied. In this,

---

847 *Cf.* Jn 15:15.

848 Jn 3:7-8.

the promotion reaches its culmination, because "No human can see [God] and live."[849]

Thus, through the concept *râzâ,* the intention is to see Christ as a revelatory manifestation ('-phany') and not as a veil, and in this case, 'to live' means to continue enjoying one's freedom, sacred freedom. Was it not the case with Jacob's adversary, this some-One Who is God and not-God at the same time, for Whom, nevertheless, 'freedom' is a necessity which pushed Him – that 'Unknown'– to ask Jacob, at dawn, to give Him back His freedom?[850] And, concerning Moses, was it not worthwhile to see God's Face and die? It seems that our freedom is dearer to Him Who entrusted it to us than our life is to ourselves.[851]

In quantum reality, the 'principle of uncertainty' allows us only to capture 'the clouds of energy' and to measure them statistically. So it is also in biblical matters. We always find a 'cloud', a 'fire', a 'storm', a 'breeze', a 'voice', an 'Angel' that accompanies His presence to prevent the inhabitants of this world from having any certainty and making any accurate prediction. This is so, it seems, out of respect for the effort they must exert to recover the phase left behind the 'Flaming Sword'. Yet, they are not left without gauges and waves guiding them toward certainty.

Already by planting Eden and before placing humans there, God has prepared for them a 'field of gauges'. By His Incarnation, He has diffused His 'guiding waves' throughout the universe so that those who want to find once more the Eden of Childhood do not get lost, but reach it with great simplicity.

Mystics testify, thanks to this promotion, that it remains essential for humans to know that transcending to Eden, to the Kingdom, also means in a very determinative sense to go back to the source by climbing a ladder that has been well described in Jacob's dream at *Beth-El.*[852]

Transcendence, in this *râzâtic* sense, is a key to general relativity when the latter would be considered a probable link between the different 'milieus' and 'phases' mentioned above. Moreover, and relatively speaking, while Science has discovered today a vast, invisible world beyond the reach of physicists, as has been expected since Ephrem's time, the same logic and the same effort would not be sufficient to

---

849 *Cf.* Ex 33:20.

850 Gen 32:23-33. Here it goes without saying that releasing the aggressor is equivalent to freeing oneself. It was not appropriate for Jacob to go on holding his aggressor when he was alone to fight and when he was suffering greatly.

851 As long as we enjoy freedom, God can restore us to life, even if we die. But once we die without freedom, God will not know what to do with us anymore.

852 Gen 28:19. Often written as Bethel. This format highlights the name of God, 'El', embedded in the place name.

transcend the quantum reality of the created invisibles to the reality of the uncreated ones. Even more, they would be insufficient to transcend to the reality of the One Who can never be seen or heard. So the question to ask is the following: where, then, can we find this 'milieu' of the uncreated where the encounter between Man and God can happen and where the discovery of the partnership, the nucleus of this universe, can be fulfilled?

3. The Kabbalist theory of *zimzum* (צמצום) and the creation of the Cosmic Milieu: The hidden God *"Deus Absconditus"*, En-Sof (אין סוף), Unlimited, *En-Sof'Or* (אין סוף אור) Unlimited Light [853]

It is thanks to the Little Prince's mind that we can discover this 'milieu'. Humanity must tame God as the Little Prince tamed the boa, the elephant, and later, the fox, thus establishing a new era of theology called the theology of Nature.[854] For this innovation, Saint-Exupéry makes himself deserving of the title of a real saint. On behalf of his Little Prince, he made an appeal, long before Moltmann, to tame what is invisible in nature. He made this appeal intending to be able to identify objectively the difference between what is physically invisible like the elephant behind the skin of the boa, and what is metaphysically invisible like the boa itself, which the 'fully-grown' took for a hat. While for his part, Moltmann appealed to the Kabbalah School, Saint-Exupéry, on behalf of his Little Prince, copies the Gospel by inviting us to accomplish the necessary transmutation of outdated paradigms. He does so to safeguard the genuineness and ingenuousness of Childhood and to be able to enter the inexplorable 'milieu' where an encounter between God and Man, Faith and Reason, could be realized. Jesus, the Son of God, invited Man to such a meeting and determined its 'place'. We read His invitation: "Truly I tell you, anyone who will not receive the kingdom of God like a little child, will never enter it."[855] It is clear and determinative, and out of respect for this invitation, we put aside any other issue. We invite each other to embrace childhood and to walk with Moltmann and the Kabbalists toward its source, the ineffable milieu, mystic for some, *râzâtic* for us, to enjoy the encounter. Will we also be able to discover the origin of spin and gravity there?

We deem it crucial to compare Luria's *zimzum* (צמצום) theory and the Little Prince's boa to see if this theory could shed light on this 'mystical' milieu and the two 'mysteries' of spin and cosmic gravity.[856]

---

853   *Absconditus* = unintelligible; imperceptible, inaudible, incomprehensible, indistinct.
854   *The Little Prince,* Chapter 21.
855   Mk 10:15; Lk 18:17.
856   Moltmann, p. 57: "Luria's answer is that God freed... a kind of original mystical space." If we go on using the concept 'mystical', it is because it is of universal use. We keep

We have seen that Luria's approach has come as an answer to a question considered unanswerable: how can Creation be *extra Deum* (outside of God), while nothing can be *extra Deum*? Professor Evgeni Tortchinov says in his electronic article *Studies in Sabbatian Kabbalah: Isaac Luria's zimzum*:

> The Lurianic tradition (based on the book of Zohar and the intuitions of Rabbi Isaac Luria Ashkenazi of the Tzfat / Safed school of Kabbalah) teaches that the imperceptible transcendent Absolute, the *Deus Absconditus* (*En-Sof*, Unlimited), or (*En-Sof'Or*, Unlimited Light) should contract to make room for the Creation. If the *En-Sof* is all, if He is unlimited and infinite, there can be no room for Creation. Consequently, *En-sof'Or* contracted from His center to 'empty' space for the world.[857]

First, it is a real linguistic masterpiece, especially for us who have an affinity for the Arabic words *akla* (أَخْلَى), *takalla* (تَخَلَّى), *kala'* (خَلَاء), *koulwat* (خَلْوة) that all orbit around what is intimate and personal. The meaning of this family of words is "to leave", "to free" (*akla*); "to deprive oneself of" (*takalla*); "open-air, vacancy" (*kala'*); "place of meditation" (*koulwat*), etc. They serve properly to express kenosis much more than the following words *faragh* (فَرَاغ), *afragha* (أَفْرَغ), *farragha* (فَرَّغ) which all revolve around the meaning of "empty" *faregh* (فَارِغ) as "to empty some container, even oneself", which leads to emptiness (*faragh*)... It is indeed the concept 'empty' that we have already critiqued in front of the tomb as inadequate, since, in the absolute, it matches absurdity, and Science, too, proves this absurdity.

In the previous chapter, we have succinctly developed our point of view on the phenomenon of contraction supported by this theory, and how it should happen without ever transgressing the *šekinah* or the morphology required by the concept 'God'. Hence, while affirming our acceptance of a first kenosis ontologically prior to the *Arche* of Genesis, which does not interfere with a succession of kenoses to come, especially that of Incarnation, we will emphasize our contribution to the detail that has emerged as a consequence to our

---

the adjective *râzâtic* for the places where it will be necessary to distinguish between the secular mystery, like the reductionist religious one, and the trinitarian and 'incarnational' Christian Mystery.

857 *Cf.* http://www.kheper.net/topics/Kabbalah/Tzimtzum-ET.htm [Accessed December 5, 2017.] "It is a very interesting approach based on etymological Hebraic background. Tortchinov continued analyzing: There were different interpretations of *zimzum* from literal and mythological (mytho-poetic) to the philosophical (the last approach was represented first of all by Italian Kabbalist Israel Sarug and his disciple Azariah Fano). And, as you have known from Prof. Avraham Elqayam's post, some Kabbalists (Moshe Hayim Luzzatto) interpreted *zimzum* as God's self-limitation for the benefit of the Creation."

questioning. How could this type of kenosis have happened, and what is the reason that must have motivated it?

Like good children, let us dive without scruples, with all of our senses and our common sense, into this 'milieu' previously unveiled and confirmed as impervious to any emptiness. With our contribution which asserts that God's omnipresence is fulfilled by His quality of Love that is of the same nature as the *En-Sof'Or,* we will come to the help of the Kabbalah School.

To get out of the embarrassment of a contradiction in the Creator, this School fell into another contradiction. This time it is about the Will of this Being Who seems to reconsider His decision in order not to allow the ontological vacuum to take place. Is it allowed to point to some hesitation or uncertainty in the concept of the Being Himself?

One of the leaders of this traditionalist School, Nathan Gazati, acknowledged this contradiction between "being" and "not being" at the same time. To alleviate this contradiction, he called the vacuum *Tehiru* (חהירו) "primordial space", and he supposed that following the contraction of the Infinite Absolute *En-Sof* (אין סוף), some residues of the Absolute Light *En-Sof'Or* (אין סוף אור) remained in this *Tehiru,* forming a *Reshimu* (רשימו). These residues later became the material substrate *Keilim* (קלים) and *Kelippot* (קליפות) – instruments, receptacles of Creation.[858]

If we apply the diagram of the 'milieu' of Creation, seen in chapter VI, to this Kabbalist case where the meaning of the contraction of the center is not well specified, we might have the following design:

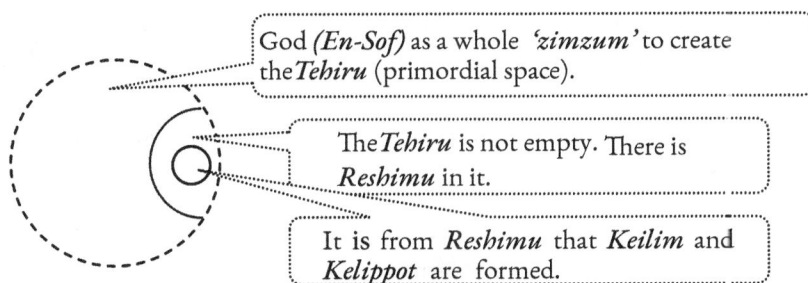

God *(En-Sof)* as a whole *'zimzum'* to create the *Tehiru* (primordial space).

The *Tehiru* is not empty. There is *Reshimu* in it.

It is from *Reshimu* that *Keilim* and *Kelippot* are formed.

Figure 55: Kabbalist theory[859]

---

858 Moltmann, p. 57; *Cf. E.U.,* "Cordovero, Moses (1522-1570)". Article written by Roland Goetschel concerning the problem of the nature of the *sefirot* (divine substance) or *Keilim* (instruments, receptacles).

859 NB: It is said that Luria wrote nothing; therefore it is even less likely that he would he have drawn diagrams.

Even if the Kabbalists insisted on the fact that the contraction spoken of by Luria would be a centrifugal *zimzum* (צמצום) as we described in the previous chapter, it would be difficult for them to determine the nature of any of the *Tehiru,* the *Reshimu,* the *Keilim* or the *Kelippot.* Anyway, according to them, all come from God and are supposed to be out of God without ever being outside of Him. They all come by emanation and enjoy at the same time the free ability to change.[860]

In our opinion, all this is just a perfect example of nominal creationism. Would it be sufficient to give names, even if in Hebrew, and functions, even if based on *gēmaṭriyā* (גימטריה)[861] for reality to be what it is? Jacques Derrida invented, not to say created, the concept "differAnce", to destabilize all the certainties of the nominalists.

To limit the list of questions without answers and avoid adding complexity to complexity, we draw the attention of scholars from among scientists and theologians to our way of interpreting this Kabbalist theory that remains, for us, very interesting. We can only excuse its founders since their era is not like ours, and the *râzâtic* spirit that we enjoy as Christians, that is more than mystic, could not be theirs, since the Church and her *râzê, i.e.,* the Body and Blood of the *Râz* Himself, is something they forbid themselves to share. The proof is that Ephrem (IV century AD), who belonged to a much more primitive era than theirs, much more primitive than their ancestor Maimonides (XII century AD), was able to comprehend the depths of Creation and kenosis much better than they were. That is why we resort to him to prove that the contribution of modern physics orients the human mind effectively toward their proper *En-Sof, En-Sof'Or, Tehiru, Reshimu, Keilim,* and *Kelippot.* It also supports in an extraordinary way the Christological truth, as well as the correlation between the different realities that until now have presented themselves to the human mind.

While using their concepts and their words, we are going to reinterpret the *zimzum* theory - as we have already planned - by applying it to the very substance of the realities, their *yât,* and not to their 'materiality':

---

860 For more details on this theory, see the above-mentioned references.

861 A Kabbalist method of interpreting the Hebraic Scriptures by calculating the numerical *value* of the words on the basis of the values of the letters that constitute them.

Râzâtic view   Kabbalist view

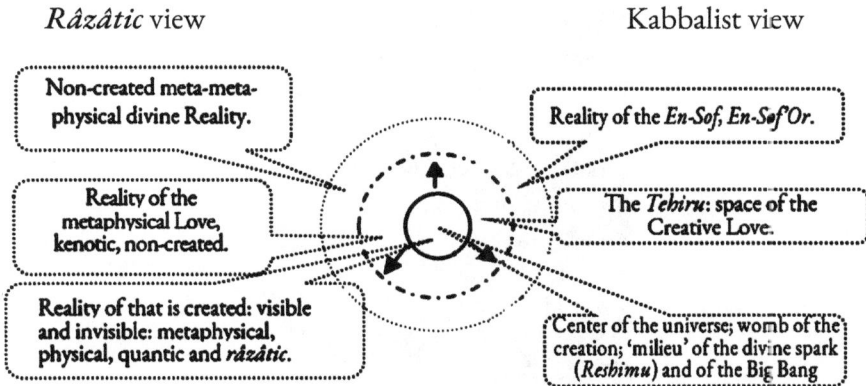

Figure 56: The *Big Decision* and its consequences
While for Luria the contraction is intrinsic, centripetal and assymetric, for
us, it is intrinsic, centrifugal and symmetric, from the central point toward
the infinite. The driving force is the Triune-Fecund Love named 'God',
and the motive is Creation.

In this last case, we read the following description made by the Lurianic School
concerning the divine Will: "It [*sc.* Will] is completely immanent to the Absolute
like the crown to the sun or heat to fire."[862] Accordingly, we can say that this
theory respects the notion of "Light from Light" of the Christian Creed, except
that the immanence of the crown, this time, like that of heat, is *ad intra* to the
Being. Therefore, the phenomenon space-time which seems not already created,
not to say that they will never be created, is the result of a centrifugal *zimzum* that
will never affect the *šekinah* of the Unlimited. However, we use the verb "create"
here, out of respect to the distinction between time and eternity. Moreover, this
phenomenon corresponds with Motherhood that we can, in its turn, consider as
Light from Light, Motherhood from Motherhood.[863]

Psychologically speaking, when the Little Prince draws an elephant in the
womb of a boa, he can be considered to be making a projection of the womb of
his pregnant mother on the boa, knowing that there is 'someone' like him there.

---

862 Tortchinov, Evgeni. *Studies in Sabbatian Kabbalah: Isaac Luria's zimzum.* *Cf.* http://
www.kheper.net/topics/Kabbalah/Tzimtzum-ET.htm [Accessed Dec 5 2017.]

863 Moltmann, p. 57. "The trinitarian relationship Father-Son-Holy Spirit is so vast, that all
Creation can find in it space, time and freedom. Creation as action of God in God and
as of God, would rather be a feminine conception: God creates the world by letting a
world become and be 'in Him': *Let it be.*"

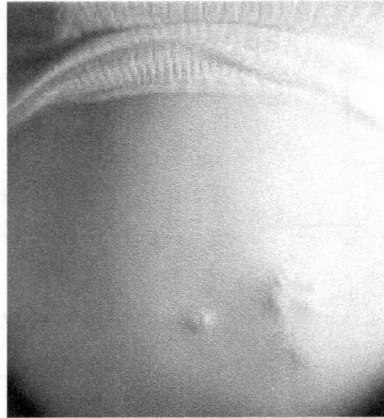

Figure 57: The 'Little Prince' on his Planet...

Nathan Gazati, therefore, introduced a second version of the Kabbalah *zimzum* theory after having criticized the first, pointing to "the existence of some contradiction in the Will (of God)" because It leaves 'a vacuum' in the place resulting from the contraction. It leaves a 'non-being' in the being. Then, to save the principle of non-contradiction in God, he speculates that, "Some residues or traces of the Absolute Light (*En-Sof'Or*) remain in this vacuum, – the so-called *Reshimu* – residues that later became the material substrate *Keilim* (קלים) and *Kelippot* (קליפות) – instruments, receptacles of Creation."[864] But with this speculation, Gazati contradicts again one of the fundamental divine qualifications in the Old Testament, the *creatio ex nihilo*,[865] without which God could no longer be called 'God'.

With Saint-Exupéry, we wonder whether there is not too much anthropomorphism in these Kabbalist speculations. Indeed, the reader is filled with the feeling of standing in the operating room of a gynecologist who performs a Cesarean section on the maieutics of ideas. It seems, according to the Lurianic School, that to solve the problem of *nihil*, God should serve as a stopgap. Once again, God forbid, we find Him obliged to save the situation in which, "The *En-Sof* projects rays of light straight into the abyss of the *Tehiru* and the process of creation is triggered by this."[866]

The more we reflect on the Lurianic *zimzum* creation theory, the wider the gaps appear and the more acute become the conflicts that could emerge between Science and Religion as well as between scientific logic and religious logic. The

---

864   Tortchinov, *Cf.* footnote 862.

865   *Cf.* 2 Macc 7:28.

866   *Ibid.*

zimzum theory, in itself, is indeed beneficial for Religion and Science, but all it is missing is the sprinkling of a little Gospel-inspired Childhood. As described by Evgeni Tortchinov, this theory seems to be the product of 'fully-grown' people.[867]

We are quite satisfied, for the moment, to have shed new light on this 'Cosmic Milieu' as being included in the 'Divine Milieu' without mixing, confusion, overlap, division, or exclusion. Their relationship can be said to be in line with our new concept, *synesseration,* in a way that if one is absent, the other is absent too.[868] Consequently, let us get back on track to try to answer the questions concerning the source of cosmic gravity and spin that were raised earlier.

4. The source of cosmic gravity and spin

Given the seriousness of this title, we will tackle it under its two fundamental components: gravity and spin, each one separately.

4.1. The source of cosmic gravity and gravitational force

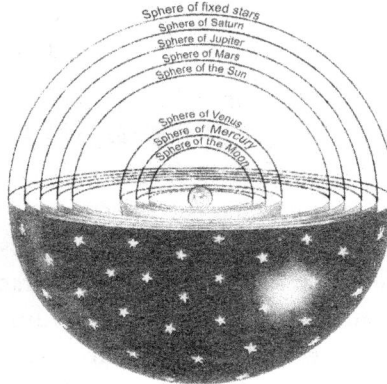

Figure 58 : From Big Bang to black holes

---

867 *Ibid.* (It seems a product of very well grown-up people).

868 It may be that a person zealous in defending the philosophical principles and the theosophical dogmas is alarmed to defend the substantial self-sufficiency of the existence of God (*Theos*), despite the absence of any creation. We reply that this is true in the case of the God, One and Absolute, of the philosophers, but not in the case of the Triune God of the Incarnation. In our case, we cannot in any way consider a substantive as 'creator' without a verb such as 'create' from whose relationship flows, by default, the substantive 'Creation'. This is due either to the principle of exhaustiveness or to the principle of holism and, in the worst of cases, to the principle of non-separability, or out of respect for the linguistic theory of paradigms. The same is said about *generatore, generare, genito; amante, amare, amato* etc. Man decides, says, and writes this, either after deliberation or before it. Everything depends on the starting point of intuition.

As the elephant cannot be outside of the boa nor the fetus outside of the womb, so this sphere cannot be outside of God.[869]

This assertion of ours cannot be considered scientific until proven. It could be regarded as childish or even pathetically religious. But even if, for a monk like us, it is obvious to begin our investigations in an aprioristic way, we are going to try to prove that this point of view is also objective, and quantum-wise, tolerable.

We have already seen that, in accordance with the *zimzum* theory, the symbol of the spherical universe, as embraced by Hawking, could exist under three possibilities: entirely outside of God as the creation theorists claim, or in God and thus risking pantheism, or in God and outside of Him at the same time, as backed by the previous analyses.[870] Thereupon, if we admit what has come to us through the Kabbalah School with its conception of 'residues', we also have to admit that the cosmic gravity of the created universe must be the result, *de facto*, of interactions between these residues as charged particles. That is why although even today all physical sciences prove that, in the vacuum, gravity – Earth's attraction included – does not have any effect, scientists continue to wonder about the cosmic gravity that the supposed vacuum does not even succeed in disturbing. The proof is that the universe has never curled up, or crumbled, or scattered.

The difficulty of this Kabbalist theory lies, as we have just pointed out, in its critical position toward the principle of *creatio ex nihilo*, since it thwarts both the Bible and the dogma of the Church. It is because of the inability to discern the nature of the so-called vacuum that the question about the source of cosmic gravity has remained unanswered until today. Could this gravity be the result of the attraction between the fullness of the *En-Sof*, contracted in a centrifugal way according to our point of view, and the so-called vacuum, created by kenosis as an absence of this fullness? If we can admit this, an attraction that affects the whole as it affects the part would then be present on both sides, on the side of fullness as well as on the side of the vacuum, because here it is about substance. There can be no

---

869   *Religion and Science*, p. 124, sends us back to the second part of Sallie McFague's book *Models of God* entitled "God as Mother, Lover, and Friend". And, on page 242, it refers to Arthur Peacocke, saying: "Peacocke also mentions briefly the alternative model of a pregnant mother bringing a baby into being *within* her body." Then the author comments saying: "I am inclined to favor the growing child analogy."

870   Aiming at resolving this Lurianic dilemma, Moltmann resorted to the dialectics of Saint Paul to say: "Only a return of God in Himself frees the space for that *nihil* in which God can then exercise His *creatrix* activity. But then, is Creation really a being and an another that exists 'outside' of God? Mustn't we say that this Creation 'outside of God' is, at the same time, 'in God', namely in the space that God has prepared for it in His omnipresence? As a result, hasn't God created the world 'in Him', hasn't He granted it time in His eternity, finitude in His infinity, space in His omnipresence and freedom in His selfless Love?" Moltmann, p. 57.

more separability, and the part enjoys the same power as the whole, especially that it is of opposite vector direction. It will be an attraction of the same nature as the one supposed to be between the rays projected by the *En-Sof'Or* in the developed vacuum and their source. That is what cosmic gravity should be, in our opinion, and not the result of the attractions among the 'residues' in the *Tehiru*. Every residue, mainly if it is multiple, is not suitable for the simplicity of the solution. From this perspective, what would happen to the symmetry so firmly adopted by Bergson and Guitton philosophically and by Ledermann and Hill scientifically?

Let us go back to the diagram of the Kabbalah theory under its first conception to compare it with our centrifugal *zimzum* one.

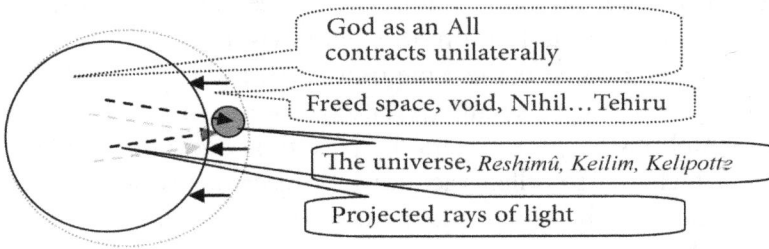

God as an All contracts unilaterally

Freed space, void, Nihil...Tehiru

The universe, *Reshimû, Keilim, Kelipotte*

Projected rays of light

Figure 59: Kabbalah understanding

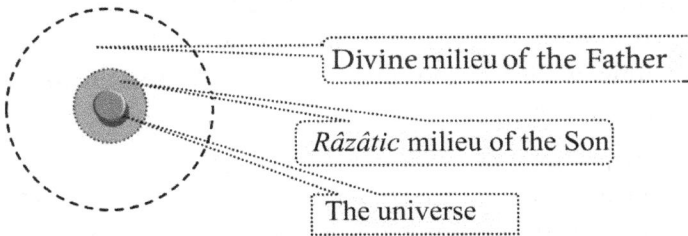

Divine milieu of the Father

Râzâtic milieu of the Son

The universe

Figure 60: Our understanding: ratio and proportions not respected

The symmetry would be a space-time marked by an imminent gravity, which would seem to be a natural and directional tension – nostalgia – of the space liberated by the liberating Act toward the liberating Will that produced it. In other words, a tension of the new state of being toward the original one, in an 'all' immersed in a commutative Goodness.[871]

---

871 "Or like the attraction among the Trinity. If the new Space-time is within the body of God, then it should reflect the principles that are true to God's essence, including a reciprocally loving communion, a communal attraction or gravitation among the members of the community." [Personal communication, Prof. Dennis O'Hara, University of Toronto, Canada].

Even Aristotle's entelechy would find a better echo here. Uncertainty and unpredictability become a fact between the two types of 'milieu' that originally shared the same nature. Even the interconnection and the non-separability theorems become so. The two 'milieus' go on sharing what a fetus and its mother would share, as well as what the 'visible' and the 'invisible' that belong to the same 'All' would share.

In spite of all the transmutation undergone by the paradigms that shift from the invisible to the visible, as from unconsciousness to consciousness, the 'All' is always 'made' of the same 'Word', the same 'Verb' and more precisely, the same letters, syllables, sentences, sounds, and generally, the same 'Wisdom'.[872]

The output of such a creation 'language' pushes us to consider that its grammaticality depends on the coordination between our tenses and the unique Tense of eternity, namely, its Tension since tension is the result of a Present Tense continuous and progressive at the same time. This coordination should have as its bases the rules of synergy, symmetry, and harmony that explain the beauty of this divine Symphony, *i.e.,* the Creation. Space-time cannot be subjected to the obstacles to which the so-called vacuum is subjected, just as the imagination of a child cannot face the same constraints as those of 'fully-grown' people. In other words, the dichotomies like time–non-time, space–non-space, become applicable to gravity too, gravity–non-gravity. In this sense, and from what we pointed out in previous chapters about the Derridean concept "differAnce", the negation 'no' or 'non-' does not have here anything to do with absolute denial, the 'nihilistic' negation. 'Non-', here, points to a different way of being and not to the inexistence of a being. A non-being can also mean a being in potency, in the process of becoming, while the absolute 'no', the 'not', and the *nihil* irrevocably mean nonexistent, and this is absurd.

We dare not say we have found the exhaustive solution to the problematic of the ubiquity of the force mentioned by Ledermann and Hill. Still, we do hope the readers will consider us to have pointed it out, and we hope it will satisfy some of those who are interested in it. Yet, to provide more support to our way of seeing the divine act of kenosis that goes all along with the act of creation, we want to delve more deeply into the search for evidence of the role *zimzum* could play in both cases, in regard to the symmetry mentioned above. For this purpose, we prefer to draw, again, some intuitive diagrams.

In the first case represented by the Figure 58, and per the Lurianic School, we cannot know from which side Creation was realized, nor whether it is *ex nihilo,* providential, or through the Big Bang.

872　*Cf.* Jn 1:1-3 and Prov 8:22-32.

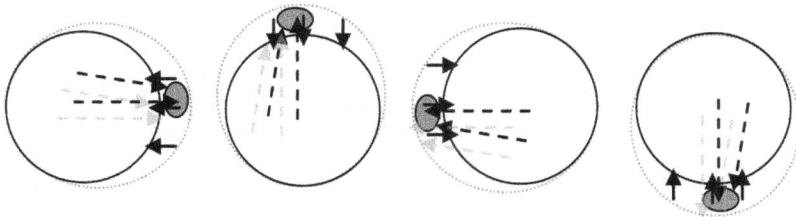

Figure 61: All the positions around the sphere are probable

The *En-Sof zimzum*s Himself voluntarily,[873] from any side, to leave a void (*Tehiru)* with some residues or traces of His essence *Reshimu*, from which are formed the *s* and the *kelippots* useful for creation. Then, He, under the qualifier *En-Sof'Or,* and going contrary to His first step, projects direct rays of light into the abyss of the *Tehiru,* to trigger the process of creation.[874]

In the case where the small grey spheres in the diagrams represent the created universe, we notice that, wherever the universe might be, the rays of light projected directly and supposed to trigger the process of creation must bombard it laterally, therefore asymmetrically. There could not be any symmetry except in the unique case where the contraction is itself centripetal and symmetrical. Then, the supposed vacuum will have the shape of a spherical membrane that encompasses the divine sphere. The figurative cross-section of this set allows us to imagine the membrane of the vacuum, like a crown of light around the sun as Tortchinov explained it. Analogically speaking, it is like saying that, if creation does not circumscribe God as the ring of light circumscribes a cross-section of the sun and includes it, God cannot enjoy symmetry, and thus He would be imperfect. The previous figures would, therefore, have the following shape:

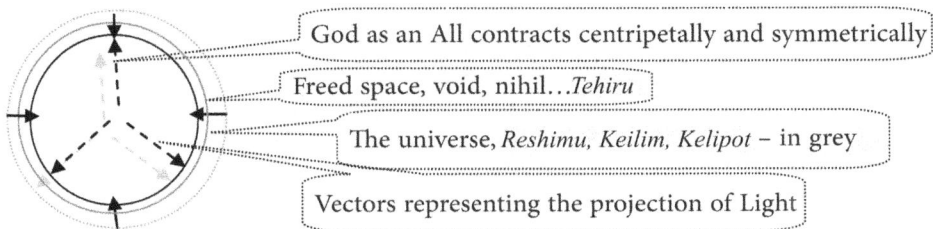

God as an All contracts centripetally and symmetrically

Freed space, void, nihil...*Tehiru*

The universe, *Reshimu, Keilim, Kelipot* – in grey

Vectors representing the projection of Light

Figure 62: Would the intermediary ring be a circular universe or rather a small sphere that revolves around God as the Earth revolves around the sun?

---

873   To avoid any idea of unintentional emanation as well as of any conditioning.

874   The arrows in the figure represent the contraction *ad intra* and the projection of the rays *ad extra*.

Consequently, we have the right to comment saying: if, on the one hand, we have a scientific proof that a vacuum contains no gravity, and on the other, that gravity detected in the small sphere is due to the large rays projected laterally in its atmosphere, what would the result be? The principle of symmetry would be jeopardized, and the small sphere, supposed to be the universe, should not be drawn like Hawking drew it, but rather similar to the following shape.

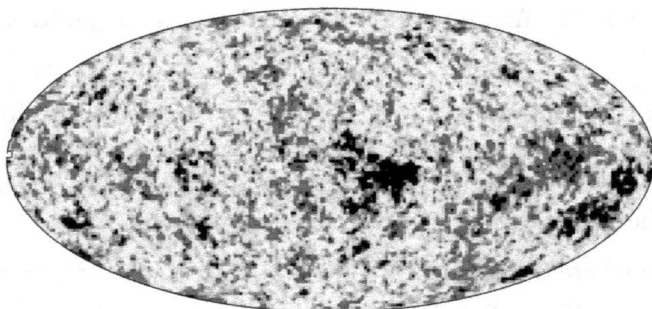

Figure 63: Cosmic microwave background radiation (NASA)

In this case, nothing in Creation would be spherical, therefore able to rotate or to have spin, which would go against Aristotle's theory![875] Nevertheless, in the following case, that of the centrifugal contraction, symmetry reigns and points to the 'King' – respect for the Magisterium – in an exhaustive and naturally harmonious way while maintaining the data of quantum physics, mainly the principle of non-separability.

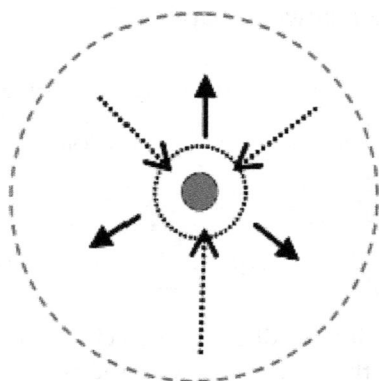

Figure 64: Centrifugal and symmetrical *zimzum*

---

875 *The Mind of God*, p. 36. "Moreover, Aristotle's universe… contained objects – the heavenly bodies – … moving forever along fixed and perfect circular orbits."

The '*zimzum*-ing' just out of the 'milieu', the center, managing the space by centrifugal contraction (which is equivalent to a retraction toward the infinity of God), allows no accusation of contradiction. Given that the kenosis is perfect, neither remnants nor residues would be necessary to accomplish creation, nor a demiurge who would be compelled to project, reversely, direct rays of his light to where he has withdrawn. On the contrary, because the kenosis is perfect, since the Will that fulfilled it is perfect, it leaves a perfect symmetry.

A natural tension accompanies this symmetry. This tension is a kind of 'present tense', continuous and progressive, or a form of 'Passion' already mentioned in Chapter VI with the name Love. This Passion is associated with an intrinsic desire of reconnection and reunification, in phase, between the divine wave-particles, which form the interference fringes of the universe, and their Source. That is because kenosis must have allowed the formation of these fringes through the double-slit of the divine Love and Will.

This is perfectly expressed throughout the Old Testament, then by the Incarnation event as Ephrem noted:

> Sun is one. It has a unique nature.
> And yet, in it, three are mixed, distinct, without division.
> Each of them is perfect, and all are perfect in the One;
> Praise is one and multiple,
> This admirable nature creates in unity,
> 'Contracts' in the One and unfolds in Trinity.[876]

Augustine also testifies for that desire of reconnection and reunification by writing, "My heart will never find rest but in You, my God."[877]

And in addition to all this, what is considered by the Lurianic School as a projection of direct light gains here much more meaning. In the present case, all exchange or all osmosis of light between the two 'milieus' is nothing but a sign of a fair balance and has allowed a perfect reflection. God and humanity are going to be at a certain distance that will enable Karl Rahner to write what we have already quoted twice: "When God wants to be what is not God, Man comes to be." This assertion is a postulate that does not seem to come from a 'fully-grown' person. It rather seems to be evoked by the mind of a 'Child', the mind of an inspired person illuminated, as a saint, by the Holy Spirit. This postulate swings between anthropomorphism – from the human side toward God – and 'Theomorphism'

---

876   *H Fid* 40, 5; *CSCO* vol 154, p. 131; *Cf. Les Origines,* pp. 107-109; *Cf. La Pensée,* pp. 124-125. Translated by the author with adaptation to the last sentence.

877   *Cf.* http://www.fordham.edu/halsall/basis/confessions-bod.html [Accessed Dec 5, 2017.] Book 1, Chapter 1.

– from the divine side toward Man. We wonder if this is not the balance that Jesus pointed to when He delivered His speech about children and everything related to childhood, as in the Beatitudes.

Moreover, this postulate leads us to deduce that the two 'milieus', the divine as much as the human, share the same central point. This unifying keystone that 'All', visible, invisible, meta-invisible revolve around, could be the holder of the *râz* of unity in diversity, *i.e.*, of the unifying dialectic. But, it could also be the holder of the *râz* of the 'Grand Unified Theory' (G.U.T.), the dream of the most learned physicists.[878]

This keystone could be smaller than any boson,[879] smaller than the ultimate imaginable *minima naturalia*. This center of all 'milieus' could be said to contain, in its essence, all the substrate of gravity that acts between these two 'milieus', the divine and the human, not to mention, between the three 'milieus', once quantum reality is considered as a 'milieu' in itself.[880] The large number of cones drawn to explain the behavior of time, as well as that of the propagation of light (particularly the quantum spectra as drawn in the following figures), gives two impressions: one, that all the rays of light cross this central point of convergence; and two, that in it takes place the essential exchange between what is created and what is not.

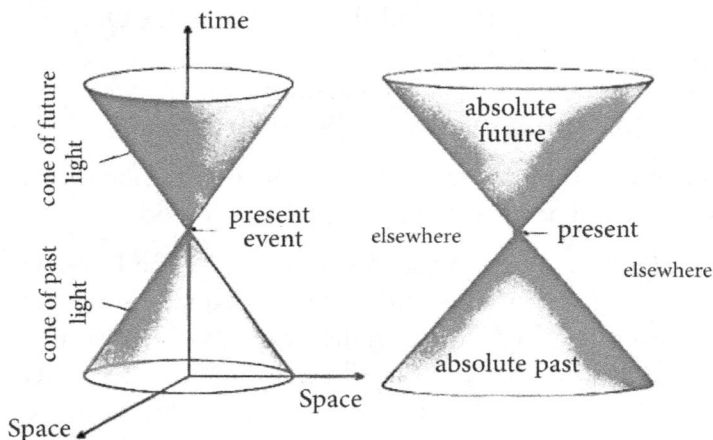

Figure 65: Hawking: cone of light, cone of time[881]

---

878   The theory through which physicists hope to build a *rapprochement* between the weak atomic force and the strong atomic force.

879   "Boson": any particle obeying Bose-Einstein statistics (mesons, photons, etc.).

880   Multiplicity in unity finds here an unlimited expansion since we can consider, likewise, an infinity of centers.

881   Hawking, pp. 124-131. The following figures are also taken from the same reference.

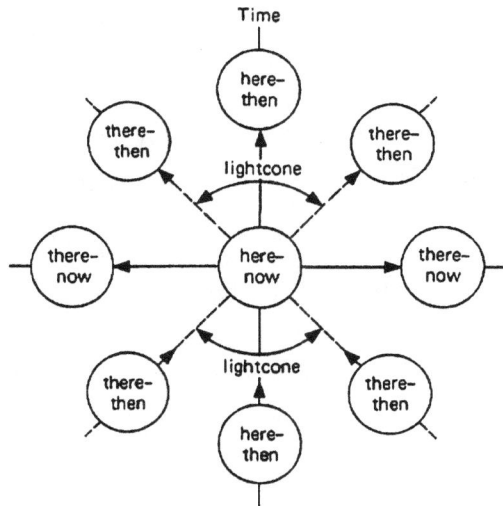

Figure 66: Illustration and explanation of the cones of space-time.

The figure reveals the difference between the traditional conception of space and time, and that of space-time considered under the shape of cones. According to the first conception, between the (here-now) or the (there-now) and the (here-then) or the (there-then) there is a measurable distance, whereas, for the conception of space-time cones, the distance remains zero. Every 'there' keeps its unity with the 'now' from where the cones start.

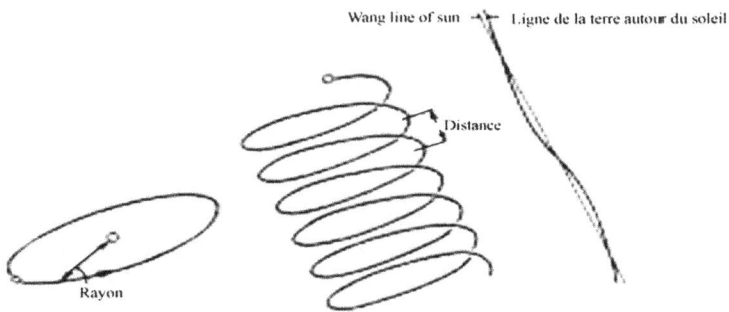

Figure 67: a) The orbit of the Earth around the sun is closed as defined in ordinary space. b) In space-time, the orbit would look like a helix. c) due to the enormous speed of light, the helix is so widely opened that the trajectory is almost straight – the pitch of the helix equals 63000 times the radius.

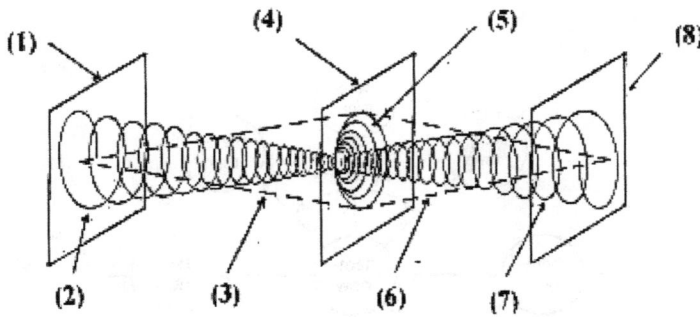

(1) Positive Projection Plane   (5) Fibonacci Spiral (Growth-Death Aspect

(2) Positive Constructive Cone   (6) Negative Destructive Cone

(3) Positive Destructive Cone   (7) Negative Constructive Cone

(4) Fibonacci Expansion Plane   (8) Negative Projection Plane

Fig. (11) Dynamic Life-Death Cycle, Which shifts from Pyramid to Cone geometry

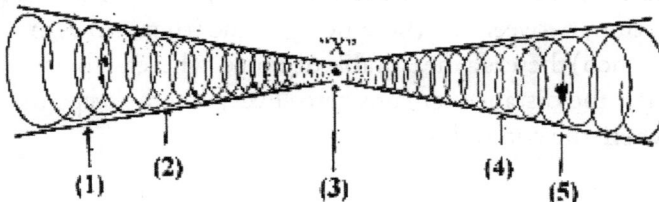

(1) Path of Past Space-Time events

(2) All Mass in this cone spirals upwards to point "X"

(3) Universal point of time – **Here Now**

(4) All Mass in this cone spirals downwards to point "X"

(5) Path of Future Space-Time events

Fig. (12) Cone of accelerating Space-Time point-events (*General Relativity*)

Figure 68:   The dynamic cycle represented by a pyramid to classically
        schematize the life course of a human being is now transmuted into
        cone geometry. Note that the direction of the vector before and after
        the 'here-now' contrasts. This theory is better suited to the Economy of
        Salvation, whose point '*x*' is the Incarnation.

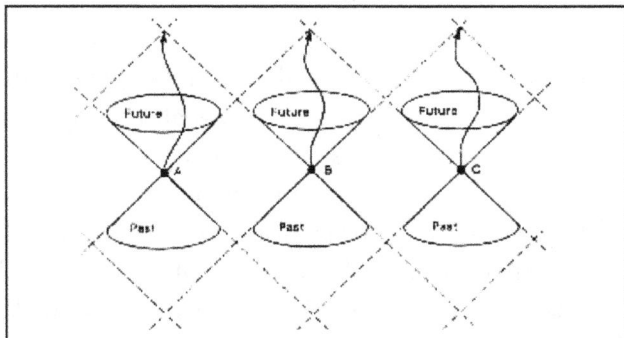

Figure 8 : A set of three light cones "belonging" to spacetime events A, B, and C. It is impossible to travel from any of these events to any of the others.

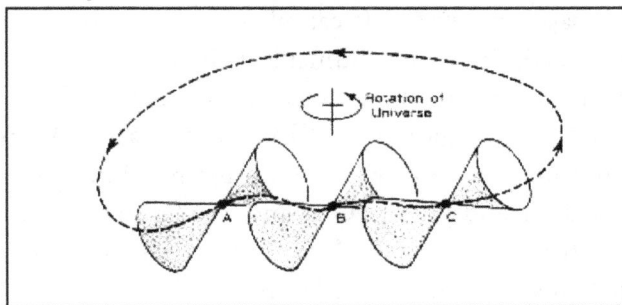

Figure 9 : If the Universe is rotating, the light cones may be tipped so that you *can* travel from A to B to C—and on around the Universe, back to event A. That is, back to the same place *and the same time* that you started from—and all without ever traveling faster than light.

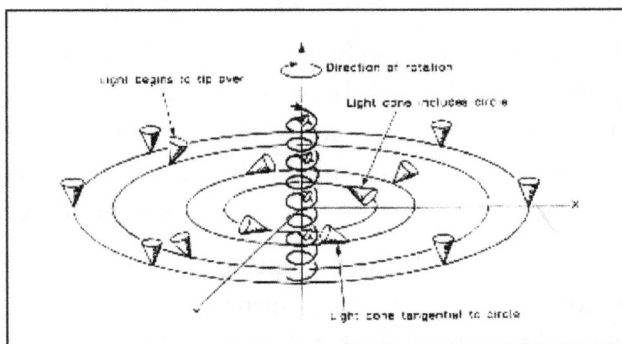

Figure 10 : A massive, rotating cylinder will also drag spacetime around with it and cause the light cones to tip over in the region of the strong gravitational field. This is the basis of Frank Tipler's design for a time machine. By traveling in a tight orbit around the rotating cylinder, you would travel backward in time, as represented by the central helix in this diagram.

Figure 69: Design of space-time cones. Note that if we drive at the speed of light, we go nowhere. And if we go forward on an orbit very close to the rotation cylinder, we will instead be going back in time.

We can find these diagrams in the majority of books and specialized websites.[882] Let us look at them with good use of our eyes and our imagination to see the elephant inside the boa. If we do so, we will be able to imagine that through this point, the 'center' of inter-trinitarian space, pass all the rays of light, and that the fundamental shift between what is created in act and what is still in potency happens. It can be said to be the point at which light and energy are condensed, the starting point of the Big Bang, the 'processor' of the immeasurable motherboard that forms the visible, invisible, and the meta-invisible of the whole universe. We can also consider it the core of God the Father, His Womb, the Uterus of divine Motherhood that generates through the Father, in eternity, through Mary in our space-time, to continue generating through the Holy Spirit and/or the Fire by means of the baptismal water. It can also be said to be the bridge between the two 'spirits' (*noûs*), divine and human, or the intersection between the two 'highways' that include, at the same time, the two paths of the Divine and the Human, of Science and Religion, *i.e.*, of Faith and Reason. An exhaustive cross-shaped 'road sign' could also be provided at the intersection of the two 'highways', with this signboard "A and Ω" or "Olaf and Tao" . Toward this 'road sign', all the vector arrows head, and from it they all start again after having been either transubstantiated, or simply transmuted or renewed.[883]

Small in an unimaginable way since the circle of which it is the center is at the same time ineffably huge, this point is the same culmination of the 'cross-highways' we have just described. It can only be the 'dew point' of the energy mentioned in the theory of the Big Bang, of the light spoken of in the Bible, and, therefore, of all kinds of quanta discussed in this book. It can only be the focal point of the convergence of the rays of divine Love and of their divergence, the causative Cause of space-time and of creation. It can only be the 'center' of cosmic gravity before which the whole of Science remains, to this day, in profound prostration.

It suffices to place one of the aforementioned cones of time, or one of the cones in helical motion, in a globe representing what is visible, what is invisible, and what is meta-invisible to admit that "this is it'. Symmetry and harmony are here, and "this is good", to repeat the divine expression in the Book of Genesis.

---

882   The most efficient diagrams with videos and animated graphics can be found at: http://visualrelativity.com/LIGHTCONE/minkowski.html [Accessed Dec 5, 2017.]

883   Notice the shape of the Tao (+) of the proto-Canaanite language: does it not fit well in the context? (We believe that it is providential and not a simple coincidence.)

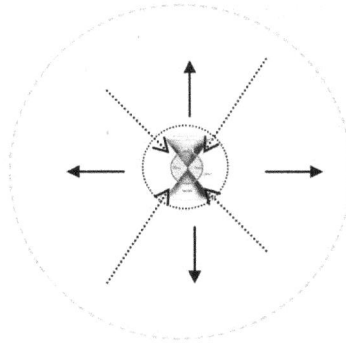

Figure 70: Integrated Space-time

Everything revolves! Thus, space-time occupies all of the 'globe' left free by kenosis till the point where divine wisdom deems it necessary to allow a porous membrane, like an enclosure, between the Creation and the Creator, between the fetus and the Mother (*Khôra*).[884] This enclosure will, then, allow osmosis between the 'Divine' and what 'He' would be, in case 'He' would desire not to be divine, never to be blocked. Nothing, according to the formula *'Maranatha'*, can be considered to have taken place, once, in a kind of past that infinitely moves away, while the God of Genesis continues daily to go through Eden, to await the return of Adam and Eve. Love is incompatible with the idea of the past. Divine kenosis, whether the 'first' or the 'last', and including the intermediaries, goes on expressing itself ceaselessly, in a continuous and progressive present that creates a time-tension, a tonus without which the universe will collapse and the divine will be lost. Yes, the divine will vanish like all 'love' that fails its subject-object, the 'hearth' of its heat and light.

At this point, what can we say about spin that, according to the *Farlex Free Encyclopedia*, is considered "the intrinsic angular momentum of an elementary particle"? Is it behind every motion that makes all things turn: quanta, Earth, stars, and the whole cosmos? Where does it start, and what will its limits be? That is what we will tackle under the next subtitle.

---

884   *Khôra* (also *Chôra*; ancient Greek: χώρα) "... the *chôra* does not simply correspond to the empty expanse where the objects are located, but also to the law of their inevitable instability and to the source of the becoming: it predetermines everything to change and corruption. In this sense, the *chôra* is a property of the perceptible, halfway between being and non-being. It is not something but the condition of possibility of all things, and ultimately the reason why no science of the perceptible nor of the particular exists, the tangible world constantly deviating from the rational web of Ideas that underlie it...We can also note Chantal Jaquet's statement that 'the *chôra* must be thought of as a receptacle capable of receiving all forms precisely because it has none'." *Cf.* https://fr.wikipedia.org/wiki/Chôra [Accessed Dec 5, 2017.]

4.2. The source of spin and all rotation

4.2.1. What is spin?

Figure 71: Somewhere, somehow, spin has started

Spin is something very specific: it is a fundamental property (quantum number) associated with each type of particle. Spin is the amount of intrinsic angular momentum possessed by a particle of any given type. The spin of a given particle is unchangeable – as long as the particle exists, it will possess its characteristic amount of spin angular momentum – no more, no less. Depending on conditions, it may also possess some amount of orbital-type angular momentum (which can be changed) about some other object. Whether or not it has orbital angular momentum, the spin will always be there, adding its fixed amount of angular momentum into any system containing theparticle.[885]

Klein makes a more in-depth analysis: he affirms that spin is an intrinsic property of particles, analogous but not identical to the general concept of self-rotation, and he adds:

The spin of an electron, when measured along an arbitrary direction, can take only two values: either $h/4\pi$ or $-h/4\pi$, where h is Planck's constant. If we imagine the electron as a

---

[885] Schumm, Bruce A. *Deep Down Things, The Breathing Beauty of Particle Physics*. John Hopkins University Press; Baltimore and London. 2004. p. 178.

small charged sphere with a radius of the order of 10^{-15}m (the radius for which the electrostatic energy of the charged sphere is equal to the mass-energy mc^2), and if its spin corresponds to a rotation of this sphere, the speed at the surface of the latter should be greater than that of light. The very existence of the spin compels us, then, to give up with building a model of the electron, and what is more disturbing, to admit the existence of a kinetic moment which does not correspond to a rotation of matter.[886]

These descriptions already provide an idea of the importance that scientists give to this fundamental element of matter that affects the movement of the rotation from the quantum level to the cosmic one. That is, what has inspired us and has continued to inspire many authors, particularly leading physicists such as Hawking and Klein, philosophers such as Guitton and Derrida, and Protestant theologians like Moltmann and Davies. This is what should from now on inspire Catholic theologians.

One of the best books we could consult is Leon Ledermann and Christopher Hill, *Symmetry and the Beautiful Universe.* Its authors use a real laboratory to describe an experiment entitled *Rotating the God Particle* and to highlight the phenomenon of symmetry.[887]

We will content ourselves with this amount of information concerning the spin phenomenon because, as it turns out from what we have said so far, our first goal is not to develop a part of the theoretical physics analysis, but rather to answer the question: where does spin come from?

4.2.2. The source of spin and every rotation: the Trinity

Let us go back to our conception of the rabbinic theory, *zimzum*. Our task for the moment, would be to specify what type of God would, precisely, reveal Himself through this conception of creation and, by the same token, how He would help to solve the 'paradox of spin', as it pleased Schumm to call it.[888]

Moltmann joins this task through an obvious and precise question: "What intra-trinitarian figure can be recognized in the creation of the universe, by God, in God?"[889] An extremely significant answer, spread over almost two and a half pages of the one hundred fifty pages of the book, follows this question. Moltmann's

---

886 Klein, p.199. (Quote translated by the author.)

887 Lederman & Hill, p. 303.

888 Schumm, p. 177.

889 Moltmann, p. 60.

words detail very well the core of this answer, as follows:

> If we understood Creation solely as an *opus trinitatis indivisum ad extra*, we could only go from a decision made by the Will of One God, and we would not be able to define the creative act more precisely.[890]

It is evident that when Science had to do with 'God', it approached Him as the first concept recognized by philosophers, the primary Cause, the Aristotelian Mover that moves everything – makes everything rotate – without anything moving Him. Much later, an accent began to appear, indicating a Trinitarian God, which, according to Moltmann, inspired a kind of intra-trinitarian movement, intrinsic to the divine milieu.[891] It is also evident that the art critics of the Rublev icon are unanimous in indicating the importance of the 'movement' of the heads painted therein.

Without stopping at the colors of the clothes as required by iconographic art, it is this movement, among other elements, that helps distinguish slightly between the three Persons and to determine intimately Who, among Them, is the Father, Who is the Son, and, therefore, Who is the Holy Spirit.

Figure 72: A reproduction of the Rublev icon of the Holy Trinity

Moltmann has not mentioned this icon, but he has described it in a way that reveals, from the elements necessary for the Theology of Nature, what has remained

---

890   *Ibid.*

891   *Cf.* O'Donnell, John. *Trinity and Temporality*. John Polkinghorne wrote another *Science and the Trinity,* after having launched its idea in the book that preceded it, *Science and Providence*. (By the way, we would like to mention that we are devotees of the Holy Trinity according to the Rublev icon.)

hidden. For him, and under a specific epistemological dialectic, considering God as an absolute Unity offers no assistance for a quantum approach to nature and reality.

As it was proved that created matter is made of quanta of energy in a continuous state of spin and rotation, the paradigm 'God' must, from now on, admit a particular transmutation that will make it more open to new challenges. 'Unity in multiplicity' becomes a condition *sine qua non* for every process and development. But it is especially necessary for understanding the particular and universal 'whirl' of creatures (from the smallest to the whole universe), their spin, and any rotational movement.

We would not have dared to raise this issue if we had not been encouraged in our design as a 'mystic' by reputable sources such as Saint John Paul II. We found this encouragement in the encyclical *Fides et Ratio* §48, in Paul Davies' *Mystical Knowledge*,[892] in Ledermann's philosophical commentaries,[893] in Herbert's meditations,[894] Polkinghorne's discourses,[895] and William Worthing's ponderings.[896] They all have admitted that 'mystics' should have something to say about it. They all respect the specific orbit on which the 'mystics' whirl, as *Sufi dervishes* do on theirs. This orbit would resemble the reef mentioned by John Paul II, a ridge where Reason and Faith can meet to dialogue and not to exchange mutual condemnations. We call on Providence for help, and we continue our analysis.

Moltmann described the intra-trinitarian movement by the following:

> But if we go from the intra-trinitarian relations of the Triune Persons, it becomes obvious that the Father creates what differs from Him, in virtue of His love for the Son, and that, consequently, Creation doesn't only correspond to the Will, but to the eternal Love of God... Creation is, on the contrary (unlike Saint Augustine's theory), a product of the Father's

---

892 *The Mind of God,* pp. 231-232. "We find ourselves barred from reaching the ultimate knowledge, the ultimate explanations, by the very rules of our reasoning that alert us to seek them in the first place. If we wish to go beyond these bars, we must embrace a different concept of apprehension from that of rational explanations. Most likely the mystical path is a face of this apprehension."

893 Lederman & Hill, p. 287.

894 Herbert, p. 250.

895 Polkinghorne, John. *Science and The Trinity, The Christian Encounter With Reality.* Yale University, New Haven and London, 2004. pp. 94, 176.

896 Worthing, M. William. *God, Creation and Contemporary Physics,* Theology and Sciences Series. Fortress Press, Minneapolis, USA, 1996. p. 1.

Love, and is then attributed to the whole Trinity.[897]

And, concerning the Holy Spirit, he adds:

> He (the Father) creates by the power of the Holy Spirit. He creates, whatever is meant by these words, as of the forces and energies of His own Spirit. Thanks to the forces and energies of the Holy Spirit, the difference is overcome between the Creator and the creature, between the Agent and the action, between the Artist and the work.[898]

And, to remove from his description any shadow of pantheism, as much as to define in a trinitarian way his point of view, he continues by saying:

> Creation indeed does not become, by that, divine in itself, but it is introduced into the Spirit's field of forces, and gets to take part in the intimate life of the Trinity's self-being... God's creation in God implies the following order of the Trinity: The Father creates the world from His eternal Love by the Son with a view of a temporal response to His Love in the power of the Holy Spirit, Who unites what is different... Creation should be attributed to the unity of the Triune God. In a creative Love, God unites with His "other" of creation and grants him/her space, time, and freedom within His own infinite life.[899]

The successive quotations, borrowed from Moltmann, are somewhat lengthy and may transgress the professional rigors of writing. Nevertheless, we have preferred to join Saint Augustine in his *felix culpa* rather than to commit the opposite by interpreting what is very clear and precise and which builds the best basis for dialogue with the sciences of today. These statements, which Moltmann wanted to be indisputable, sow the best predictability and certainty in both quantum and *râzâtic* realities, as well as in the continuous and progressive present tense of our human state and redemption. Besides, from a mystical 'angle of view', being worshipers of the Holy Trinity, we are assured that this description, so symmetrical and coherent with the data of quantum reality and the 'Mystery' it reveals, raises us to a more advanced heaven than the one on which we believe ours to be based. We find here, in the icon of Rublev, an entirely convincing proof of this intra-trinitarian ontological movement where the direction of the spin, counterclockwise, appears clearly, thanks to the

---

897   Moltmann, p. 60.

898   *Ibid.,* p. 61.

899   *Ibid.*

inclination of the heads of the three Persons. They seem to revolve around a well-defined axis, the one that goes through the Wine Cup of Love.

Figure 73: The Start of Spin

The intra-trinitarian movement expressed by the icon seems, at first glance, to belong only to the Holy Trinity *in Se* and not to be transmitted to Creation. In fact, the Nicene Creed does not highlight this dimension of Love as being one of the divine acts: generation, spiration, and procession.[900] For example, we do not read in the Creed, "True Love from True Love". It is, at least, what the text of the Creed inspires, whereas, in our opinion, the icon of Rublev hides a *pre-creational* scene, seen by its 'mystical' icon writer, and through which the three Persons ponder the consequences of Their *Fiat.* It is at this moment of consultation and of care around the Sacrificial Cup of Love that we feel them in the state to take the 'Great Decision', which means, using our human vocabulary, to 'dare', together, to say "Yes" to the ontological 'Big Kenosis'. Love, represented by the symbol of the Cup that seems to be the icon's triune center of gravity, is an 'inconvenient' Love because it has to communicate what is intrinsic (*ad intra*) to a given extrinsic (*ad extra*) that could never be, in turn, outside this center of gravity, outside the same Love.[901]

---

900   *Spiration:* Here, we are again in front of the problematic of the *Filioque* that continues to weigh, even today, on the unity of the Church: Orthodox East / Roman West. According to Bonaventure, the two Persons, the Father and the Son "spire" the Spirit (love, gift and link) and constitute Him a Person. According to Aquinas, in God, procession is intellectual emanation like the intelligible verb emanating from the one who speaks and yet remaining inside him/her. *Cf. The Holy Trinity*; Ictus Win. Version 2.7, "Procession". Aquinas's point of view, that in fact backs Augustine's, is more convenient to our description of spin in the Trinity. We will attempt to discuss this issue of the *Filioque* in the light of our contribution in a separate work. We think it is time to review it after almost twelve centuries of silence, for the sake of the unity of the Church and the dignity of the *râzâtic* Body of Christ.

901   Jesus says to John, "You will drink my cup...", to make them understand that what

Analogically speaking, we can conclude that starting from this point of sharing and pondering around the Cup of Love (that is translated by generation, procession, and spiration at the divine triune level, and consequently, by kenosis, creation, and incarnation at the human level) the 'divine spin' began, *ad intra* as *ad extra*. It is the phenomenon expressed in the *Pater Noster* by the words, "on earth, as in heaven".

The 'spinning' starts with or without 'Bang', as described above, and causes the formation of the spirals and helicoids of eternity and time as well as of spatial fluctuations.

All this starts, following a given 'mystical' good sense, from the same central point, going in all conceivable directions, making everything rotate, everything being spherical or almost, thus causing that everything moves in a spiral, from time itself to history,[902] to the tree of the DNA of every living creature, in a space managed by the gravity of Love in perpetual kenosis, and an eternal attraction between the central *râzâtic* particle and the *râzâtic* periphery in *šekinah*.

This movement can only be an eternal cycle that humans describe with verbs of fecundity, parenthood, rational-affectivity, emanating-willingness, all synthesized by one imperative tense: Love.[903]

By the fact of the 'Big Kenosis', the *Fiat* of creation, a new dynamism is added to the intra-trinitarian movement of the icon, that of the spiral seen in the previous figures:

Figure 74: The spiral of the cone of Eternity, prior to the 'Big Kenosis'

comes from Love is inevitable. *Cf.* Mt 10:35-40 and Mt 20:20-28.

902   *Cf.* Our Arabic book, *Taḥawol al Mafahim Fi Bina' Al Jumhuria*, pp. 24-26.

903   "Love and do what you want," said Saint Augustine.

But, if, analogically speaking, the actions 'to generate', 'to breathe'[904] and 'to proceed', just like the general spinning movement of divine Love, are all considered to be the properties of the 'God' particle, they will require, in line with Feynman's diagram, the emission of a new particle outside of them, by kenosis, as alteration effect (Feynman says "by decay").

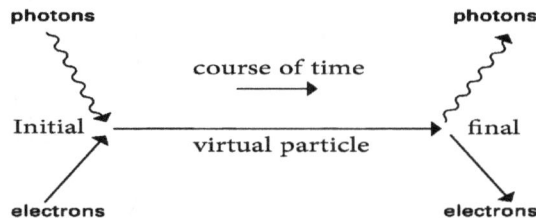

Figure 75: Feynman's graph

Feynman interprets his graph after redesigning it, by the following schema:

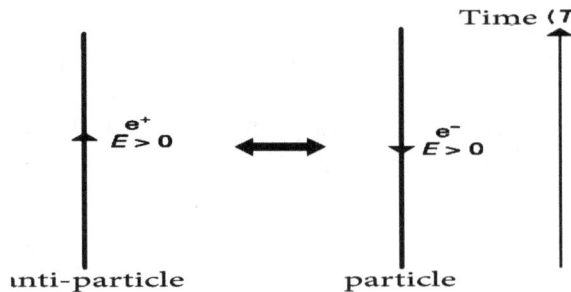

Figure 76: New quantic conception

He explains:

> "An antiparticle of positive energy that propagates in the same direction as time is equivalent to a particle of negative energy that goes back in time."[905]

---

904 "Spire the *Ruah*": We stress that there is a real difficulty in translating the Latin verb *spirare* proper to the ontological relation between the Father and the Holy Spirit – the Father 'breathes' the *Ruah*. Is it not strange not to find the verb *spire* in English, when we have verbs such as *'respire'* (even if not very common) and *'aspire'* and the corresponding substantives 'respiration' and 'aspiration'? *Cf.* General Audience of John Paul II, Wednesday November 7, 1990: *The Spirit Who "proceeds from the Father and the Son"*.

905 *E.U.*, "Quantum (Mechanics) – Fundamental Properties". Article written by Alain Laverne, Jean-Marc Lévy-Leblond.

And going back in time means, for us humans who can only understand time in a movement toward future, is to go backward. This will lead, while exhaustively considering the movement of both the particle and the antiparticle, to expect a Particle with a capital P, capable of moving in the aforementioned cones of time on either side of 'time present', or time zero.

This amazing contribution of Feynman, which has essential repercussions on all quantum physics, would be a lot more surprising once applied to the scene of Christ's tomb, as described at the end of Chapter V. Therefore, a question brushes against our mind, the same question that puzzled Guitton, who wanted to know what existed before the Big Bang, in the cone symmetrical to that of the time of creation.[906]

This Particle could be called *Verbum,* '*Olaf* String' or '*Tao* String', 'Singularity' or 'Cloud', it does not matter. What matters is that from each of these last concepts, the Big Bang could have happened.[907] It could have occurred from all of them, exhaustively, being included in each of them, as in Ephrem's 'Pearl'. In both cases, the spark of the Big Bang from which Creation is supposed to have reached what it is today must have kept throughout the process of development the tonus of the first spin, the divine one. Then, gravity would seem to be this entelechial nostalgia, or this nostalgic entelechy, from one 'milieu' to another, from the 'milieu' of the Cup of the icon to that of the Last Supper, the Paschal Meal. It would be a 'milieu' of the same nature but "differAnt".[908] And the spin would be this activity resulting from the correlativity and entanglement between the Particle (A) and the Particle (B), that exert their activity on both sides of space-time zero as described by the EPR paradox theory, while respecting the contributions of Böhm and Bell at the same time. It is what guarantees the activity of guidance for a safe return to the first state of being left behind the 'Flaming Sword', in the *maškanzabnâ* of the origins.[909]

---

906 *God and Science,* pp. 45-48.

907 *Ibid.,* pp. 29 and 51. "Little by little, we begin to understand that reality is veiled, inaccessible, that we hardly perceive its shadow in the temporarily convincing form of a mirage. But, what is there, then, *under* the veil?... And I believe that long before the physicists, Bergson grasped something of the mystery of Creation: he understood that the world we know today is the expression of a broken symmetry. And if Bergson were still among us, I'm sure that the last conquests of physics would make him add that it is from the very imperfection that life could arise."

908 Jesus indeed used a verb that expresses the passion of missing, of desire: "I have eagerly desired to eat this Passover with you..." (Lk 22:15) Neither Jesus nor the text determine the moment when this strong desire started to develop in Jesus' consciousness. Is it since the Last Supper described in Rublev's icon or since the Wedding at Cana?

909 This conception leads us to say, Christologically, that our avidity for the Omega is intrinsic, and that it is present throughout the universe as a proper characteristic.

Again, we wonder where this entelechial 'milieu' is situated that receives the 'God Particle', and from where the primordial spin of the created universe is triggered in complementarity with the spin of creative kenosis. And again, we reply that from our *râzâtic* interpretation of the *zimzum* theory, it is in the Uterus of God. This answer removes any shadow of the contradiction in the divine *mdabrânûtâ* (divine Economy of Salvation) where the kabbalist theory fell short.

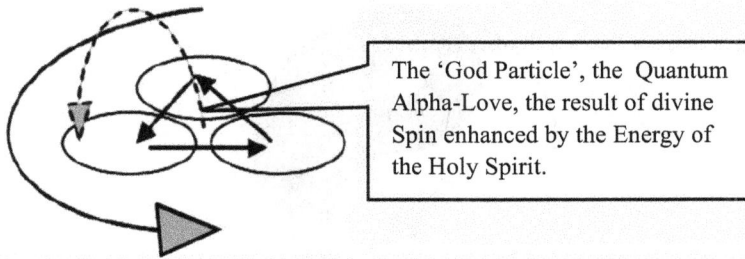

> The 'God Particle', the Quantum Alpha-Love, the result of divine Spin enhanced by the Energy of the Holy Spirit.

Figure 77 : The outcome of the trinitarian spin

This figure symbolizes a kind of spin in the divine milieu, first of all, between the three Persons Who cannot be said to be One unless the movement of generating, breathing, and loving – *i.e.*, giving life – starts. All, as aforementioned, is done only out of Love. Love is the divine Flame that warms, inflames, illuminates, provides energy without ever consuming, and that, at the same time, like a magnet, attracts all toward the point Omega. All of this is taking shape in the Cup of salvation placed at the center of the Space developed by the centrifugal *zimzum* of Trinity and meditated on by the Son, the emissary Person. It is the Cup of successive kenoses that accompanies the continual creation, and, therefore, the constant *Mdabrânûtâ* that accompanies the expansion of the universe, its conservation, its directionality and especially its redemption.

All of this is difficult to be swallowed by non-Christians. That is why, for more universal requirements that help the dialogue between civilizations and religions, while respecting the Science–Religion relation, we sought non-mystic indications (signs and symbols). These indicators are meant to support the idea of the existence of the universe in the 'Womb' of its Creator, even His 'Uterus', and consequently, to help us forge a universal awareness of the 'Place' where we are and from which Ptolemy sent us his famous message. Thus, once this awareness is possessed, cosmic gravity and the source of the dominant movement of spin and rotation will find a more open way to the unique Truth in which they are rooted and toward which they are deemed to guide humanity. What can these signs and symbols be?

5. Spin and the cosmic Trinity

Having read, decades past, the various philosophies of East Asia, we borrow from them the famous sign of Yin-Yang, the symbol of the principle of an apparently contrasting duality, but according to the many explanations that accompany it, nothing else than a version of Derrida's "differAnce".

Figure 78 : The Yin-Yang

This "differAnce", in the present case, wants black to be simply a non-white, and vice versa, neither the opposite nor the anti.[910]

It turns out that this symbol goes back to a civilization older than that of the Middle East, to one almost contemporary to the Pharaonic civilization. We find it even among the basic symbols of the Kabbalah School. Nowadays, in the modern sense, it becomes a point of interest for the supporters of naturalistic movements, of mysticism, of dialogue between civilizations as well as between religions. Given the universality and the seriousness of this symbol, we had the intuition that it could well satisfy our openness toward those civilizations of cosmic religions, whose center is human values, especially self-consciousness, and whose followers exceed in number half of humanity.[911] How will it help us do this?

Until now, we have seen that Einstein's relativity has affected the interpretation of the two concepts space and time. It transmuted them from the blocked Newtonian model to the relative, statistical, quadri-dimensional

---

910   *Cf. E.U.*, "Chinese (Civilization) – The Chinese Thought". Article written by Claude Gregory.

911   Bergson defines the spirit as follows: "By spirit, we should understand a reality capable of drawing from itself more than it contains." Bergson, Henri. *Writings and Words.* Message at the Descartes Congress (P.U.F). *Cf.* "Spirit" http://www.philagora.net/ philo-agreg/corps-esprit.php [Accessed December 5, 2017.]

space-time one, unavoidable to the human mind. Well, no doubt, this must have had its repercussions on all religious phenomena that had resisted it, even Christianity. To gain necessary information able to prove the universality of our point of view about the Creation accomplished by God, in God the Triune, we began our research on the internet and were lucky. We fell on what can come to our help, in support of our openness to the universal.

First, we focused our interest on the traditional Yin-Yang symbol with its two dimensions, religious and philosophical. This symbol seems to refer to an apparent unity of two contrasting principles, 'symmetrical' and 'complementary' at the same time, the 'such' and its 'different', in intertwining and without elimination, confusion, or separation. It is a very common duality among the religions of East Asia. But, we would say, that this is not yet unity in plurality since plurality begins with three. This symbol could also represent, once it turns on itself, according to a given cosmic spin, 'infinite movement'. However, being based solely on a bipolarity, seen under the two colors black and white, similar to a certain extent to one of the dichotomies of the Bible ('darkness' and 'light' of Genesis), it falls into the category of reductionist and harsh theories that leave no space for a spirit of tolerance, forgiveness or redemption. Despite this, for the believers of the religions that adopt it as a symbol, it only covers, with its two "philosophical colors" and the perfection of its spherical shape, tangible reality as it appears. It does not refer in any way, as a universal sign, to the dance of the waves and fringes of the interference in Young's experiment. All this is very important because it proves, as we have already seen, that 'black' is nothing but the absence of 'white', *i.e.,* it does not exist in itself and that dark is only the background left by the reluctance of light to shine.

Wave interference and fringe formation now concretely symbolize the difference between the cosmic religions and the Trinity-based Christian religion. The two principles, that of non-separability and that of exclusion, find a better application in the latter, and the paradox EPR finds satisfaction in it too.

So, in spite of all the spirituality, the wisdom, and the mysticism of East Asia, we notice that the "All", in the universal sense, belongs to a unity which is natural, cosmic, three-dimensional and repetitive, at the rhythm of all tenses of conjugation. Thus, though this symbol is beautiful, symmetrical and perfect, it has little to do with the continuous, progressive present tense and even less with metaphysics, eschatology, and everything about the 'Omega' dimension and eternity.

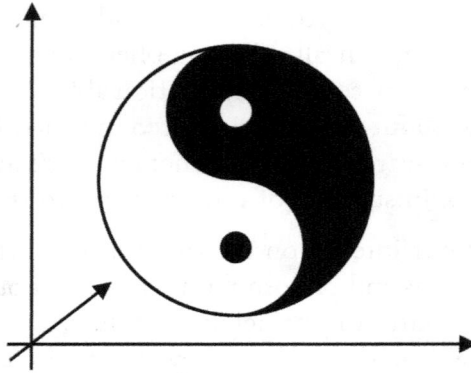

Figure 79: The three-dimensional Yin-Yang

Despite this, our intuition assured us that this symbol would add some novelties to our project. It would do marvels if a tri-polar version could exist. In fact, and much to our surprise, we found a version that satisfies our requirements.

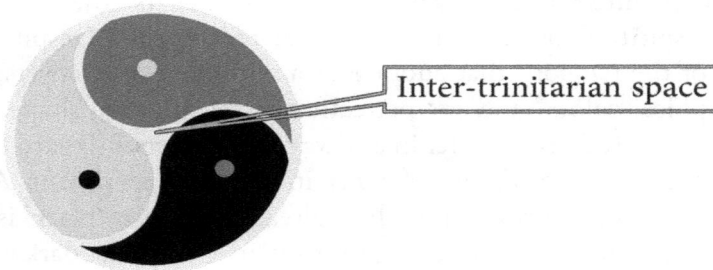

Inter-trinitarian space

Figure 80: Three poles and four dimensions based on spin

The colors given by the unknown author to this tripolar Yin-Yang are, in origin, blue, red, and yellow. They could well symbolize the three Biblical virtues Hope, Faith, Love (ἀγάπη), as well as the symbolic colors that physicists give to the theoretical particles of the quark. This 'tri-polarity', caught up in a rotational movement managed in space-time by its inner spin and the ambient gravitational force would explain the creation and the expansion of an intra-trinitarian space, by a centrifugal *zimzum*. The 'all', in line with the figure, rests on a background of a different color (cream color in the original schema). It would represent, on the one hand, cosmic gravity, and on the other, *i.e.*, according to the *râzâtic* theory, the 'gravity' caused by kenosis. We have described the latter as 'the tension of attraction' – the space of creation between the 'three milieus'. At the level of the concept 'God', this color would represent the exhaustive Love. Outside of this Love, nothing could exist, and it plays the role of gluon that attracts the 'bosonic' parts to each other, following a 'gauge field' that would maintain the elements of creation in perfect order of balance, harmony and tripartite symmetry. For philosophers, they might well call it meta-gravity. And scientifically speaking, for Laplace's 'celestial mechanics', it would represent the simple newtonian gravity, whereas, for quantum physics, it would represent the enigmatic gravity physicists are looking for even today and will not discover unless they can see the wave and its particle at the same time and calculate its four-dimensional coordinates. Finally, for the *Mdabrânûtâ*, the Economy of Salvation that reached its apex with the Incarnation and Resurrection, it represents the *rāzâtic* gravity.

Surprised and delighted by this discovery, we can only pay homage to the perceptive mind that conceived it, because it seems to have been in search of the 'mysteries' of the same Truth.

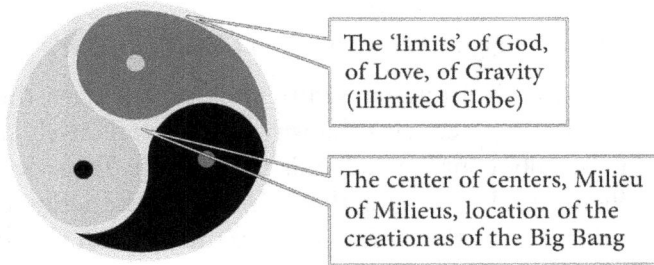

The 'limits' of God, of Love, of Gravity (illimited Globe)

The center of centers, Milieu of Milieus, location of the creation as of the Big Bang

Figure 81: The hemispherical section of the tripolar Yin-Yang. It could have the three colors of the Persons of Rublev's icon. Does it reveal the Crucifixion?

For more precision, we highlight the fact of having discovered this symbol in a website specialized in plastic arts, accompanied by the following description taken from the Taoist philosophy: "From One comes the Two, from Two, comes the Three and from the Three all the shapes of the universe develop."[912]

Being one of the solar symbols, what is wanted from this one is a representation of energy in its multiple forms, constantly changing, always in balance, and never motionless. This symbol is more than impressive to us — it is inspiring. It perfectly symbolizes, among other things, Ephrem's 'sun', 'amber', and 'pearl'. It covers also all the trinitarian symbolism and refers, in a more precise way, to the creative dynamic hidden behind its form (*Gestalt*). When seen turning, it can be taken for a fireball, a sun, or a quite impressive bush reminiscent of Mount Sinai.[913]

---

912 *"Yin and Yang"*, http://www.iep.utm.edu/yinyang *Cf.* also "Fibonacci sequence", https://www.livescience.com/37470-fibonacci-sequence.html [Accessed December 5, 2017.]

913 Recently, on the 17th of March 2017, at the feast of Saint Patrick whose symbol is the

It is genuine and awesome at the same time. This symbol wonderfully expresses the origin of spin in God-Trinity and the Space that He formed in His Womb for Creation. The 'milieu' of Creation finds its natural place, as indicated by the white sphere in the following figure, in the intimacy of the three wings that intertwine.

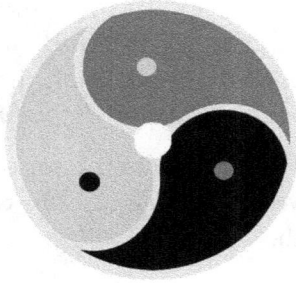

Figure 82: Centre and Milieu

However, as long as this trinitarian symbol is in a static state, this 'milieu' can be neither conceived nor perceived exhaustively. It reveals itself spontaneously to the eye that mystically contemplates this tripolar Yin-Yang when all rotates, which analogically means, by bringing the divinity into the action of generation, procession, breathing, and by the same token, into this visceral and fecund kenosis of creating.[914] This divine, intra-trinitarian and exhaustive action expresses itself, most of all, as said before, by retraction and liberation of the 'milieu' of Creation where the nostalgia for the entelechy of Love will cause, at its fullness, the silent Big Bang. The vibrations of this Big Bang and its waves are neither audible nor visible but life-giving since they remain the means of communication and exchange of information between the visible created, the invisible created and the Invisible uncreated. Even the Creator is indiscernible to those who do not have the Luminous Eye, either of the mind or of the heart.

As regards eternity, it is that trigger of exchange of information between incorporated human Love and divine Love, through entelechial Love (Fire, Light, and Heat), called *râzâtic* due to the Incarnation that allowed the temporal era to start. It is the same Big Bang that triggered the stopwatch, and so, the realization of space-time. Instead of being deemed reciprocally transcendental, this exchange, whose point of departure is only the intra-trinitarian divine activity, deserves

---

heart-shaped three-leaf clover, Pope Francis, so close to childhood, used a toy called a "Fidget Spinner", now in the hands of most children in the world, to give an idea of what a triunity could be.

914 What is to be noted is that the isolation of DNA, as every kind of genetic fertilization, must go through a stage of 'spinning'. What about the 'spinning' phenomenon of the chain of DNA itself?

then to be qualified as immanent. It is so because it is caught between that same point of departure (center of centers) and the same indescribable confines of the *En-Sof,* with due respect for the difference between the diverse 'milieus', though considered not just different, but "differAnt".

As Ephrem says, any determinative definition or description becomes, at this level, reductionist. The white spot we named above the 'center of centers' marks well the 'point' of convergence and retransmission, with or without objectification of 'light' into 'matter' and vice versa, according to the design of the Cones described above. The two key symbols, Alpha and the Omega (*Olaf* and *Tao*), can marvelously be included in it, as it pleased the Son of Man to refer them to Himself. It also includes, in fact, the Love symbolized by the 'Ember' that transmutes the human bread (*laḥmâ*) into 'Ambrosia' (divine bread), darkness into Light, and death into Life.

This central white spot also refers to the primordial 'singularity' that obsesses scientists, chasers of constants and principles hidden behind experimental reality. It is a singularity – unique particle – that is supposed to be the basis of every universality and that announced the genesis of the phenomenal world. In Christian vocabulary, this fits perfectly well Jesus Christ's assertion that He made everything – divine-human singularity – and that, apart from Him, nothing could have been, neither spin nor cosmic gravity.

Figure 83: The Spin of the Crucified by the sculptor Rudy Rahme.[915]

915 Internationally renowned Lebanese sculptor and painter expert in lines and shapes that designate the unifying dialectic of the Incarnation between God (the Creator) and Man. We call this sculptural masterpiece *Essentials of the 'Big Kenosis'.*

Let us imagine what this will teach us once this sculpture is placed on the Little Prince's planet.

We wonder what the scientists' reaction would be toward the Little Prince, if he hid the universe, represented by the following gyroscope, in the white spot of the previous figure, with the divine Word just at the center of its central ball.

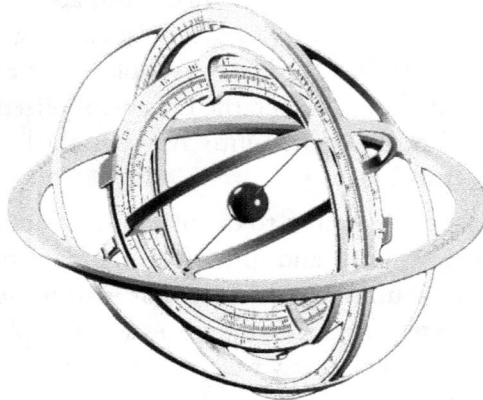

Figure 84: Gyroscope (Photo: Courtesy of Mike Sims.)

We have deliberately dwelt on the origins of universal gravity and spin, the basis of all circular, wave and spiral movement, and we believe we have succeeded in supporting the key idea that time and space are not entirely created realities or not created at all. Instead, they are made out of the nostalgia between the creating Love and humans created out of that Love. To satisfy our sense of responsibility toward the Triune God Who entrusts intuitions, we still have to diagnose the nature of cosmic gravity, its relation with the emitted 'God particle' (that we have renamed the 'Love particle'), its spin, and the way this gravity affects humans.

6. The nature of cosmic gravity and its relationship with Love

As we have mentioned before, if God is Love, everything that proceeds or comes out of Him must be so, or at least must be part of everything that comes from the concept 'Love' and the declension of the word 'Love'.[916]The same can be said if He is considered Light, Life, etc. What, then, makes cosmic gravity,

---

916   N.B.: In this assertion of ours, the expression 'proceeds' must be differently understood, *i.e.,* in line with "differAnce".

which generally influences the entire creation as a whole or as a part, stand behind so many limitations, sufferings, and sorrows? What makes it stand behind the entropy from which Science and the physical world suffer, on the one hand? And why is it the basis of Religion through its ethics and morality, on the other? We see the dilemma each time the two branches of our subject, Science and Religion, are challenged by its practical applications.

We can list traditional answers at the spur of the moment, such as that the universe is in a continual becoming and is therefore imperfect, and entropy emanates from this imperfection, or that Love is not loved in this world,[917] or that humans prefer walking in the darkness rather than in the light. Furthermore, egocentrism and selfishness break any harmony and any symmetry, and humans, created to cultivate and keep the universe, are no longer up to this task.[918]

However, as we have pointed out several times, these answers do not satisfy us anymore since they belong to the field of resignation and withdrawal, or to the field of simple refusal to bear responsibility toward this universe. At the religious level, these stressful answers cause 'mutual blame' between the human and the divine, focused on anthropomorphism from our side – the accusation of God – and on 'Theomorphism' from God's side, according to Genesis, by the condemnation of Adam and Eve as guilty. The proof is Jesus' reply to those who wanted to know who to condemn for the blindness of the young man born blind. Jesus explained, saying, "Neither this man nor his parents sinned; he was born blind so that God's works might be revealed in him."[919]

Brian Greene, one of the new quantum catechists, expounds in his work *The Fabric of the Cosmos, Space, Time, and the Texture of Reality* what we consider to lead to definitive answers to our questions. He provides drawings that we deem inspired by the Master of every science and of every religion.[920] These figures back up our hypotheses. We are even tempted to say that they were made only to serve them. We borrow them from him out of academic courtesy, since both of us are at the service of the same *râzâtic* reality, the same creational truth, that of the Creator, Love. Let us, then, reproduce them right away:

---

917 Chrysostom, *Maronite Breviary (Ar)*, Easter Time, USEK, Jounieh Lebanon, 1987.

918 Gen 2:15.

919 Jn 9:3.

920 Greene, Brian. pp. 70-71.

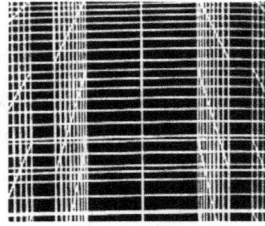

Figure 85: Flat space; (*version 2D*)   Flat space; *(version 3D)*

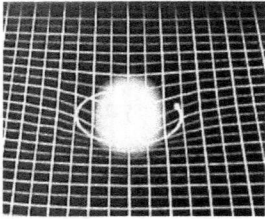

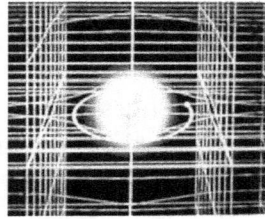

(a) version 2D   (b) version 3D

Figure 86: The Earth maintains its orbit around the sun because it follows
the curves caused by the presence of the sun in the space-time fabric:

What would the commentary of our conscious reader be, "conscious" and "conscious of being conscious", if we consider these nets as a representation of the divine Laws that rule the universe, the Big Bang included, and that even God forbids Himself from disturbing? Or, what would his/her commentary be in the light of the Ephremian intuitions and of our contribution, as to the location of the universe and the nature of gravity? What would the reader's reaction be if these filaments, which bend and unbend to surround in order to incubate the universe, are taken for rays of the divine Fabric that spreads in the 'milieu' of creation as a spider's web? It is a spherical web resulting from the osmosis between the two 'milieus'. It is grafted to the walls of the divine 'Uterus' created by Its continual kenosis. The purpose of the divine Fabric's existence is to maintain symmetry, harmony, and therefore predictability, as prescribed in the hidden Laws of the universe. These divine Laws, transmitted from the divine Mind to the human one through constants, axioms, principles, theorems, etc., as mentioned earlier, corroborate the existence of the Eternal 'Watchmaker' and the durability of His 'Watch'.

These figures inspire for us the bases, the foundation stones, and especially, if we dare to say it, the "gauge field" of the gravitational force, as well as of the human language. They also inspire for us the rays of divine Light (*En-Sof'Or*) where the universe accomplishes its expansion, as a fetus would do in the womb. And as we mentioned above, if the womb of a mother is limited, that of God 'Father–Mother' is not at all. These threads of Ariadne are the radii of the circles, their axes, and diameters. They also draw the diagrams of Love and have a role in keeping the concentric order of the spherical layers which represent the diverse realities and direct everything toward the purpose for which they came out of the *En-Sof.*

What would happen if the lines drawn in these figures also symbolize the rays of the Fabric of God, the proof of the order of His retraction, and that the Kabbalah School has taken for **Reshimu**? What if they represent the radii, axes, and diameters of the concentric spheres of Light and of Love, similar to the layers of the 'Ember' of Chapter IV, which are also part of the particles and waves and which likewise submit themselves to Planck's constant?

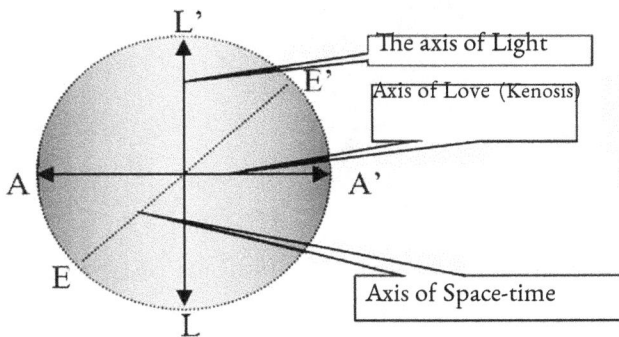

Figure 87: The axes of the Creation

All rotate in perfect order and harmony, as suggested by the tri-unitarian symbol of the Yin-Yang. Thus, the waves of the *En-Sof,* emitted from the three Persons by kenosis and correlatively, 'create' the divine Womb that becomes a free 'space' for the Big Bang with a particular gauge field for every reality. These gauge fields will conserve harmony, symmetry, evolution, and expansion, without confusion or any other incompatible action (especially without contradiction), between a visible 'all' and an invisible 'all'. It is once more expressed in the words of the *Lord's Prayer* "... on earth as it is in heaven" and their *râzâtic* exchange.

All of this invites us to glimpse an unimaginable pattern of interference between creative Waves and created waves, between divine Will and human will, through the conservation of life and of the universe, as through the inevitability of death and decay imposed on everything. A pattern of interference between the divine and the human qualities appears. Freedom, in action as in reaction between divine gravity and human one, between space, time, and movement on both sides, is respected. It is respected in a way that freedom of conscience of humans is kept safe and sound to the end, even if, as 'fully-grown' people, they make the decision to destroy both the waves and their interference. That translates into the fact of killing each other, of committing genocides, particularly the 'geno-suicide' committed by slaughtering the environment while thinking they are killing their God. Without this freedom, no verb, language, grammar, no redemption or reconciliation would have been imaginable.

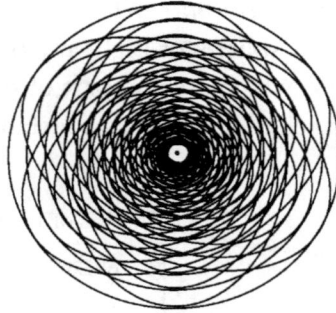

Figure 88: Intra-trinitarian interlacing and cosmic interference
inspired by Brian Greene's figures.[921]

The figure above represents the hemispherical cut of the sacred enclosure, the 'milieu' of creation, and its central spot where the Big Bang was triggered following the 'Big Kenosis'. This enclosure, which is in continuous expansion under the effect of the continual kenosis since creation would always need more space-time, is formed of a three-dimensional circum-spherical *râzâtic* fabric made of three types of thread, all called Ariadne's. They are the threads of Love, Light, and eternal Space-time. Together, they ensure the gravitational background and the gravity necessary for the good development of creation and its purpose. And as we have suggested above, these threads emanate from the interlacing of Love within the Trinity, in Its Eternal Movement that comes from Its Will. This proper perichoresis permits the verb 'to create' to complement the verbs 'to generate', 'to breathe'(*spirare*) and 'to emanate', no matter the kenosis that this will cost. It is in this 'milieu' that the universe will continue its expansion with a progressive dialectic that recognizes no limits except the ones that God alone imposes on Himself as He does on the Creation, with the promise to respect those limits at all costs. The Flood episode guarantees this promise, thanks to the regret that God has expressed there.[922] This limit which the *En-Sof* has imposed on Himself by accepting the kenosis principle can be revealed again in the New Testament in the Gospel story of "The Rich Man and Lazarus". In this reference, we read what Abraham tells the rich man, "Between us and you a great chasm has been set in

921 Figure drawn by the author.
922 Gen 9:11-17.

place so that those who want to go from here to you cannot, nor can anyone cross from there to us."[923] We note that the location of the rich man did not hinder the communication with Abraham and that expressions such as "father", "my son", and "have pity" continue to be exchanged there.

In this *râzâtic* atmosphere, divine and non-divine at the same time, the universe 'revolves' in full 'symmetry' and 'harmony', relative to the divine trinitarian Source Who makes its spinning possible. Suffering is a normal consequence of the continuation of kenosis, on the one hand, and of the expansion of the universe, on the other. So if God has admitted the principle of suffering and assumed it until His death on the Cross, why would man, whether on the scientific side or the religious one, not admit it? We will not engage here in the controversy of the Suffering of God as such, that is, the Father's and Holy Spirit's participation in the Suffering of the Son before, during, and after His earthly mission. However, it will be enough to suggest that the entropy phenomenon accompanies Love in all Its metamorphoses: attraction, fertilization, birth, separation, longing, nostalgia, encounter, unification, creation, expression, disintegration, etc.[924] All this would be absurd if God were not Triune. Quite plainly, He (Being) could not have been and could not be, as nothing could have existed if language itself and writing did not.[925] We even dare to say, according to the seven chapters that we make available to scientists, theologians and mystics, – Christians and non-Christians alike, – that God would be Trinity or would not be at all.[926]

Suffering being a normal consequence of the act of creating, the word 'evil' would have served only as a descriptive concept for every type of extinction of Light and Love, whether it be through any activity of exclusivism or egocentrism.[927]

---

[923] Lk 16:26.

[924] *God and Science*, pp. 98-99. Quoting Bergson: "The considerations in my essay 'Matter and Memory' make us realize, I hope, the reality of the spirit. From all this, naturally comes out the idea of a creating God, Who freely generates matter and life at the same time." How did Bergson get to such certainty? Simply by relying on the idea that, at the origin of the universe, there was a surge of consciousness, a rise upward that, at a certain moment, stopped and "fell". It is this fall, this relapse of the divine consciousness that engendered matter as we know it. No wonder, then, that this matter has a 'spiritual' memory, linked to its origins.

[925] **Alain** (Émile Chartier) wrote in his *About Education:* "All the means of the mind are locked in language; and the one who has never thought about language, has never reflected at all." *Cf. Digital Dictionary Bibliorum Larousse*, "Mind".

[926] *Cf. The Thought,* pp. 125-126:
Look at the images *(demwâtâ)* in creatures
And doubt not of (the reality) of the Three
So that you won't be lost. *(H Fid 73, 20)*

[927] Rabindranath, Tagore. Poem entitled *"Why?" Cf.* http://danslombreduvent.over-blog.

These are here the bases of the loss of the 'Little-Prince type' of childhood. Ephrem condemns reducing faith to definitions, and Ian Barbour, with other epistemologists, condemns every reductionism. The Creation and all that flows from it can only be described by the two keywords very dear to computer programmers: "open source".[928] Thus, Adam and Eve can assume the responsibility God entrusted them in the first two chapters of Genesis.

We will have to identify in the 'milieu' of creation the two 'milieus' borrowed from Saint-Exupéry and highlighted by the introduction: the 'milieu' of the 'grown-ups', even of the 'fully-grown' ones, then the 'milieu' of the 'kids', the 'little ones', even Childhood itself. Is it not true that Jesus preferred the latter 'milieu'? 'Children', once taken into the trap of the world of the 'fully-grown', would be healed only by a flow of 'Maternal Love'. This Love is unconditional, and it breaks down any dam, any reservation, any enclosure, etc. Hence the necessity to be born again by taking the unique 'Luminous Bridge' back to the Eden of ontological nostalgia of all human beings. In our opinion, this is what Jesus meant by his reply to Nicodemus.[929]

This unconditional Love that was revealed by the Son as being the Father's joins in Himself 'Fatherhood', 'Motherhood', and all the 'Feelings' that come from them, if one can use the concept 'Feelings' analogically to refer to what governs the intra-Trinitarian activity.

It is this Love that, on earth as in heaven, represents the singularity that initiated all types of gravity and spin. It is what was revealed to us, quite humbly, by the relation between spin and Trinity, as well as between gravity and Love. Thereupon, we ask, with all those who are concerned with the Environment and with the Theology of Nature, what would sin be if not to break natural symmetry and harmony, and by the same token, the symmetry and harmony of Mother Nature (*Gaia*)?

Lebanon's beloved son, Khalil Gibran, is one of those who understood, even if in his own way, this *râzâtic re*ality. We quote him through his *Prophet*, describing Love, saying: "Love possesses not nor would it be possessed; for love is sufficient unto love... When you love you should not say, 'God is in my heart', but rather,

---

com/article-23860113.html [Accessed December 5, 2017.]

928 One of the best examples we can give of this activity is the Koha Library which is becoming the library of the world.

929 Jn 3:1-9; *Cf. God and Science*, p. 90. Guitton says: "Sixty years after the big discoveries of the quantum theories, are our beliefs in the 'spirituality' of matter, or in the materiality of the spirit, objectively founded?" The jump is so extreme, we reply, that it remains in the realm of metaphysics, *i.e.*, that it surpasses the palpable human intellection. There must be something between the plow that Teilhard's hand touched and Jesus Christ about Whom his mother spoke. Inevitably, there is the Motherhood from which Teilhard's spirit has come and still comes. He has described it as *"Eternel Feminin"*.

'I am in the heart of God'."[930] Taken by His perichoresis, by the gravity and the spin of His Love, at the rhythm of His infinite kenosis, we ask: do we see, through Gibran's words, a transfiguration of the *yât* of Childhood?

Figure 89: Divine Realm: the spin of Creation according to Khali Gibran

7. The *yât* (ܚ) of Childhood and the Eternal Motherhood

We would have liked to tackle at greater length the relation between Cosmic Gravity and Love. However, it would have weighed heavily on our book. What comforts us is that Love has infiltrated everywhere in this concluding chapter. This is what it does whenever we open the heart's door to it, be it a heart passionate for Science, for Religion, or simply for exhaustive Truth. Through this last chapter, the reader must have grasped the relationship between Love and the Gravity of our cosmos, the big cosmos as much as the small one. It remains that if Love is indefinable, uncontrollable, and unavoidable, its *yât*, which is Childhood, is too. We will continue to deliberate as a 'Child' until we leave this chapter with a definite conviction about the indispensability of the Childhood state of mind. We call it, from now on, the *maskanzabnâ* common to God, Science, and Religion, so that the 'Luminous Bridge' can become practicable in all directions. Finally, we will reveal what it is about, or rather who it is about, when we speak of the 'Luminous Bridge'.

God is Love. Childhood is His perfect expression.[931] The deprivation of

---

930 *Cf.* http://www.lebanonpostcard.com/en/gibranpaint/2-divine-world.shtml [Accessed December 5, 2017.]

931 1 Jn 4:16 and Lk 8:16-17.

Childhood from which Adam and Eve suffered, as we have pointed out above, was filled by Christ – the New Adam – Who entered our world as a fetus, then, a child, through a family, via Mary – the New Eve. Mary too was able to be a fetus, then a child, and was able to know the tenderness of a mother, of her mother, a feeling that the first Eve missed. That is what, among other things and from the *râzâtic* angle of view, the Gospel expression "the fullness of time" would mean. It means the 'time' when humanity was going to make space for its fullness and see in itself the fulfillment of the phase that was lacking in the perfection of its Genesis, and which is indispensable to its humanization, Childhood.

This way of reading the reality in which we bathe, without being sufficiently aware of it, is supported by the stanza of the hymn written by Jacob of Sarug mentioned in the first chapter. It is also supported by the evangelist Matthew's depiction of the fierceness of Herod, one of the 'fully-grown' of history, against the infant Jesus, the newborn, and the massacre of the 'children of Bethlehem'. Matthew's chapter ends with, "Then what was said through the prophet Jeremiah was fulfilled (Rachel weeping for her children and refusing to be comforted, because they are no more), as if it were Childhood itself, that is no more. By the way, for such people, this Childhood, so much dreaded by them, should never exist. Is it not also the prophecy as said in Genesis: "I will put enmity between you and the woman, and between your offspring and her Offspring; He shall bruise your head, and you shall bruise His heel"? So, it is by Childhood, whose cradle the New Eve will rock with her left hand, that She will overcome the Evil One and will restore the reputation and the dignity of the terrestrial Adam and Eve. That is why we would say that Bethlehem, the King's city, refused to make space for the birth of this Childhood and that Herod was so relentless against the sole probability of its existence. He would not have committed this horrible act, the slaughter of the children, if he had not felt his throne shaken by the mere fact that this promised Childhood, the *yât* of every childhood, was there. It would fulfill the prophecy of Genesis as Mary sang it before Elizabeth, saying, "...He has scattered the proud in the thoughts of their hearts; He has brought down the mighty from their thrones and exalted those of humble estate."

We recall here what distinguishes the concept '*yât*' of Aramaic origin, from the concept 'self' or '*se*' of Latin. To say the *yât* of Childhood is not quite like saying 'the self' of Childhood as one could recognize it in Cain and Abel, Eve's offspring by Adam whose name indicates the clay from which he was modeled (*Adîm* = dust). This distinction goes hand in hand with the one mentioned earlier between creation out of *Nihil* and creation out of Love. If the "self" of Cain and Abel's childhood goes back to the *Nihil* and to the *Adîm*, the *yât* of Childhood which arrived in our world through the Incarnation goes back to the divine Love, especially the one that the Son and His Mother constituted 'in the beginning', when Creation was still in the "Silence of the Memory of God".[932] What do we mean by this?

To be able to answer this question, we have to replace the concept 'God' by the concept 'Love' to indicate, from the *râzâtic* perspective, the Supreme Being. It is only in this case that one will find in the following paragraphs what we mean by what we have said.

7.1. Creation of the alphabet and the first syllables

The letters of the ontological alphabet conceived by divine Wisdom and presenting themselves as the DNA of the creating Intellect began 'interlacing' in the 'Silence of the Memory of Love'. As soon as this 'interlacing' began to form the first syllables, the first words and the first divine name, "Love" (*H'AHaBaH* הָאַהֲבָה), appeared. The two fundamental letters A and B constituted the beginning of this divine alphabet, the letter H being the Breath, the Wind, even the Spirit between the two, as in *EHieH* or *YHWH*. From these two letters, A and B, came simultaneously the two Names "Father" (*AB* אב) and "Son" (*ABN/BN*), the Names most expressive of the divine Love. Why do we insist on simultaneity?

It is evident that in this 'milieu' of divine Wisdom Who knows everything and Who is, in the *hic et nunc,* conscious of everything, these last two Names made of the same two letters Olaf (א) and Bet (ב) could not but appear together at the same moment of Eternity. No lapse of time should have separated their appearance, for the "cause" of the existence of one is in the "*raison d'être*" of the other. Neither of the two could have a "*raison d'être*", a directional orientation, without the object-subject of His purpose, the other. And, united in perfect equality by the letter H (ה) of the Hebrew verb 'to love' in a Trinity that revealed Itself at the Epiphany, both share, in spite of the "differAnce" between the two Names, the same final cause of their Being. This cause is to objectivize, at the service of the divine perfection, the Fatherhood-Sonhood creational dimension of the exhaustive Love indescribable and indefinable in Itself.

Consequently, the verb "to love" (*AHaBaH* אֲהָבָה) should be, by the Spirit (*H* ה), the first Driving-Force that appeared at the service of the same perfection, and so too the verb "to create" (ברא), simultaneously. divine Love cannot be but Creator. No lapse of eternity could have passed between "to Love" and "to

---

932 An expression used by Khalil Gibran in his *Prophet* to figure out the eternity of the matrimonial link: "You shall be together even in the silent memory of God."

Create". At the level of this divine milieu, things do not happen as in our space-time where there is contingency that inspired for Aristotle, and then for Thomas Aquinas, the philosophical principle of 'Act' and 'Potency'. At the level of that milieu, there is no way to think about the phenomenon of waiting, not even for a 'blink of the eye'. In fact, the kerygma does not define for believers and is not able to identify for them, in the absolute, Who of the Three divine Persons is the Father, Who is the Son, and Who is the Spirit, as Ephrem suggested and as we have noted from the legible faces of Rublev's icon. According to Rublev, this is only feasible through some signs and symbols, among others the 'movement' of the heads and the fingers, exclusively recognizable by experts and left to interpretation.

Consequently, a question remains unanswered: Who among the three Persons has primacy over the Others?

If we insist on designating a primacy, it should be the Holy Spirit's, the Love Who unifies (as gluon) the Two Others. That is why Saint Augustine insisted on clarifying that the works of the divine Trinity 'outward' (*ad extra*) are categorically indivisible (*indivisa sunt*), even if the concept 'outside' presents a dilemma. These works are the product of the Three unified. Yet, this unity of Action should by no means lessen the primacy of the first Driving Force, which is the divine Love from Whom Paterno-Filial Love originates. For this reason, the only Person of the Trinity Who has not been objectified by a familiar figure is the *Ruah* (רוּחַ). The *Ruah* is the Paterno-Filial Love Who, being the very essence of the absolute Love, allows us to say that the Father and the Son also proceed from Him just as Their two names proceed from the Alphabet of divine Love, and not only that He, the *Ruah*, proceeds from Them.

In the *râzâtic* meaning, it is Love that gives both Fatherhood and Sonhood their *raison d'être*, and not the opposite. It suffices to recall, in this view, the parable of the Prodigal Son to ascertain this truth.

The Maronite Liturgy offers through the words pronounced during the Fraction of the Host, a proof of the active role of the Holy Spirit Who controls causation in itself, from the pre-creational causative cause, to the final post-creational cause, qualifying Him as "...Principle, End, and Perfection of everything that was, is and will be in Heaven as on Earth." This makes understandable the comparison made by Christ between the Spirit and the wind, when He said to Nicodemus, "Wind blows where it wants; you hear its voice, but you know neither where it comes from, nor where it goes," (Jn 3:8) and furthermore, the supreme dignity which He conferred on the Spirit and which surpasses His own: "...Anyone who speaks a word against the Son of Man will be forgiven, but anyone who speaks against the Holy Spirit will not be forgiven, either in this age or in the age to come." (Mt 12:32)

The Holy Spirit, according to our *râzâtic* point of view, is at the origin of Paterno-Filial Love, without exclusively being the Paterno-Filial Love. He will play the different roles referred to Him in the Gospel, yet without ceasing to play the role of Revealer that He has fulfilled throughout Salvation History. He has been the agent of diverse hierophanies in numerous forms, of which the two most cited in the New Testament are the 'dove' of the Epiphany and the 'tongues of fire' of Pentecost. The Holy Spirit has enjoyed, throughout the Bible, the highest authority in matters of Faith so that whenever it is said that the Spirit of the Lord "came powerfully upon" (1 Sam 10:6-10), "seized" (1 Sam 19:23), "has come on" (Lk 1:35), "has filled" (Acts 2:4) someone, one is faced with an indisputable divine Authority. It is the strength of this prerogative of the Holy Spirit, so deeply rooted in the collective consciousness of the Jewish people, that Christ is going to emphasize. He said to His disciples: "Nevertheless I tell you the truth; it is expedient for you that I go away: for if I go not away, the Comforter will not come unto you; but if I depart, I will send Him unto you." (Jn 16:7) Moreover, "...He will teach you all things and bring all things to your remembrance, whatsoever I have said unto you." (Jn 14:26) "He will guide you into all truth...and He will show you things to come." (Jn 16:13) It is on this prerogative that Paul of Tarsus will also base his vocation, to make his authority equal to that of the apostles, especially Peter. (Acts 9:17)

Thus we are, up until now, in the "Silence of the Memory of God", the Memory of Love, with a dichotomy Father–Son that objectivizes this Love, especially since the Person of the Holy Spirit remains the guarantor of the equality between the two other Persons and of the unity of the Action accomplished in their Trinity. This Unique and Absolute Love, incomprehensible in Himself, is conceivable only through this living Trinity and in all that will happen through Him, specifically the Eternal Motherhood, Creation and Redemption. This form of 'objectivization' is reflected in the Creation by all that is triune, particularly, at the quantum level, by the wave-particle and what holds them together, by two particles and the gluon that holds them together, by correlation, by entanglement, by non-separability. It is reflected by all the components of the cosmos and the Cosmic Gravity that keeps them in order and synchronization, as said above.

Sincerely, we cannot hide our conviction that it is the Holy Spirit Who teaches everything and Who is the Person Who reveals these truths upstream to the *râzâtic* eye of Childhood that probes Science deeply, so that downstream the kerygma of the Church comes to approve them with a certain "differAnce" which underlines that **as far as classical theology is concerned, primacy is for God Who is Love, whereas for râzâtic theology, it is for Love Who is God.** From Love Who is God proceed, in the absolute, the other concepts such as Paternity, Maternity, Filiation, Triunity, and all types of love like *caritas, agape, eros,* etc. Even the distinction which Plato made between the good *eros* and the evil *eros,* once applied in the world of moral and ethical relativities conditioned by materiality, applies equally to all other emotional categories. This last phenomenon, materiality, produces the veils, the aforementioned layers of ash that hinder clairvoyance and which are excused by the classical axiom re-invoked by the apostle Paul to the Corinthians, "For, who has known the mind of the Lord that he may instruct Him?" (1 Cor 2:16) It is an inevitability to which we react in the manner of the Virgin Mary at the Annunciation, by questioning, "How then can we love Him, or rather, how can we love in general?" Paul, being attentive to this dialectic, hastens in the same verse to open the 'door' for himself and for us all by adding, "But we have the mind of Christ."

Through this difficulty of knowing the mind of God, Paul wanted to support every believer matured by faith, confirming, "But he who is spiritual judges all things, yet he himself is rightly judged by no one." (1 Cor 2:15) This means that every person born of the Spirit and Water, or of the Spirit and Fire, can reach the knowledge to which Christ's thought guides, the thought that leads to the depths of Love as Paul describes it in the same epistle. (1 Cor 13)

Since God is Love, or rather Love *par excellence* as Pope Benedict XVI highlighted in his encyclical *Deus Caritas Est,* let us try, at least, to reduce everything that subtracts Him from us and "live of Love" with Saint Thérèse of the Child Jesus. Didn't Jesus, after all, resurrect to perpetuate Childhood which sees behind the skin of the boa of the Little Prince?

Let us continue to read from the same perspective, as 'children', the scene which according to Sarug's prayer mentioned above preceded creation. We have to do this in search of the *yât* of Childhood, which is from now on the Tent of Meeting under which Love invites Science and Religion to meet. Let us furthermore highlight the 'Luminous Bridge' supposed to lead us there. With the view of doing things for the best, we reproduce the translation of Sarug's prayer in the following section.

7.2. Maturation of the alphabet: the beginnings of communication and kenosis

> The Father took in His holy hands the dust prepared for the creation of Adam, called His Son and said to Him: "Here is the one, who on the Cross will lift You and mock You; here is the one who to the tomb will lead You, and by him, You will be humiliated and despised. If You agree, I will create him. If You do not, I won't." The Son said to Him: "Create him because it is from Mary that I will 'carry' him, I will endure suffering, and will save the world.

The divine Alphabet had already matured so to allow the Father and the Son to communicate. The Word, even the Verb (*ha melah* המילה), was already present in the 'Silence of the same divine Memory of Love'.

It is necessary here to clarify a crucial point that would hinder the growth of children's faith. A young boy would spontaneously make the link between the Son–Father relation and his own with his dad. He would do the same for the bond between himself and his mom and Jesus' bond with Mary. For him, both sets of bonds should be similar. To find out if this natural way of children's thinking makes sense a big question comes to our mind: has the Mother of the Son existed already, before Creation, somewhere, and with the name of Mary? Would God be a family as this young boy's?

Overt masculinity characterizes the Old Testament that recognized Israel only as a 'son' of God, with God characterized as male in His gender, without this 'son' even being worthy of pronouncing God's name. Yet, we note that this God put in the prophets' mouth maternal attributes, some of which we mentioned in the course of our work, that refer to Him as a conceiving Mother. The divine Maternity was well spread among the peoples neighboring the Hebrews, especially the maternity of Astarte who even Solomon venerated in the apogee of his glory. (1 Kings 11:5) Yet, it turns out from what we have just said about the Old Testament that the collective consciousness of the Jewish people, through all that was written in their sacred texts, has not accepted to substitute this 'maternity' either by Wisdom (חכמה) or by Word (המילה), both of which are feminine terms in Semitic languages. Nostalgic for a Childhood that they never tasted, the Jewish people longed for a Mother.

It is from this point that Sarug's text draws its inspirational force, from this nostalgia, this desire for a Savior Who will break the circle of masculine rigidity, Who will fecund the sterility of the Law and Traditions, Who will sow everywhere the maternal tenderness palpable in the heart of the just and good Father of the

Prodigal Son missing a mother.

This pre-creational scene of Sarug implies, from the *râzâtic* point of view, the following interpretation:

A- Since Creation, the first "concretization" of divine Love, is a necessity for the perfection of divine Love, the Trinity, endowed with all Will, all Wisdom, and with an "alphabet" open to every letter and syllable that will be needed, should plan a vision and a design for this Creation. It is this supposition that inspired Saint Jacob Sarug.

Let us then admit a kenotic 'understanding' within the Trinity in which, according to the order of the letters of the alphabet, one of the Persons would be '*AB*', Father, another '*ABN*' (with muted A), Son,[933] with a "differAnce" between the two Names and not a difference, and the third, the Spirit, the Love that binds both of them, so that these three designations serve 'language' during the creation and, afterward, human language.

B- As far as the Father is concerned, to create Man (the male Adam and his race) or not to create him was the question, and this was in view of the consequences that the Son will undergo. Sarug's text suggests an attempt of 'de-responsibilization' of the Father toward the Son. This should never be understood this way; it is rather the perfect sharing in the eternal *hic et nunc,* as is the case with everything that this question of the Father evokes. The proof is that the Son did not react as He did at Gethsemane when He said, "Abba, Father, everything is possible for You. Take this cup from me. Yet, not what I will, but what You will!" (Mk 14:36); neither did He undertake an analysis of the feasibility of the project before replying. Rather, in a perfect Equality with the Father, His Partner, and in full consciousness, and conscious of being conscious of His role, He surprises Him by putting His Word in the design of which It became the Key: "Create him, for it is from Mary that I will carry him. I will endure suffering and will save the world," wrote Sarug. Mary became the "differAnt" partner, the Mother.

C- If creating Adam (temporal fatherhood) was the question for the Father, Mary (Eternal Motherhood) was for the Son, and not only the temporal motherhood represented by Eve that will become deified and so eternalized. The text stresses in due form the urgent order of the Son to the Father: "Create him." The hearts of humans vibrate at the sound of this order, for it guarantees the self-

---

933 It should be noted that the first letter of the alphabet is not the alpha (A) pronounced as such, but rather the guttural sound "ah" normally symbolized by an apostrophe (') as in ('h) and which with the diacritic (a) becomes ('a), and with the diacritic (e) becomes ('e) etc. Consequently, in the syllable 'aBN', this sound is silent so as to be pronounced 'BN', which makes it as if the name of the Son starts with B.

esteem of every son, every child, just as it provides a 'mother' in the subconscious of all children. It may be for poetic requirements that the author put it this way: no matter! It is also a means which the Holy Spirit uses to convey a message. What message is it in our case?

D- Here, the anthropomorphizing necessary to answer this question is more than interesting: it is fundamental. It already represents Mary as the fourth concretization of the exhaustive, absolute Love, after the Trinity, through the Trinity, and thereby makes her a necessary condition between the male, paternal Trinity and the creation of Adam, without disturbing the order of the letters of the divine Alphabet. It is the role of this name 'Mary' to objectivize the Motherhood of the absolute Love, then to incarnate this Love in the world through the Mary of the Gospel, the Virgin predicted by the prophet Isaiah, so that from her the Son-Verb takes flesh. By this, the Son draws the Father's attention to two facts:

> 1- That the Eternal Fatherhood alone is insufficient for the perfection of creation.

> 2- That an inter-Personal synergy is essential for the Eternal Motherhood to have its part and to be the result of their Love, in the service of creation and its perfection.

Is it not for the completion of this perfection that the Son will incarnate to save the world, by restoring to humans through Mary's umbilical cord the lost Edenic Motherhood, and by His birth from her, Childhood which they did not know before Him? And all this through the same Holy Spirit, "Principle, End, and Perfection of everything that was, is and will be in Heaven and on Earth"? This is what stands, from a *râzâtic* point of view, behind the dogma of the Immaculate Conception, for if it were not so from the beginning, it would not have been realized in the fullness of time. It is the message we were expecting: the 'immaculating' relationship, Son-Mother, designed for the Creation.

Therefore, nothing hinders us from saying that the Son has voluntarily limited Himself. Effectively, He has limited Himself at two levels: on the one hand, at the level of the Love He shares with His Father, and on the other hand, at the level of the Love He shares with His Mother and humanity, all of it for the sake of the Creator-Love. It is in this act of limitation that the beginning of the Son's kenoses takes root. Yet, this also leads to saying *râzâtically* the same thing about the beginning of the Father's kenoses. Why? Because of the Love He shares with the Son and His acceptance that the Eternal Motherhood shares with Him the Love of the Son, of Creation, and of the totality of the Economy of Salvation. It is from this angle that we understand Mary's intervention at the Wedding of Cana and her adherence to her Son until the Cross, until the institution of the Church, and then, her perfect return, without dying, to the perfection of divine Love.

As to the question of the kenosis itself of divine Love (as we pictured it by the intermediary of the *zimzum* theory) from Whom comes all deification, creation, existence, incarnation, and redemption, as well as to the application of the Fatherhood-Sonhood's 'limits' over Him, the case is undoubtedly "differAnt". Yet, it is because, as we said, absolute Love admits no limits and, even less, contradiction. Therefore, the limitations of kenosis came to be addressed by Love's unlimitable gravity. *Râzâtic* gravity crosses any obstacle by osmosis. As long as 'slits' and 'holes', even 'pores', exist in the 'skin' of all created beings, as between the 'membranes' of their different milieus, 'Love-God' crosses them in the way seen in Young's double-slit experiment.

Since the act of creation is an expression of divine Perfection, so is the act of 'objectivizing' Love by fatherhood, motherhood, filiation, fraternity, friendship, and even by the erotic bridal love. All of these emotions are aspects of this Perfection. Thereupon, we go back to the words of Khalil Gibran's *Prophet*, already mentioned above, to affirm that: "Love gives naught but Itself and takes naught but Itself. Love possesses not nor would It be possessed; For Love is sufficient unto Love." 'Love-God' possesses not, for such kind of possession enchains and limits. He is sufficient unto His exhaustive *yât* because nothing exists before Him, nor after Him, nor outside of Him to be lacking. This is what entelechy (ἐντελέχεια) has meant and will always mean. As a result, any apparent objectification of this Love, 'God' *par excellence*, to use the philosophical concept, is a *râzâtic* hierophany that does not limit and that nothing limits. However, it does not exclude the possibility of recognizing the difference between the 'milieus' as seen in Chapter IV through the symbolism of the ember.

E- Yet, this "differAnt" kenosis that happened, this time, from the side of the paterno-filial Trinity and that made Mary, as may be deduced, a partner in Creation as in Redemption, if not as a causative cause, at least as an intermediary cause, requires clarification to protect the trinitarian sacredness. There is a reason to fear the accusation of flirting with the heresy of "quaternity", even if the latter consisted in the past of considering the Son Jesus Christ to be different from the divine Word, rather than considering the Person of Mary under the wing of the Eternal Motherhood mentioned above. For this clarification, we go back to Fuller's theory, mentioned in the first chapter, and we approach it at the same time with Rublev's icon, to say:

If we consider Mary '*co-creatrix*', as well as *co-redemptrix*, it would be in the image of the fourth triangle whose existence was not expected. Let us go back to the figure of these triangles as per our observations.

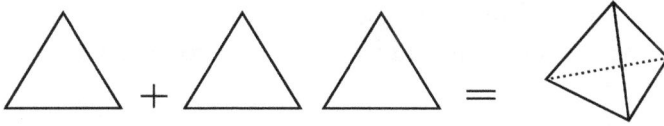

Figure 90:   Holy Spirit + (Fatherhood - Sonhood) → Trinity + Eternal Motherhood

What is "differAnt" here, whether on the side of scientific logic in general or the mathematical one in particular, is that there is no way for the triangle of the base to appear apart from the union and the synergy of the three independent triangles, or to be substantially equal to any of them. However, in the first chapter, we had given the fourth triangle the value of the Creator's act or the creation. In the present case, while we put the last touches to the *râzâtic* theological current, we have to consider it, first of all, Mary.

The figure which will represent the creation, humanity included, will be a completely new pyramid, once Mary occupies the base of the Paterno-filial Pyramid. Its Persons will be the Holy Spirit, this time under the term Materno-Filial Love, Mary the Mother of God, the Son of Man and, fourthly, the new unexpected base, the *râzâtic* 'Eternal Motherhood' and the Redemption. This is the pyramid of the Incarnation.

The movement emphasized by Rublev's icon is also vital to this new pyramid so that later, in the fullness of time, the two movements will intertwine by the encounter of the two relative centers of gravity, in the centre of gravity of the Milieu of the divine Love. The tripolar Yin-Yang might very well represent that perichoresis.

Thereupon, it becomes clear that the concept 'Mary', the Mother of God and Man, represents the 'Luminous Bridge' that ensures the communication and the passage of Religion and Science to the *maškanzabnâ* that is Childhood. It is 'Mary', bearer of the Eternal Motherhood, who has always been here, present long before creation, in the "Silence of the Memory of divine Love". We affirm it at once, *râzâtically* and objectively, leaving no opportunity for Schrödinger's uncertainty to surprise us. The Bridge is Mary, with all that this Name means.

7.3. Implications of this Name for Marian dogmas

We have no fear, by this statement, of being accused of heresy, because the kerygma of both Eastern and Western Churches affirms that the predilection, or pre-selection, of Mary, as a "concept" and as a "person" in whom the "concept" will be incarnated, has been from eternity. The same kerygma also praises chastity and virginity for their correlation with the hymen, the sign of the perfection of Eve, as seen in the third chapter. This way of seeing is at the base of the problematic of the virginity of Mary before, during, and after the birth of her Son, as taught by Saint Ephrem. Where does this lead us? It leads us to review three of the Marian dogmas of our faith:

1- The two accounts of the creation of Adam and Eve and the promised redemption.

2- The Immaculate Conception.

3- The Assumption, in the sense of the negation of Mary's death.

7.3.1. Regarding the two accounts of the creation of Adam and Eve and the promised redemption.

There is a controversy raised by the two accounts of the creation of Adam and Eve. Our *râzâtic* vision helps to put an end to it by opening fully the possibility of uniting the two with an inclination toward the first narrative (Gen 1: 27). It supports a solution through a 'prototype' much more conformed to reality than the one meant by the image and likeness of God. The Bible says: "So God created mankind in His image, in the image of God He created them; male and female He created them." From now on, we will read it as follows: "...in the image and likeness of the Son and His Mother He created them, male and female He created them..."

This new prototype of the image and the likeness brings us closer to the second account of creation. There we witness the extraction of the image and likeness of Mary (Eve) from the side of the image and likeness of the Son (Adam). This takes place for the same reason that made the prototype Mary, the Eternal Motherhood, pass from the Eternal Wound in the side of the Son to the "Silent Memory of God," *i.e.*, "of Love". Consequently, this inspires for us the following:

In Gen 2:18, we read: "It is not good for the man to be alone..."; therefore, it is also not suitable for masculinity to remain without femininity, nor paternity without maternity, nor the masculine divinity without a kind of feminine divinity. Incarnation will reveal this with a reversal of roles when the Son will come forth from the Womb of 'she' who first came forth from the eternal Wound of His Passion for 'her'.

This is what we mean by putting an end to the controversy between the two narratives. In the two cases, the one in which "God created Man in His image and likeness, man and woman, He created them", and the other where He transformed the rib He had taken from man into a woman, it is in the image and likeness of the relative prototype that both become 'created'. This also reveals that the two accounts represent the one same phase in the expectation of the Incarnation which will reflect positively on the *râz* of the relationship between 'Mary' and the Holy Trinity at the two levels of Creation and Redemption.

7.3.2. Regarding the Immaculate Conception

So, from all eternity, the being 'Mary', Mother of all living (Gen 3: 20), has existed, following a kenosis of the divine Paternity, leading to the objectivization of divine Motherhood, under the requirements of the Son, so that in the fullness of time she (Motherhood) incarnates in humanity, and He from her to fulfill the plan as described by the hymn of Sarug.

From the desire of the Son for a Mother, Mary has come. And, from kenosis to another kenosis, in the exhaustive Love, the Son will accomplish a total kenosis by renouncing the glory that was His in act, in the Paternal Milieu, to incarnate in her and from her, to enjoy motherly Love, in act also (even if with the Cross), and "save the world". He "bore the flesh" like any human being created in Their image, the image of Him and His Mother, and Their likeness to bring this image and likeness to its perfection. *Râzâtically* speaking, this is what we see behind the dogma of the Immaculate Conception because, as we have said before, if Mary, bearer of the Eternal Motherhood, were not immaculate from the beginning, she would never have been so in the fullness of time, all this happening through the same Holy Spirit in complete unity in the exhaustive Love.

Theologians found themselves embarrassed to explain how the 'immaculation' happened, whether regarding its timing or regarding the event. Did it happen at the moment Mary was conceived in the womb of her mother, or when the Holy Spirit came upon her? We will not linger on what has already been said and written on this question since we find it insufficient. According to our new paradigm, we find that it compromises Mary's perfect humanity, which directly affects Her Son's humanity. Therefore, we prefer to suggest our *râzâtic* point of view.

If we return to our conception of creation, as well as to our understanding of the Eternal Motherhood in the image and likeness of which Eve was created, a comparison between the Mother, Mary, and her Son will become more approachable. The fact is that the Maronite Liturgy presents her Son as "the single Person Who appeared on earth without sin". This statement remains incomplete unless it is also assigned to Mary because she is entitled to all that is her Son's, as her Son is entitled to all that is the Father's and vice versa (Jn 17:10). The Father and the Son are One (Jn 10: 30); the Mother and the Son are so too. Yet, we have to respect the ontological "differAnce" resulting from the fact that the Eternal Motherhood was elaborated by the wish of the Son and the kenosis of the divine Fatherhood, and not directly from the Eternal Love in parallel to divine Fatherhood. The divine Alphabet emphasizes this order, which is more in conformity with the state of the human family. So, we insist that it is a right and a duty that "immaculation" means that

Mary also appeared in our world without any sin, since she was exempted from it from the moment of her predilection. She has enjoyed her Son's qualifiers, except for what pertains to the properly divine Personality that is exclusive to the three Persons of the Trinity. This will be reflected in Adam and Eve, the image and likeness of Jesus and Mary. As Adam, from the moment of his Edenic 'modeling' in the image and likeness of the Son, will be free from all sin, so should Eve be, in the image and likeness of Mary already existing with her Son long before creation. Thereupon, we conclude that it is a right and a duty that Liturgy attributes to Mary the Mother of God the exemption from all sin that she has conferred on the Son from the moment of His conception on our earth.

Through this, the prophetic allegories of the Old Testament which refer to Mary become more comprehensible, precisely that of Jacob's Ladder which reminds us of the umbilical cord. This umbilical cord is what will bind the divine Milieu and the human one to make out of both the one and only Milieu with "differAnt" layers whose common centrality is the Childhood that Adam and Eve missed. The fact is that when our ancestors appeared in order to assume responsibility for the universe, they were adults, *i.e.*, with neither navel nor childhood. It is via the Motherhood of the Incarnation and the Childhood that accompanied it that humanity will enter the "kingdom of perfection".

So, we reiterate that if Mary, the utterly human girl in whom was incarnated the Eternal Motherhood and who is Eve's prototype before the fall, were not immaculate from any sin since the beginning, before her fertilization in the womb of her mother Anne, she would never have become so. She would never have been able to keep virginity during her pregnancy and after the birth of her Son. Even her Assumption would not have happened without death, whatever words would have striven to describe her 'immaculation' over time.

This is in principle, according to our view, what inspired the Fathers of the Church to declare her at the Council of Ephesus (431 AD) *Theotokos*. This too is what encourages us, based on what we advance in this book, especially the scientific part, to support definitively the admission by the Catholic Church of the so long expected fifth Marian Dogma, Mary co-redemptrix.[934]

7.3.3. Regarding the Assumption of Mary

The Orthodox Churches call the Assumption "the Feast of the Dormition" and the Syriac Churches "the Feast of the Rest" (*Niâḥâ* ܢܝܚܐ). In all cases, this

---

934   Cf. *"Co-Redemptrix* as Dogma?" https://udayton.edu/imri/mary/c/co-redemptrix-as-dogma.php [Accessed Dec 19, 2019.]

Dogma means Mary's passage, body and soul, into 'Heaven', Milieu of the Trinity. This dogma was promulgated by Pope Pius XII in 1950 as follows:

> By the authority of our Lord Jesus Christ, of the Blessed Apostles Peter and Paul, and by our own authority, we pronounce, declare, and define it to be a divinely revealed dogma: that the Immaculate Mother of God, the ever Virgin Mary, having completed the course of her earthly life, was assumed body and soul into heavenly glory.[935]

Subsequently, the dogmatic constitution *Lumen Gentium* of the Second Vatican Council of 1964 stated:

> Finally, the Immaculate Virgin, preserved free from all guilt of original sin on the completion of her earthly sojourn, was taken up body and soul into heavenly glory, and exalted by the Lord as Queen of the universe, that she might be the more fully conformed to her Son, the Lord of lords and the conqueror of sin and death.[936]

The point of difference between the Catholic Church and the Orthodox Churches is the passage through death: should Mary have known physical death, even if only for a moment, to be raised by the power of the resurrection of her Son? The Catholic dogma avoids her death, which translates into an ascent in the manner of the Prophet Elijah. The Orthodox insist on the necessity of the difference between her ascent into 'Heaven' and that of her Son. Our Maronite liturgical prayers try to satisfy both sides by their description of the Assumption. It is often that the mythological style draws from the Song of Songs or the Book of Revelation to save the situation.

At a personal level, we declare our faith in this dogma as promulgated by Pope Pius XII, and our persistence, as a 'child' of Mary, to respect it. Nevertheless, we do not hide that the scientific theories that we have discussed above have helped us understand it better. Frankly, they have assisted us in supporting better the point of view of the Catholic Church. We have proved this already by tackling the problem of the hymen and the navel at the end of the third chapter. But the scientific theory that have been the most useful to us is the paradox of 'Schrödinger's cat' and its *râzâtic* application in the fifth chapter to the probabilities raised around Christ's tomb. This deliberation served us well to confirm the importance of quantum theories in proving the Resurrection of Christ. By the same process, it demonstrates the objectivity of the dogma of the Assumption, shedding more

---

935   *Cf.* Pope Pius XII, Apostolic Constitution *Munificentissimus Deus*, 1950. §44.

936   Dogmatic Constitution *Lumen Gentium*, Second Vatican Council, 1964. §59.

accurate light on **Mary's unbroken life during her passage from our milieu to the Holy Trinity's.** Thanks to the *yât* of her Eternal Motherhood that became divine Motherhood in act, through the Son, she regained her place in the divine Milieu as if she had never left it.

It will no longer be appropriate to doubt, not even one percent, of the continuity of the Eternal Motherhood that Mary represents, especially to admit a shadow of burial in this sense. It is that, once behind the 'Wall', whether Planck's 'Wall' or that of the tomb, the phenomenon becomes similar to that of the Son. But it is out of the question to suppose two resurrections that come back to Eve and her descendants. The Son had the power of resurrection in Himself (Jn 5: 26), whereas Mary, according to the consequences of Sarug's plan, had in herself only her Son, the Life and the Light. **That is why it would be inadequate to Faith to refer to the Mother, after the resurrection of her Son, the same phenomenon endured by Him. Only one, unique Resurrection from physical death, as stated in Genesis, is sufficient and unrepeatable.**

In this case, mentioning Mary as possessing *par excellence* in herself the Light arouses again the energy model of wave-particle that applies well to Mary and Jesus. If the wave-particle duality is inseparable at the physical level, how can we allow ourselves to think it is separable at the level of the divine Reality that has shaped physics from Love and by Love?

Since imagination, fertilized by the inspiration of the Holy Spirit, granted human intuition a phenomenon similar to that of the Prophet Elijah which is no longer strange to Science, we allow ourselves, with these quantum data in support, to adopt the model of this Prophet having been taken up to heaven by a whirlwind (2 Kgs 2:11). Once this is said, our *râzâtic* opinion would be the following:

It was not necessary at all for Mary to go through physical death like her Son. Her Son conquered death by opening wide the door to eternal life. Nevertheless, and according to Sarug's plan itself, Mary, bearer of Eternal Motherhood, who appeared as intermediary Cause, necessary for the Son and the Father, and by that very fact, intermediately *co-creatrix* and *co-redemptrix*, should share her Son's fate with full conformity. This statement means, in line with the data of quantum mechanics, that Mary should have already undergone the following:

a- **Death at the same time as her Son.**

b- **Remaining in Life to fulfill the Mission entrusted to her by her Son.**

c- **Conceiving in her womb, the Church, the *râzâtic* Body of her Son, and giving Her birth.**

Let us explain this shocking *râzâtic* deduction:

a- Being dead at the same time as her Son.

Mary should, in the *râzâtic* sense, have given up her soul with the soul of her Son, to the Father from Whose kenosis she has appeared and with Whom she shares a direct 'correlation' with the Son. Once "it is finished" (Jn 19:30), *i.e.*, the mission of the redemption of the human race of which she and her Son are the authentic representatives, she should either leave the world forever, disappear as in the case of the Prophet Elijah, or live another Life granted by the same Source, for a new purpose. This latter Life is what her Son gave her through His 'Testament' on the Cross. He, by the same power with which He had ordered Lazarus a few days before to come out of the tomb, He Whose power accepts neither probabilities nor uncertainty, confided to His Mother a new Mission, granting her at the same time the Life necessary to fulfill it. He addressed her in the same manner He used at the wedding at Cana: "Woman, here is your son." The same he did to the disciple whom He loved: "Here is your Mother." (Jn 19:26-27) Since then, Mary was no longer to live for her Son, but rather for those represented by the "disciple He loved".

b-Remaining in Life to fulfill the Mission entrusted to her by her Son.

In the same *râzâtic* sense, from Love to Love, which means *ipso facto* from kenosis to kenosis, the new mission of the Eternal Motherhood was a corollary to Love, the divine Verb: "...Here is your Son...Here is your Mother". Still, we insist on repeating that nothing could have been done apart from the Absolute Love, apart from the Womb of the Exhaustive Love, the Triune God.

Thus, according to the model of the wave and the particle of the Young Experiment and the theories of quantum mechanics in general, Mary (the embodiment of the Eternal Motherhood) should have shared with her Son all He underwent. She should have died with Him and accompanied Him to Sheol. There, she should have drawn strength from Him with one hand, to bring Adam and Eve out, together with Him, with the other hand.[937] By this, she should have accomplished, with Him, on that glorious Saturday, the service of the Wedding of all those who were waiting for Him and who were part of the procession of the Bride, the Church. Then she had to resurrect with Him as Mother of God.

This is on one side. On the other side, she should have remained in life, in the world, to fulfill the new mission contained in the Testament pronounced on the

---

937   "The Hand That Rocks the Cradle Is the Hand That Rules the World" is a poem by William Ross Wallace that praises motherhood as the preeminent force for change in the world. The poem was first published in 1865 under the title "What Rules the World". The refrain of the poem is a commonly quoted phrase.

Cross. It is as if she was invited to conceive the humanity of the post-resurrection and to give birth to the Church, in the world and for the world, in full union with the Spirit of Paternal Love, to continue at the same time the mission she shares with her Son, by serving the Redemption. It is under Paul's *"Maranatha"* (1 Cor 16:22) and the "Come Lord Jesus" of the Revelation (Rev 22:17-22) that this Redemption should be in continual becoming.

This is the way we understand Mary's presence at Pentecost with the disciples, while she did not share with them the Last Supper. Otherwise, what would a second descent of the Holy Spirit on her, already 'full of grace', have served, if it were not then to conceive the same Body of her Son, this time the *râzâtic* Body, the Church. Let us not forget that the *Râz,* in line with our Maronite Liturgy, is the Eucharist *par excellence, i.e.,* Christ under the appearances of bread and wine. This is the reason we would say that the teaching of the Church regarding the transubstantiation of the bread and the wine of the Last Supper, from the state of body and blood of Jesus in potency to their state in act, also applies to Mary before and after the Cross. The bread and wine that Jesus had just given to His disciples, naming them 'My body' and 'My blood', would be so in potency until becoming so in act at the moment of the Immolation on the Cross. Likewise, Jesus' words to His Mother and the disciple whom He loved would gain all their active meaning after He had given up His Soul. It was at that moment that Mary became the "Mother of God" in act.

c- Conceiving in her womb the Church, the *râzâtic* Body of her Son, and giving Her birth.

Enjoying henceforth meta-human and meta-divine qualities, Mary had to conceive in her womb on the day of Pentecost the *râzâtic* Body of her Son, the Church. With the Holy Spirit, she will give birth to the Church, feed Her, immunize Her, and accompany Her till the time to wean Her. Once this new mission had been fulfilled, her visible person that lived only for this mission, being dead–alive at the same time, had to go spontaneously to the place that had been prepared for her from eternity, in the same way as the Ascension of her Son. Thus we understand the Assumption. We can imagine how much this enhances the paradox of 'Schrödinger's cat'!

This is also what we understand by the *yât* of Childhood. It is through the Eternal Childhood that we can see the Eternal Motherhood who generates it in a quantic (discrete) and infinite way. According to the 'Little Prince', those who do not recognize this Motherhood behind this Childhood do not distinguish either the elephant behind the skin of the boa or the DNA of Mary, the Eternal

Motherhood, behind her Son, Jesus Christ.

Thus, by the fact of adopting Childhood, by its *yât*, as the *maškanzabnâ* where Science, Religion, and absolute Truth discuss together, deliberately, in search of perfection, the 'Luminous Bridge' that leads there should always be 'Mary', the Eternal Motherhood. That is what the Incarnation revealed, as it revealed the trinitarian divine Love.

Before we move on to the conclusion of this chapter, we wish to share this little Ephremian wisdom story which expresses marvelously the role of Mary who reflects her Son's face and vice versa, and the way of reading as a 'Child' the truth of the Eternal Motherhood, without sinning against Christ's absolute centrality. Saint Ephrem's Bishop had just blamed him, based on so many complaints, that he exaggerated with his Marian spirituality and that he risked falling into heresy. So he categorically forbade him to start the prayers in Mary's name, but instead, to address them directly to Christ. As a saint obeying his Bishop, Ephrem started the next Morning Prayer addressing Jesus as follows: "O Splendor of Your Mother's face".

Can we expect a more brilliant expression that satisfies the scientist and the theologian, Filiation in line with Motherhood and Fatherhood?

Conclusion

"Were not our hearts burning within us while He talked with us on the road and opened the Scriptures to us?" (Lk 24:32)

"Spin and the Trinity: the relationship between cosmic gravity and Love" is the title of this last chapter, and, from this title, the Eternal Motherhood revealed herself to us as the core of all Gravity. As for spin, it is not enough to describe it as the angular momentum of divine Love between Father, Son, and Holy Spirit, according to the classical expression. Divine Love imposed itself as a spontaneous result of the relationship Son–Mother–Holy Spirit to end up with a synergistic expression of an indispensable movement between the hereafter and here, metaphysics and physics, Paterno-Filial Love and Materno-Filial Love, the whole *râzified* by the Incarnation into one unique Love, human-divine as divine-human, on Earth as in Heaven.

Our heart was burning with the feeling that before the *B/riš/it* of the Bible, there was a previous *B/riš/it*, that of the Phoenician-Aramaic alphabet transmitted to the incarnated Word by His *Generatrix*, Mary. He used it during His sojourn in the Creation, whether to communicate with the human or with the divine. It was quite before Planck's Wall and the time zero of the Big Bang that its 'being' was, so that the Word, or the Verb, could take shape from it and let the communication take place. Here it is:

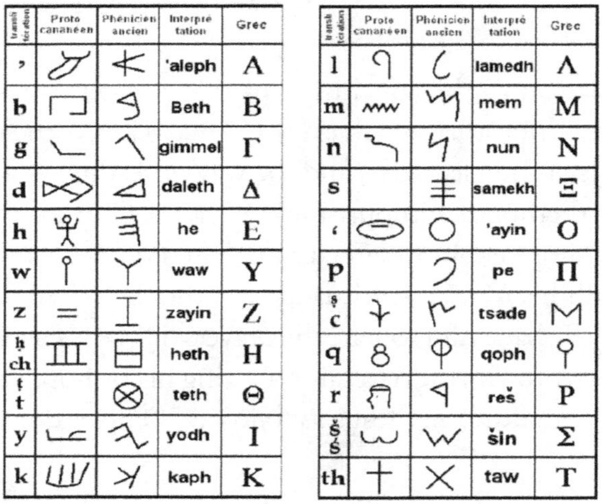

transliteration	Proto cananéen	Phénicien ancien	Interpré tation	Grec
ʾ			'aleph	A
b			Beth	B
g			gimmel	Γ
d			daleth	Δ
h			he	E
w			waw	Y
z			zayin	Z
ḥ ch			heth	H
ṭ t			teth	Θ
y			yodh	I
k			kaph	K

transliteration	Proto cananéen	Phénicien ancien	Interpré tation	Grec
l			lamedh	Λ
m			mem	M
n			nun	N
s			samekh	Ξ
ʿ			'ayin	O
p			pe	Π
ṣ c			tsade	M
q			qoph	Ϙ
r			reš	P
š s			šin	Σ
th			taw	T

Figure 91: The Phoenician-Aramaic alphabet

In the image and likeness of humankind, if this alphabet were not purposefully so "from the beginning" for expression, communication, and writing, it would never have been so – neither it nor humankind, – and consequently neither Love nor the Deity. In the image of the Eternal Motherhood, the alphabet, neutral in English but feminine in the Semitic languages, represents the 'womb' of every word, spoken or written.

In the same way in which the Ten Commandments, mother of all legislation, were written, these letters were 'drawn' in a space-time of creating Emotion in the Memory of divine Love to be able to communicate eternally with its image and likeness in the "differAnt" space-time aforementioned, in which memory is short-lived. With the alphabet, Writing (*l'Écriture*) was invented and safeguarded in the service of the successive *fullnesses* of "time" and of their respective kenoses upon which the space of "Archives" will have its roots. To become conscious of the alphabet and to succeed in conceiving, understanding, speaking, reading, and writing, humanity needed necessarily to regain its Childhood with the umbilical cord and the navel that it had missed since Genesis. It needed to do this in order to build the Bridge between the 'here' that succeeds Planck's Wall and the 'there' that precedes it, based on Gravity that joins the 'all' in a "differAnt" unity.

Once our heart was satisfied, we could not but thank the divine Love Who came to meet us on the path of our 'Emmaus'. In gratitude, we found ourselves obliged to develop our *râzâtic* inspirations, to write them down, to share them with humanity, and to preach our conviction under the aegis of Mother Church, never outside Her womb or that of Trinitarian Love.

We end this conclusion with some verses from the *Song of Songs* (8:6). Nothing better expresses this dialectic of love, which joins its perfection only in mutual, fraternal, friendly respect and particularly in Childhood which sees exhaustively:

> "Put me like a seal on your heart, like a seal on your arm; For
> Love is as strong as death, jealousy is as severe as Sheol; its
> flashes are flashes of fire; the very Flame of the Lord."

General Conclusion

Figure 92: World Centre for Love Research [CERN=WCLR]

Under the Tent of Childhood, the *maškanzabnâ par excellence*, embraced by the gravity of Trinitarian Love, first principle of every being *(Sein)* and every existence *(Dasein)*, Religion will go toward Science via Mary, Science will go toward Religion via Mary, and both will recognize each other, in the incarnation of the Light, of the Word, in the service of the unique Truth, Love.

The "Creation out of Love" (*Creatio ex Amore*) is a creation theory that intends to be an essay at the service of the improvement of the Church kerygma for the sake of humanity. We insist that it is not in opposition to the traditional dogma that wants God to have created everything out of *nihil* (*Creatio ex Nihilo*). A creation out of Love, by Love, and in the womb of Love can only be an achievement of a God Who is Love in essence.

This vision also leads to seeing, from a different angle, the dogma set out by Saint Augustine which says: "*Opera Trinitatis ad extra indivisa esse...*" This dogma, as shown above, was re-discussed by the theologian Jürgen Moltmann wondering if biblical and philosophical horizons can predict any "*extra Deum*", *i.e.*, any 'outside' to the omnipotent and omnipresent God: "Is there, after all, an *extra Deum* for the Trinity *opera ad extra*?"

From this angle of view, with the Ephremian theology in one hand and quantum physics in the other, we have been able to frame a new theological current in line with the Ephremian concept *râzâ* that has become its key. We have qualified it as *râzâtic*, *râzâ* being the Persian word borrowed by Saint Ephrem from the Book of Daniel, specifically the pages written in Aramaic. He needed it as a signifier to substitute the Greek concept 'mystery' which he found unsuitable for the correct catechism of the Church. Thus, this signifier, after we compare it to Jacques Derrida's "differAnce", means only itself and refers only to itself. It shuns any polysemy so that no lapse of time, or lambda of space or space-time, separates the signifier, the signified, and the Being to whom the signified refers. By *Râz*, we specifically recognize Jesus Christ telling the disciple Philip, "... Anyone who has seen Me has seen the Father..." Out of this *Râz*, Who by the Holy Spirit is One with His Father, flew all the *râzê* mentioned in the consecration words of the Maronite Liturgy, as well as in all types of signs and symbols, classically called mysteries but which are not mysteries at all. They are, instead, hierophanies of the divine Being, expressions of His presence in the secular reality. However, from '*râz*' as a common noun, we have forged the verb 'to *râzify*'. This verb, which is essential to bring together the different realities, is nothing but the reverse of the icon of the verb 'to love' that restores, renews, and deifies.

Accordingly, we have highlighted that it is only since the Incarnation of the divine Verb in our universe to lead Creation to its perfection, especially the descendants of Adam and Eve, that a "differAnce" started to be felt. Only Love Who created everything can, by the same Love, save His wounded creature and perfect it. In today's scientific language, we can compare this operation to stem-cell medication. That is why an 'Eternal Motherhood' was needed from the very beginning, before any creation.

Our research has helped us indeed to unveil a very delicate Marian truth. It is within the name of Mary that the Trinity made space to this Eternal Motherhood "to fulfill all righteousness". The Eternal Fatherhood allowed this, by kenosis (*zimzum*), urged by the Paterno-Filial Love, for the purpose of Adam and Eve's redemption.

Indeed, we have insisted on the predilection of a 'woman' to whom the role of bearing the divine Motherhood will be granted before triggering the creation. To this Lady, "in the fullness of time", the Eternal Motherhood will be granted by the Holy Spirit so that the Son may incarnate through Her, the Holy Spirit being the divine Paterno-Filial Love. At first sight, the purpose seems to be, as Jacob of Sarug asserted, that the Son might have a Mother from among humans and redeem them. Nevertheless, speaking more deeply, it was to restore their old motherhood that originated from the earthly Adam's rib through Mary's Motherhood which originated from the *râzâtic* New Adam's rib.

Yes, we assert that it is from the 'rib' of her Son Who was before her, her Creator and her Lord, that she was taken. According to Sarug's inspiration, the Son's desire was a condition *sine qua non* for the Creation to be, and therefore, for the Son to be born from her, enjoy a Mother, and redeem Adam's race. This is what our research has led us to, so that a new layer of the Redemption comes unveiled in favor of Mary as the Immaculate, Daughter of the God-Father, Mother of the God-Son, and Spouse of the Holy Spirit. Based on this, we strongly support the long-proposed *co-redemptrix* dogma. What has astonished us is that all that has been created spins, in Love, by Love, and for Love, as if it is an integral extension of the aforementioned divine Perichoresis.

To this point, and with full consciousness not to go beyond either the biblical or the scientific data, our perceptive mind has urged us to open the horizon of the Incarnation of God from a virgin woman (Mary by name, preselected and preserved from the Original Sin of humans) to become the Messiah. In other words, the Messiah's mission was first of all to redeem humanity through the divine Motherhood, by triggering the liberation of Eve, "the mother of all the living", from the curse imposed on her by the Mosaic Law. He did it by subjecting her to the Law of the New Eve, His Mother, who pushed Him at Cana's wedding to start the act of Redemption and who stood beside Him until He yielded up His spirit to His Father.

Our research also makes it apparent that, as asserted by the Ephremian Syriac prayers, the core of the Redemption was set as an exchange of a human mother for divinity. By becoming the Son of the preselected immaculate Mary, still a human woman, Jesus Christ extended to humanity the Fatherhood He enjoys as the Unique Son born from the Godhead in Eternity. In our *râzâtic* opinion, it is what is meant by the words "to fulfill all righteousness" of the baptismal scene at the Jordan River. It is how we see the foundations of the Kingdom of God, often declared by Jesus Christ. We want to make it clear that, above all and from eternity, it has been prepared for those who maintain themselves in Childhood.

Moreover, we have pointed out that Adam and Eve, as the Bible tells, knew neither Childhood nor Motherhood, and that the central point of the Incarnation was to perfect humanity by offering them those two essential components they were missing. It is in the act of the Incarnation of the Divinity from the 'womb' of Mary, on one side, and from the 'womb' of the baptismal water of the Jordan, on the other, that the *râzâtic* reality has begun. It is a reality where it would be challenging to separate the Creator and the creature, human thought and God's. The difficulty stands in a particular way in separating the love for the 'other' and the love for the 'All Other' since the One Who is 'All' has been caught, from that moment on, in the loving perichoresis, in the rhythm of Trinitarian Love. Paterno-filial Trinity in 'heaven', Materno-Filial Trinity on 'earth' – all is achieved by the same Holy Spirit Who is the beginning, the perfection, and the accomplishment of all that has been, is, and will be in heaven as on earth.

Quite interesting drawings have accompanied our explorations, in the manner of one of the most symbolic personalities of Childhood that we have followed, the Little Prince of Antoine de Saint-Exupéry.

It is an adventure we have embarked on, adamant to reconcile Science and Religion, as from the 'reef' Pope John Paul II spoke of in his encyclical *Fides et Ratio*. We can summarize the genuine common denominator that solicited us to do so by the word 'Light'. At the theological level, it is Christ's unshakable declaration, "I am the Light of the World", perpetuated by the formula of the Creed "Light from Light", as well as by the various hymns to the Light, especially those written by Saint Ephrem. At the scientific level, it is the information brought up by quantum physics in general, in particular about the nature of light and the photon, and especially by the physicists who revealed the wave-particle relationship, to mention only De Broglie.

Christ's challenging words about His identity as the Light of the world have obliged their aprioristic respect from traditional theology, spirituality, and the Christian 'mystics' of any given time. We have been able to detect that the case should be the same concerning the discoveries in Science. To prove it, we have dared to project these discoveries on Christ by comparing the human in Him to the 'particle' and the divine to the 'wave'. This comparison is not unheard of and is not only ours. Saint Ephrem himself inspired us. Already in his time, he compared the sun and the Trinity, as the Book of Nature and the Bible. He did it all in an apologetic purpose against the civilizations surrounding his parishioners, mainly worshipers of fire and of the sun. It is proof of the importance of imagination as an instrument available to divine inspiration, and further, to the importance of the 'synergy' between human and divine to lead Creation to its perfection.

Hence, to *râzify*, as we have defined it, means to love until causing a change in the nature of the subject-object of love. It is, from our perspective, what the words "to make all things new" of the Book of Revelation imply. The Gospel is very clear about this, and the One Who is its primary Author has signed this principle with His Blood and has encouraged us to do the same: "Love your enemies; bless those who curse you…" However, we still have to decide on the nature of this interchangeable love between Divine and Human, or more specifically, between two incompatible realities, to bring out of them a Luminous one called the Kingdom.

One verse of Saint Jacob of Sarug's hymn describing the last dialogue between the Father and the Son, just before launching the act of Creation, although not being of academic authority, has served us to the very end as a benchmark for the exploration of this new *râzâtic* discovery. This exploration has guided us to the Eternal Motherhood that we qualified as the 'Luminous Bridge'. This 'bridge' is Mary, the Immaculate Mother of the Man-God Jesus Christ. She is the 'bridge' who will join by her virginal-maternal Love, Heaven and Earth, Science and Religion.

In short, we have gradually been able to answer the fundamental questions raised by scientists, mainly the ones regarding the identity of God as absolute Love. In resolving several enigmas, we have succeeded in establishing a method to overcome any difficulty between Science and Religion, as well as between atheism, rationalism, secularism, and faith. That is because, *râzâtically* speaking, Faith proves to be the sincere expression of Love and Trust in this created world, as supported by Chapter 13 of the first letter of Paul to the Corinthians.

We have proven that the Tent of Meeting between God and the most brilliant of religious and scientific people, whether atheists or believers, is Childhood. That is also what we mean when we say that the 'Luminous Bridge' that leads to this 'place of encounter' is the Eternal Motherhood, the 'umbilical Cord' that *râzâtically* binds the Creator's Milieu to the milieu of the Creation, divinity to humanity. Further, the divine Word, with the full consent of the divine Fatherhood and the full kenosis of the divine Trinity, took human flesh through her. That 'Bridge' is Mary, descended from David, who enjoyed from the first moment of her eternal predilection the three meta-human qualities: Daughter of the Father, Mother of the Son, and Bride of the Holy Spirit. **Were it not so from the beginning, Mary would never have become what she has become.**

All the preceding has driven us to affirm our conviction regarding the long-desired *co-redemptrix* dogma. The same conviction has evoked in our spirit a new Marian theory to submit to the reflection of the Catholic Church that is about the Eternal Motherhood of Mary which makes her worthy of being considered an intermediary Cause, a partner in the act of Creation (Mary *co-creatrix*). For, to believe Sarug's hymn, the Son would not have accepted that Creation be done, if He were not sure that one day He would have Mary's own free, unconditional Maternal *Fiat*.

Have we concretely answered the scientists' question to biblical religions: What kind of 'god' is it about?

Of course. Based on our *râzâtic* current of theology, it is Love we have considered 'God'. We should henceforth adore Love and call scientists to join us in the worship of Love. Love is God *par excellence*. Between the two sayings, "God is Love" and "Love is God" sits, as we have shown, the Derridean "differAnce" in which, as we noted, the 'A' is for '*Amour*', Love. He is the Triune Love in continuous perichoresis between Fatherhood, Sonhood, and Paterno-Filial feeling called the Holy Spirit. Love created everything, specifically us, humans, out of Himself, in Himself (in His Maternal Womb), so that He Loves what He created, is loved by what He created, and that His creatures live from loving each other.

Furthermore, we pointed out that the source of all spinning (whirling) in the universe and all its effects, as Aristotle also imagined it, follows the same divine perichoresis. Even the absorbing, stunning feelings of vertigo should be seen in correlation, or better said, in entanglement with that phenomenon. Love's perfect expression in the created reality is not secular peace, which is for humans the result of a balance of terror between different striking powers. It is instead Childhood that leads to Peace by the power of acceptance of the 'other' and by the capacity for self-kenosis (*zimzum*) so that the 'other' can grow and enjoy his/her share of communal Loving Life. We have highlighted well the role it will play as the Tent of Meeting mentioned in the Bible, the *maškanzabnâ* in the Syriac language, the Environment *par excellence* of encounter between the divine Word, Science, and Religion.

However, we have asserted above that the sharing in this unique communal Loving Life within the Womb of the Trinitarian Love can only be realized under two sources: on the one hand, the watchful Eye of the divine Fatherhood of which each earthly father is a representation, and on the other hand, under the Eye of the Eternal Motherhood of which the New Eve, Mary, became by predilection the unique permanent partner in the created world. We have based our opinion on the words of the Son warning us "not to call anyone on earth 'father' for we have one Father, and He is in heaven," and Who avoided saying the same about mothers. It is that *râzâtically* speaking 'motherhood' is granted as an original spark from Mary's Motherhood to each genuine earthly woman (mom) by the power of the womb granted to her by birth. That is why, in our opinion, Jesus did not warn us not to call any woman on earth 'mother – mom'. He enjoyed, personally, this appellation and lived it in significant harmony and synergy with His Mother until His death on the Cross, and even after it. We have insisted on this "after" in the last chapter in order to take part in the solution of the polemic about the Assumption of Mary with or without having known death.

Science was a great help to us in laying the foundations of *râzâtic* theology. It was possible thanks to the incontestable similarity between the divine Triunity and the wave-particle whose third integral element is the electromagnetic – gravitational force which ensures the non-separability between the two, their directionality and their spin, without any risk of mutual elimination, confusion, or separation. Then came the two phenomena of correlation and entanglement supported by the EPR paradox. For us, this is only a quantum simulation of the nature of the "god" in question. The perichoresis specific to the particle-wave moved by its energy to the value of Planck's constant, in perfect interweaving and harmony, looks like that of the divine Trinity moved by the proper altruistic Paterno-Filial Love. The fact is that, in the two cases, it is impossible to distinguish between the two principal components, as Young's double-slit experiment proves for Science and as Rublev's icon, which occupies a good part of our analyzes, shows.

The *rázátic* theory does not limit itself to the comparative expression "on earth as in heaven" of the Lord's Prayer. The latter leaves an impression of separation and exteriority between the two "realms". Basing itself on the dialogue between Jesus and the disciple Philip, it insists instead on the intimate correlation between earth and heaven, the basis of their partnership in the same realm, the Kingdom. Otherwise, the Incarnation and the Resurrection would not have accomplished their primary goal of removing the Flaming Sword of Genesis, making the 'Mosaic veil' of mystery and mysteriousness fall, and re-opening wide the communication between Father and children. In other terms used by Jesus, we would not have become friends of the Son and, therefore, of the whole Trinity.

The theory does not, however, deny the need to continue to distinguish between the holy, the sacred, and the profane, but categorically refuses the perpetuation of the oppression caused by the 'veil' of Moses (as Saint Paul asserts in Chapter 3 of his second letter to the Corinthians) which our theory considers a mortal sin against divine Love.

In our exposition, what helped to develop the "structure" of the *rázátic* theology are fire, light, heat, energy, synergy, quantum physics, mathematics, and theological currents such as the Trinitarian, the kenotic, the environmental, and the apophatic ones. Christology from above, from below, and from the *"Core"* and Saint Ephrem's Mariology also joined as essential 'packets'/quanta of epistemological luminous waves. We have not even spared the Kabbalist *zimzum* theory, the Taoist Yin-Yang, western philosophy, and precisely the deconstructionist one, as well as linguistics, sciences, etc. Through their best representatives, all these disciplines have contributed to explore the phenomenon of the tomb of the Man-God, said to be empty.

The culminating point of our contribution was the comparison done between Jesus Christ in His tomb and Schrödinger's Cat in its box. We had prepared the scene to this matching in probability and certitude with a prophetic deliberation *post mortem* between seven personalities who found themselves on board our explorer ship. They are Moses, La Place, Planck, Einstein, Schrödinger, Guitton, and Ephrem. This deliberation happened on a *râzâtic* platform based on the extreme data of Science, Philosophy, and Religion.

The group came out of it convinced that based on R. Buckminster Fuller's synergistic equation $(1+2=4)$, reality must be seen exhaustively. According to the two principles of correlation and inseparability, the state of things that reigns behind the 'Wall' of the tomb should be the same as that which prevails before it: death/death; life/life. Consequently, the exhaustive optimism offered by the Resurrection of Christ (the *Râz* of Love) better suits the coordination between Science and Religion. It is because those who stand behind the two disciplines are all mortal humans who read and write using the same Alphabet. Likewise, both form the two wings of the Absolute Truth. Otherwise, everything risks falling into absurdity. This way, we came out with the unavoidable importance of recognizing that Creation exists not in some space, outside of its Loving Creator, *ad extra*, but rather in Him, just at the center of the 'milieu' He, in the Old Testament, called "His Womb".

Being convinced of the results of our contribution, we have invited every human being to remove any anxiety about the unpredictability and uncertainties that discoveries may arouse, especially those at the level of quantum mechanics. We have been able to prove that what has been created "out of Love" and "by Love" cannot be but good ontologically, and shall definitively end in Love, in continuous mothering, in eternal regeneration. This is what Eternal Motherhood means. The total destruction and annihilation of the universe are out of the question. The future of humanity, managed by unshakable natural laws set out of Love, by Love, for the conservation of the universe, is protected from any annihilation or absurdity.

All we have written after almost eighteen years of research, meditation, sharing, and consultation, especially after the publication of our first book in 2014 under the French title *La Création, de l'Ère Patristique à l'Ère Quantique / Essai en Théologie Râzâtique*, seems today to be nothing more than a small step in our route toward our *râzâtic* Emmaus. However, since we started our work in 2002, our terrestrial globe has known, unfortunately, alarming epidemics like AIDS, SARS, EBOLA etc., and in the present time, the pandemic of COVID-19 that is shaking the whole of humankind. Our globe has also experienced devastating natural events like the tsunami of 2011, massive fires that devoured immense forests around the world, and mainly the dangerous depletion of the vital ozone layer. Furthermore, this world of humans has gone through the worst fanatic religious wars and oppression of all kinds. The race toward the domination of energy sources, rare materials, and market created and nurtured these. It has proved, once more, the criminal egocentric tendency of the human being toward possessing and dominating. Consequently, global anxiety is growing. The questioning of the role of loving Providence is surfacing, raising the challenge between Science and Religion, highlighting the necessity of our optimistic *râzâtic* theory based on God-Love.

All the preceding keeps the 'door' open to any perceptive mind and any creative imagination motivated to contribute to reassuring humans about the future of the earth and the universe. It is a strong invitation for worshipers of the divine Love to support our *râzâtic* theory, providing that Love is God, that "Love is stronger than death", and that Life is worth living under the Tent of Childhood and the care of the Eternal Motherhood.

Appendix of the years 2020, 2021

We are writing these last lines during Holy Week of the year 2020, the epicenter of the COVID-19 pandemic. As of today, this virus has quarantined fifty percent of the entire population of the globe for almost three months. Since the beginning of February 2020, it has affected more than two million people around the world and killed almost 135 000 persons. The most significant number of affected persons belong to the most developed countries such as the U.S.A. and the European ones.

We wonder, in awe, why that the countries most affected by this plague are the ones that have been involved the most in the application of the theory of "Constructive Chaos" throughout Arab countries, calling it sarcastically the Arab Spring?

The ravages consequent to this have been done to subdue the Arab countries to a forced peace with the bizarre colonial country Israel. It is indeed bizarre, because it substitutes for the Chosen People of the Old Testament a geopolitical entity contrary to the biblical God's Will. They caused without scruple millions of deaths, millions of displaced persons, and a general state of misery and humiliation of peoples who are now subject to the new slavery of globalization. Would this reading of the present dramatic events of COVID-19 be necessary to consolidate a universal conviction that personal wellbeing, and not the individualistic one, begins with the welfare of the 'other', starting with neighbors?

Is it not fair to deduce from these painful circumstances that the universal Golden Rule, underlined by Jesus Christ in the Gospel, remains the best advice for being together and not for "having together"? It is a frequent philosophical distinction that we highlighted by our new concept 'synesseration' forged from the Latin verb *esse* which means "to be"? We insisted on this new concept to distinguish between "withing, to have together" and "withing, to be together". It is supported by the Cartesian saying "I think, therefore I am", refuting the globalization standard of "to have", or more critically, "to possess". An invisible virus derailed all dreams of dominating, possessing, controlling, substituting the Loving Creator with a financial, human dictator, as has happened at different times in history, as the Bible itself witnesses.

Wisdom has repeated since Horace: "When what is natural is driven off, it returns at a gallop." All our pages have supported this saying and have proved that triunity is the most natural component of Creation, at the level of Religion as well as of Science, of which Love is the most profound proof and expression. To love the other, the Whole-Other, the creature and its Creator, to love and to be loved in a healthy environment, inspires more certainty and predictability.

To love as a 'child' under the aegis of the Eternal Motherhood is the foundation of the Kingdom of the Trinitarian understanding between Science, Religion, and divine Love, the Unique Truth that the two are supposed to serve as being their Unique Reference.

We close our book by paying tribute to Pope Francis for his first encyclical letter entitled *Laudato Si', On Care for Our Common Home*. In this document, as if by prophecy, the Holy Father took over the wishes of Pope Paul VI and warned the economic and political leaders of the world not to force the sacred natural law of the Creation. Under paragraph 18, he declares:

> The continued acceleration of changes affecting humanity and the planet is coupled today with a more intensified pace of life and work which might be called '*rapidification*.' Although change is part of the working of complex systems, the speed with which human activity has developed contrasts with the naturally slow pace of biological evolution. Moreover, the goals of this rapid and constant change are not necessarily geared to the common good or to the integral and sustainable human development. Change is something desirable, yet it becomes a source of anxiety when it causes harm to the world and to the quality of life of much of humanity.

Based on this warning, we wonder: could humanity have been able to avoid the COVID-19 disaster? Do not the tears shed by Pope Francis during Holy Week of this year 2020 not recall the tears of the prophet Jeremiah in deep lamentation over Jerusalem, demolished then by Nebuchadnezzar, and today, treated by a modern Nebuchadnezzar as real estate? Or, to each era its need to hear, like prodigal sons and daughters, the warning of the Creator found in the same prophet Jeremiah, saying:

> Hear this, O foolish and senseless
> people, who have eyes, but see not,
> who have ears, but hear not.

Do you not fear me? says the Lord;
Do you not tremble before me?
I placed the sand as the bound for the sea, a
perpetual barrier which it cannot pass;
though the waves toss, they cannot prevail,
though they roar, they cannot pass over it.
But this people has a stubborn and rebellious
heart; they have turned aside and gone away.
(Jer 5:21-23)

Yes, indeed, Lord, and we have made out of "Love" a consumable package valid for all kind of commercialization and abuse. We did not grasp that "fearing You" means fearing to lose Your Paternal Love. Forgive us as You forgave the Prodigal Son. We are coming back to You with more awareness and less stubbornness. We have sharpened our skills so that we can pretend no more to have blind eyes and deaf ears. We confess that we humans are made out of Your Fabric, and no matter what non-loving people have done to us, we will, by Your Love, overcome their evil, even death, and grow until the fullness of the size of Jesus, the Son Who founded our Trinitarian relation on sonhood, fatherhood and motherhood. We confess – we, the fruits of the womb of fatherhood and motherhood – that we are made out of Your love, by Love and for Love, and that our mission is to keep Love running in this world, blessing forgiving and giving life abundantly. Our hearts can never rest except in You. We sincerely love You, Love!

BIBLIOGRAPHY

Actes du Colloque XI-Alep 2006; ensemble de Conférences données au 17ème centenaire d'Éphrem. Alep; CERO, 2007.

Babu, Paul. *Veni, Vidi, Vici*. Trivandrum, India: Rabban Benjamin Joseph Publisher, 1982.

Bachelard, Gaston. *La Psychanalyse du Feu*. Folio/essays, 25. [Paris]: Gallimard, 2008.

Barbour, Ian G. *When Science Meets Religion*. [San Francisco]: HarperSanFrancisco, 2000.

————. *Religion and Science: Historical and Contemporary Issues*. A Rev. and Expanded Ed. of *Religion in an Age of Science*. [San Francisco]: HarperSanFrancisco, 1997.

————. *Myths, Models and Paradigms In Science and Religion*. N.Y.: Associated Press, 1960.

Beck, Edmund. *Des Heiligen Ephraem des Syrers Carmina Nisibena*. CSCO, Corpus Scriptorum Christianorum Orientalium, Vol. 218; Scriptores Syri, Tomus 92. Louvain: Secrétariat du Corpus SCO, 1961.

————. *Des Heiligen Ephraem des Syrers Hymnen contra Hæreses*. CSCO, Corpus Scriptorum Christianorum Orientalium, Vol. 169; Scriptores Syri, Tomus 76. Louvain: Imprimerie Orientaliste L. Durbecq, 1957.

————. *Des Heiligen Ephraem des Syrers Hymnen de Ecclesia*. CSCO, Corpus Scriptorum Christianorum Orientalium, Vol. 198; Scriptores Syri, Tomus 84. Louvain: Secrétariat du CorpusSCO, 1960.

————. *Des Heiligen Ephraem des Syrers Hymnen de Fide*. CSCO, Corpus Scriptorum Christianorum Orientalium, Vol. 154; Scriptores Syri, Tomus 73. Louvain: Imprimerie Orientaliste L. Durbecq, 1955.

————. *Des Heiligen Ephraem des Syrers Hymnen de Nativitate (Epiphania)*. CSCO, Corpus Scriptorum Christianorum Orientalium, Vol. 186; Scriptores Syri, Tomus 82. Louvain: Secrétariat du CorpusSCO, 1959.

————. *Des Heiligen Ephraem des Syrers Hymnen de Sermo de Domino Nostro*. CSCO, Corpus Scriptorum Christianorum Orientalium, Vol. 270; Scriptores Syri, Tomus 116. Louvain: Secrétariat du CorpusSCO, 1966.

————. *Des Heiligen Ephraem des Syrers Hymnen de Paradiso und Contra Julianum*. CSCO, Corpus Scriptorum Christianorum Orientalium, Vol. 174, 175; Scriptores Syri, Tomus 78, 79. Louvain: Secrétariat du CorpusSCO, 1957.

————. *Des Heiligen Ephraem des Syrers Hymnen de Pascha: de Azymis, de Crucifixone, de Resurrctione*. CSCO, Corpus Scriptorum Christianorum Orientalium, Vol. 223; Scriptores Syri, Tomus 94. Louvain: Secrétariat du CorpusSCO, 1962

_____. *Des Heiligen Ephraem des Syrers Hymnen de Virginitate*. CSCO, Corpus Scriptorum Christianorum Orientalium, Vol. 248; Scriptores Syri, Tomus 108. Louvain: Secrétariat du CorpusSCO, 1964.

Bennington, Geoffrey et Jacques **Derrida**. *Jacques Derrida.*Série les Contemporains. Paris: Seuil, 1991.

Benoît XVI [Joseph Ratzinger]. *Jésus de Nazareth, Deuxième partie*. Rocher, Groupe Parole et Silence. Librairie éditrice vaticane, 2011.

Blondel. *Les Exigences de la Pensée Contemporaine*. Œuvres Complètes, Tome II. PUF, 1997.

Böhm, David; Stanislav **Grof**, préf. et Tchalaï **Unger**, trad. *La Plénitude de l'Univers*. [Monaco]: Le Rocher, 1987.

Bou Mansour, Tanios. *La Pensée Symbolique de Saint Éphrem le Syrien*. Kaslik, Liban: *PUSEK, 1988.*

_____. *La Théologie de St. Éphrem*. Kaslik, Liban: PUSEK, 1993.

Brock, Sebastian. *The Luminous Eye: The Spiritual World Vision of St. Éphrem*. Kalamazoo: Cistercian publications, 1992. [Rev. Ed.].

_____. *Syriac Studies: A Classified Bibliography (1960-1990)*. Kaslik, Liban: Parole de l'Orient, 1996.

_____. *L'Œil de Lumière. La Vision Spirituelle de Saint Éphrem*, suivie de *La Harpe de l'Esprit, Florilège de Poèmes de Saint Éphrem*. Trad. de l'anglais et du syriaque par Didier Rance et Dom Louis Leloir; Préf. de Dom Louis Leloir. *Spiritualité Orientale, N° 50*. Godewaersvelde: Abbaye de Bellefontaine, 1991.

Capra, Fritjof. *The Tao of Physics: An Exploration of the Parallels Between Modern Physics and Eastern Mysticism*. 4th ed., updated. Boston: Shambhala, 2000.

_____. David Steindl-Rast and Thomas Matus. *Belonging to the Universe: Exploration on the Frontiers of Science & Spirituality*. N.Y.: Harper San Francisco. 1992

_____. *The Turning Point*: Science, Society, and the Rising Culture. USA: Bantam Book; Canada: Simon and Schuster, 1983-1988.

Caputo, John D. et Michael **Scalon** J.; eds. *Augustine and Postmodernism: Confessions and Circumfession*. Bloomington, Ind.: Indiana University Press, 2005.

Cauvin, Patrick. *E=MC², mon amour: roman*. Paris: J. C. Lattès, [1977].

Collins, Francis S. *The Language of God: a Scientist Presents Evidence for Belief*. New York: Free Press, 2006.

Davies, Paul. *God and the New Physics*. New York: Simon and Schuster, 1983.

_____. *The Mind of God: The Scientific Basis for a Rational World*. New York: Simon & Schuster, 1993.

_____. *The Fifth Miracle: The Search for the Origin and Meaning of Life*. New York: Simon & Schuster, 1999.

De Lubac, Henri. *Teilhard et Notre Temps*. Aubier, Série Foi Vivante. Paris: Aubier, 1971.

Derrida, Jacques. *Circumfession*. London: The University of Chicago Press Ltd, 1993.

Den Biesen, K. *Bibliography of Éphrem the Syrian*. Giove in Umbria, 2002

De Chardin, Pierre Teilhard. *Le Phénomène Humain*. Série Points, Essais. Paris: Seuil 1955.

──────. *Le Milieu Divin*. Série Points-Sagesse, 1. Paris: Seuil, 2003.

──────. *Mystique Savant*. Série Points-Sagesse. Paris: Seuil, 2003.

Eliade, Mircea. *Le Sacré et le Profane*. Paris: Gallimard, 2008.

Éphrem, de Nisibe. *Les Chants de Nisibe*. Trad. de Paul **Feghali** et de Claude de Navarre. Antioche Chrétienne III. Paris: Cariscript, 1989.

──────. *Célébrons La Pâque*. Les Pères de la Foi, Migne. Paris, 1995.

──────. *Hymnes sur le Paradis*. Ed. et Tr. En français par Lavenant, René, Sources Chrétiennes N° 137. Paris: Cerf, 1968.

──────. *Saint Éphrem: un Poète pour notre temps*. Patrimoine syriaque; Actes du colloque XI. Paris: CERO.

Éphrem (saint). *Hymnes sur la Nativité*. Trad. du syriaque par François Cassingena-Trévedy. Paris: Éd. du Cerf, 2001.

Éphrem le Syrien. *Hymnes sur l'Épiphanie: Hymnes Baptismales de l'Orient syrien*. Intr. Trad. Du texte syriaque, notes et index par François Cassingena. Spiritualité, n° 70. Godewaersvelde: Abbaye de Bellefontaine.

Feghali, Paul. *Les Origines du Monde et de l'Homme dans l'Œuvre de Saint Éphrem*. Paris: Cariscript, 1997.

──────. et C. **Navarre;** trad. de Paul **Féghali** et Claude **Navarre**. *Saint Éphrem, les Chants de Nisibe*. Antioche Chrétienne, 3. Paris: Cariscript, 1989.

──────. «Commentaire de l'Exode de l'Orient par Saint Éphrem. Introduction, traduction et notes», *in Parole de l'Orient 12* (1984-1985) 91-131 [→ § 123, § 260, § 274].

──────. «Les premiers jours de la création. Commentaire de Gn1:1-2, 4 par Saint Éphrem», in *Parole de l'Orient* 13 (1986) 3-30 [→ §121: § 260].

──────. «Un commentaire de la Genèse attribué à Saint Éphrem», in *Kh. Samir (ed.), Deuxième Congrès International d'Études Arabes Chrétiennes* [Orientations Christiana Analecta 226], Roma, 1986, 159-175 [*cf.* Tabet, title 154, and Samir, titles 901 & 907 → § 13].

──────. «Influence des Targums sur la pensée exégétique d'Éphrem?», in *H.J.W. Drijvers*, R. Lavenant, C. Molenberg & G.J. Reinink (eds.), IV Symposium Syriacum 1984. Literary Genres in Syriac literature (Groningen-Oosterhesselen 10-12 September) [Orientalia Christiana Analecta 229], Roma, 1987, 71-82 [→ § 259].

Ford, Kenneth W. *The Quantum World: Quantum Physics for Everyone*. Cambridge,

Mass.: Harvard University Press, 2004.

Gibran, Kahlil. *The Prophet*. New York: Alfred A. Knopf Inc., 2001.

——————. *Le Prophète*. Trad. Jean-Pierre **Dahdah.** Paris: J'ai Lu, 1993.

Girard, René. *La Violence et le Sacré*. Paris: Hachette littérature, 1998.

Greene, Brian. *The Fabric of the Cosmos: Space, Time, and the Texture of Reality*. New York: A. A. Knopf, 2004.

Guitton, Jean. *Mon Testament Philosophique*. Collection Petite Renaissance. Paris: Presse de la Renaissance, 2007.

——————. et Igor et Grichka **Bogdanov**. *Dieu et la Science: vers le Matérialisme*. Paris: B. Grasset, 1991.

Haight, Roger. *Jesus Symbol of God*. Maryknoll, N. Y.: Orbis Books, 1999.

Harrison, Edward R. *Cosmology*. London: Cambridge University Press, 1981.

Haught, John F. *Science and Religion from Conflict to Conversation*. New York: Paulist Press, 1995.

Hawking, Stephen William. *A brief History of Time: from the Big Bang to Black Holes*. Toronto, New York: Bantam Books, 1988.

——————. *Une Brève Histoire du Temps: du Big Bang aux Trous Noirs*. Trad. de l'anglais par Isabelle **Noddeo-Souriau**. Paris: Flammarion, 1991: 2004.

Herbert, Nick. *Quantum Reality: beyond the New Physics*. Garden City, N.Y: Anchor Press; Doubleday, 1987.

Jaki, Stanley L. *Road of Science and Ways to God*. Chicago; London: The University of Chicago Press, 1978.

John-Paul II. *La Foi et la Raison*, Lettre Encyclique Fides et Ratio. Prés. par Michel Sales, sj. Ed. Paris: CERF/Bayard, 1998.

Jung, Carl Gustav. *Psychological Types*. A revision by R.F.C. Hull of the translation by Helton Godwin Barnes. (Collected Works of C.G. Jung Vol.6); Bollingen Series XX. Princeton (N.J.): Princeton University press, 1990.

Klein. *Petit Voyage dans le Monde des Quanta*. Paris: Flammarion, 2004.

Lalande, André. *Vocabulaire technique et critique de la Philosophie*. Paris: PUF, 1988.

Lederman, Leon **Max** et Christopher T. **Hill**. *Symmetry and the Beautiful Universe*. Amherst, N.Y.: Prometheus Books, 2004.

——————. et Dick **Teresi**. *The God particle: if the Universe is the Answer, what is the Question?* New York: Dell publishing, 1993.

Lewandowski, Raymond C. *The Imprint of God: Secrets in Our Genetic Code*. New York: Morgan James Pub., 2008.

Mathews E.G. and J.P. **Amar**. *Saint Éphrem the Syrian, Selected Prose Works: Commentary on Genesis; Commentary on Exodus; Homily on Our Lord; Letter to Publius. Translated by E. G. Mathews Jr. and J.P. Amar; Edited by K. McVey.* The Fathers of the Church, 91. Washington (DC), 1994.

Moltmann, Jürgen. *Le Rire de l'Univers*. Paris: les Éd. du Cerf, 2004.

Noujaim, G. *Anthropologie et Économie de Salut chez S. Éphrem autour des Notions de 'Ghalyata'*. Thèse non publiée. [Rome, 1980].

Olthuis, James H. *The Beautiful Risk*: *a New Psychology of Loving and Being Loved*. Grand Rapids, Mich.: Zondervan, 2001.

Otto, Rudolf. *Mystique d'Orient et Mystique d'Occident (Distinction et Unité)*. Traduction et préface de Jean Guillard. 278, Petite Bibliothèque Payot. Paris: Payot et Rivages, 1996.

Polkinghorne, John Charlton. *Science and The Trinity: The Christian Encounter With Reality*. New Haven; London: Yale University Press, 2004.

Rahme, Georges. *Teilhard De Chardin: Mystique Savant*. Beyrouth: Publishing and Marketing House, 1984.

Rahner, Karl. *Foundations of Christian Faith: an Introduction to the Idea of Christianity*. Translated by William V. **Ditch**. New York: Crossroad, 1978.

«**Revue Internationale de Philosophie**». *La Mécanique quantique,* n° 2, juin 2000.

Robertson, Donald W. *Mind's Eye of Richard Buckminster Fuller, New York, St. Martin's Press, 1974.*

Shumm, Bruce. *Deep Down Things: The Breathing Beauty of Particle Physics*. London: John Hopkins University Press, Baltimore, 2004.

Smith, R. Payne. *Thesaurus Syriacus*. Oxford: Clarendon Press, 1957.

Vico, Giambattista. *A cura di Nicola Abbagnano*: *La Scienza Nuova e Altri Scritti*. Torino, Italie: Unione Tipografico-Editrice torinese, 1952.

Vorreux, Damien. *Un Symbole Franciscain, le Tau: histoire, théologie et iconographie*. Paris: Éditions Franciscaines, 1977.

Worthing, Mark William. *God, Creation and Contemporary Physics*. Minneapolis: Fortress press, 1996.

أوريجانس، في المبادئ. تعريب الأب جورج خوام البولسي. منشورات المكتبة البولسية، 2003. العنوان الأصيل بالفرنسية:

(Origène, Traité des Principes, Sources Chrétiennes NN 252, 253, 268, 269; par Henri Crouzel et Manlio Simonetti; Editions du Cerf.)

أفرام السرياني (مار). منظومة الفردوس. ترجمة الأب روفائيل مطر. مجموعة أقدم النصوص المسيحية. الكسليك: جامعة الروح القدس، 1980.

ــــــــ. أناشيد الصوم والفطير والصلب والقيامة. قدم لها ونقلها إلى العربية وكتب حواشيها الخوري بولس الفغالي. سلسلة ينابيع الإيمان، 8. الحدث، بعبدا: منشورات الجامعة الأنطونية، 2004.

ــــــــ. بين مائدة وبين مائدة. ينابيع الإيمان، 12. تعريب الخوري بولس الفغالي. الحدث، بعبدا: منشورات الجامعة الأنطونية، 2007.

. في الكنيسة أو الجهاد المسيحي. تعريب الخوري بولس الفغالي. ينابيع الإيمان، 14. الحدث، بعبدا: منشورات الجامعة الأنطونية، 2007.

ــــــــ. أناشيد في الإيمان، الجزء الأوّل 1-40. تعريب الخوري بولس الفغالي. ينابيع الإيمان، 15. الحدث، بعبدا: منشورات الجامعة الأنطونية ، 2007.

ــــــــ. أناشيد في الإيمان، الجزء الثاني، 41-80. تعريب الخوري بولس الفغالي. ينابيع الإيمان، 16. الحدث، بعبدا: منشورات الجامعة الأنطونية، 2007.

بسترس، فاخوري عبسي. تاريخ الفكر المسيحي عند آباء الكنيسة. جونية: المكتبة البولسية، 2001. الشحيمة المارونية (معرّبة). بيروت: المطبعة الكاثوليكية، 1981.

الجميّل، بطرس. صلاة المؤمن: زمن الفصح. جونية: مطبعة الرسل، 1987.

رستم، أسد. آباء الكنيسة. ط. ثانية. جونية: منشورات المكتبة البولسية، 1990.

روحانا، ميخائيل. تحوّل المفاهيم في بناء الجمهورية، رسالة لبنان من أجل السلام في الشرق والعالم: نحو جمهورية لبنان الخامسة. بيروت: منشورات صادر الحقوقية، 2006.

السروجي، يعقوب. رؤى دانيال. قدم لها ونقلها إلى العربية وكتب حواشيها الأب إميل أبي حبيب. الحدث، بعبدا: منشورات الجامعة الأنطونية، 2006.

طوق بولس. النار والنور في الفكر العالمي: مجموعة الوجدانيات وشخصية جبران. بيروت: دار نوبيلسي، 2000.

الفغالي، بولس. يعقوب السروجي: الأيام السبعة. مجموعة ينابيع الإيمان، 11. الحدث، بعبدا: منشورات الجامعة الأنطونية، 2005.

الفغالي، بولس. المحيط الجامع في الكتاب المقدس والشرق القديم. جونية، لبنان: جمعية الكتاب المقدس،

المكتبة البولسية، 2003.

قرداحي جبرايل، *اللباب*، كتاب في اللغة الآرامية السريانية الكلدانية، الجزء الثاني، المطبعة الكاثوليكية، بيروت، 1891.

كامل المنجد أبجدي، مترجم قاسم بوستاني، إنتشارات فقيه، تهران، 1373هـ. 1953م (قاموس فارسي - عربي).

كتاب القداس الماروني. بكركي: [د.ن.]، 2005.

كتاب رتبة المعمودية. بكركي: [د.ن.]، 2003.

كتاب التشمشت، الليتورجية المارونية، أحد الموتى، صلاة الصباح، باعوت مار يعقوب (مخطوط سرياني).

كتاب زمن الدنح المجيد، جامعة الروح القدس- الكسليك، لبنان 1978.

Electronic references

Encyclopædia Universalis, 2009, (DVD)

Kostaz, dictionnaire syriac-anglais-syriac (File) 2008 (PDF)

Bouchez, Arnaud. Ictus Win, version 2.7, 1994-1999 (CD)

Dictionnaire numérique Bibliorum Larousse (CD)

Petit Larousse des Symboles, Larousse, 2006 (CD)

Internet

Barbour, Ian G.

——. *Christianity and the Scientist*. Site: Religion Online by Ted and Winnie Brock. https://www.religion-online.org/book/christianity-and-the-scientist/

——. *Graduation Speech*. http://www.swarthmore.edu/past-commencements/graduation-speech-ian-barbour-44

——. https://en.wikipedia.org/wiki/Ian_Barbour ;

Benoît XV, Pape.

——. *Sancto Éphrem Syro* http://w2.vatican.va/content/benedict-xv/la/encyclicals/documents/hf_ben-xv_enc_05101920_principi-apostolorum-petro.html

Benoît XVI, Pape, ;

——. *Deus Caritas Est;* 2006 http://w2.vatican.va/content/benedict-xvi/en/encyclicals/documents/hf_ben-xvi_enc_20051225_deus-caritas-est.html

——. *Homélie Du Pape Benoît XVI. http://w2.vatican.va/content/benedict-xvi/fr/homilies/2006/documents/hf_ben-xvi_hom_20060615_corpus-christi.html*

Bible; Logos Library System, Ver 2.1, DVD, *The New Revised Standard Version*, 1997. Faithlife Corporation.

——. https://www.logos.com/

Derrida, Jacques. *Déconstruction et 'différAnce'*

Comme élément constructif d'une philosophie déconstructive.

——. http://www.signosemio.com/derrida/deconstruction-et-differance.asp

Déconstruction et 'différAnce', Art, écrit par Lucie Guillemette et Josiane Cossette;

——. http://www.iep.utm.edu/derrida/

Éphrem: "My bones shout from the grave that Mary has given birth to the son of God";

——. http://www.aramaic-dem.org/English/History/Mor Ephrem is an inspiration of our time. pdf

الفغالي، بولس. *وحضور المسيح*، مقال منشور على الصفحة الإلكترونية التالية:

——http://boulosfeghali.com/boulos/index.php/site/content?ID=3325&Cat=368

——http://boulosfeghali.org/boulos/index.php/site/content?ID=3089&Cat=368

Fibonacci sequence

——. https://www.livescience.com/37470-fibonacci-sequence.html

Fiddes, Paul S. *The Creative Suffering of God* [book on-line] (Oxford: Clarendon Press, 1992, accessed 14 December 2005), iii; available from Questia,

——. https://www.questia.com/read/13914671/the-creative-suffering-of-god

Jean-Paul II, Pape;

——. *Fides et Ratio*, (Vatican: 14 Sept. 1998): http://w2.vatican.va/content/john-paul-ii/fr/encyclicals/documents/hf_jp-ii_enc_14091998_fides-et-ratio.html

Minkowski, Light Cones,

——. http://visualrelativity.com/LIGHTCONE/minkowski.html

Rabindranath Tagore, Poème intitulé «Pourquoi?»

——. http://danslombreduvent.over-blog.com/article-23860113.html

Audience Générale de Jean-Paul II, Mercredi 7 Novembre 1990:

L'Esprit qui «procède du Père et du Fils».

——. http://w2.vatican.va/content/john-paul-ii/it/audiences/1990/documents/hf_jp-ii_aud_19901107.html

Saint-Exupéry, Antoine de, *Le Petit Prince*; disponible à :

——. http://gutenberg.net.au/ebooks03/0300771h.html#ppchap1

——. http://www.cmls.polytechnique.fr/perso/tringali/documents/st_exupery_le_petit_prince.pdf

Schaeffer, John D. *Sensus Communis: Vico, Rhetoric, and the Limits of Relativism* [book online] (Durham: Duke University Press, 1990. iii; available from Questia.

——. http://www.questia.com/PM. qst?a=o&d=97538681

Yin and Yang (Tri).

——. http://www.iep.utm.edu/yinyang

Zimzum.

——. http://islamport.com/k/ser/4405/221.htm

——. https://ar.wikipedia.org/wiki/زمزم

———. http://www.encyclopedia.com/religion/dictionaries-thesauruses-pictures-and- press-releases/*zimzum*

——. http://www.kheper.net/topics/Kabbalah/Tzimtzum-ET.htm

Other resources

http://www.astronomynotes.com/gravappl/s7.htm#A6.2

http://www.universalis.fr/encyclopedie/tommaso-campanell

http://www.universalis.fr/encyclopedie/intellect-et-intelligibles

http://www.earlychristianwritings.com/fathers/ephraim2_7_mani.html

http://w2.vatican.va/content/john-paul-ii/fr/messages/urbi/documents/hf_jp-ii_mes_20001225_urbi.html

http://remacle.org/bloodwolf/philosophes/platon/rep7.htm

http://www.willeime.com/Einstein-Kant.htm

http://www.saintcharbel-annaya.com

http://www.soufanieh.com/FRANCAIS/2014.syr.fra.soufanieh.en.syrie.et.dans.le.monde.web.pdf

https://www.universalis.fr/encyclopedie/augustinisme/3-spiritualite-et-vision-du-monde/

http://www.falstad.com/coupled/

https://physics.ucf.edu/~ishigami/Teaching/Phys4083L/labdescriptions/NETD/blackbody theory.pdf

http://remacle.org/bloodwolf/philosophes/julien/soleil.htm

https://hal.archives-ouvertes.fr/halshs- 00167290/document

http://www.albishara.net/dictionary/m/

https://books.google.com.lb/books?id=SyJi8TI5PzkC&printsec=frontcover&source=gbs_ge_summary_r&cad=0#v=onepage&q&f=false

http://www.fordham.edu/halsall/basis/confessions-bod.html

https://fr.wikipedia.org/wiki/Chôra

http://www.lebanonpostcard.com/en/gibranpaint/2-divine-world.shtm

THEMATIC INDEX

A

B

D

E

F

G

H

I

M

N

O

Olaf (ܐܠܦ) 61, 207, 283, 368, 378, 379, 384

ophthalmography 44

P

pantheism 155, 197, 325, 327, 358, 373

Paraclete 83, 100, 271, 273

paradigm 31, 62, 205, 206, 285, 301, 351, 357, 360

paradox 155, 181, 189, 224, 257, 293, 306, 327

pearl (Margânitâ ܡܪܓܢܝܬܐ) 6, 7, 35, 45, 50, 70, 71, 99, 101, 105, 117, 157, 211, 212, 243, 269, 271, 378, 383

perceptive eye 189, 201

perceptive mind 208, 269

Perichoresis 27, 235, 299, 392, 402

phlogistic 49, 79, 86, 95, 96, 97, 101, 103, 105, 202, 203, 273

photon 10, 24, 107, 192, 197, 198, 231, 232, 233, 249, 253, 265, 266, 282, 324

predictability 27, 32, 45, 72, 87, 93, 110, 112, 113, 119, 138, 140, 210, 233, 240, 242, 253, 260, 264, 297, 306, 350, 374, 388

primordial Fire 24, 118, 127, 157, 180, 192, 195

principle of uncertainty 220, 253, 255, 256, 258, 350

probability 250, 259, 261, 262, 288, 329

procession 98, 100, 299, 375, 383, 408

proton 226, 234, 242, 244, 250, 341

Providence 4, 28, 78, 311, 329, 330, 372, 373, 379

Q

qiâmtâ (ܩܝܡܬܐ) 18, 171, 184, 202, 205, 206

Quantic Energy 27

Quantic Eye 212

quantic phenomenon 7

quantum 9, 10, 24, 25, 26, 45, 58, 69, 70, 76, 83, 103, 107, 146, 150, 154, 170, 185, 198, 212, 217, 218, 219, 220, 222, 223, 224, 226, 234, 239, 240, 241, 242, 243, 245, 247, 248, 249, 250, 251, 252, 253, 254, 258, 259, 260, 263, 264, 265, 266, 267, 268, 272, 273, 285, 291, 296, 297, 298, 299, 305, 309, 327, 330, 331, 343, 344, 347, 350, 351, 358, 362, 364, 370, 371, 372, 374, 378, 387, 391, 396, 406, 407, 408

quantum reality 24, 69, 154, 170, 219, 243, 245, 252, 264, 266, 296, 298, 309, 344, 350, 351, 364, 374

quark 234, 242, 244, 245, 341

R

S

T

U

NAMES INDEX